Firefighting Operations in High-Rise and Standpipe-Equipped Buildings

Firefighting Operations in High-Rise and Standpipe-Equipped Buildings

David M. McGrail

> **Disclaimer**
>
> The recommendations, advice, descriptions, and the methods in this book are presented solely for educational purposes. The author and publisher assume no liability whatsoever for any loss or damage that results from the use of any of the material in this book. Use of the material in this book is solely at the risk of the user.

Copyright© 2007 by
PennWell Corporation
1421 South Sheridan Road
Tulsa, Oklahoma 74112-6600 USA

800.752.9764
+1.918.831.9421
www.FireEngineeringBooks.com
www.pennwellbooks.com
www.pennwell.com

Marketing Manager: Julie Simmons
National Account Executive: Francie Halcomb

Director: Mary McGee
Managing Editor: Jerry Naylis
Production/Operations Manager: Traci Huntsman
Production Editor: Tony Quinn
Book Designer: Sheila Brock
Cover Designer: Clark Bell

Library of Congress Cataloging-in-Publication Data

McGrail, David M.
  Firefighting operations in high-rise and standpipe-equipped buildings / David M. McGrail.
    p. cm.
  ISBN-13: 978-1-59370-054-6 (hardcover)
 1. Tall buildings--Fires and fire prevention. 2. Standpipes. 3. Fire extinction. I. Title.
 TH9445.T18M38 2007
 628.9'25--dc22

                                           2006101652

All rights reserved. No part of this book may be reproduced, stored in a retrieval system, or transcribed in any form or by any means, electronic or mechanical, including photocopying and recording, without the prior written permission of the publisher.

Printed in the United States of America

1 2 3 4 5   11 10 09 08 07

For my family, my mother and father, who provided me with a good foundation, excellent examples to follow, and the opportunity to succeed through education and hard work.

For my dad, who carefully exposed me to the fascinating, exciting, and brave world of firefighting from the time I was just a little boy. Dad, my success stems greatly from your years of mentoring, based on your valuable firefighting and life experience.

For my lovely wife Gina, my soul mate, for understanding the passion I have for this, the greatest of professions, and supporting me through both the good times and bad.

For my daughter Caroline, the sweetest, most beautiful little girl. I am so blessed to have you and your brother in my life. For my son Joseph, who fights everyday to overcome his physical disabilities. He represents the true meaning of courage and valor.

For my friends, colleagues, and mentors on the Denver Fire Department, and throughout the American fire service, thank you for teaching me what you know and for protecting me in the heat of battle; for those elite members of my profession who are truly passionate, from the bottom of their hearts, about firefighting; for those who continue to excel and tenaciously work to keep firefighters alive by improving the way we do business; and for those who have taken the time to read this book, I truly appreciate that you would allow me to share my thoughts and ideas about firefighting with you, my fire service family.

Contents

Foreword . . . . **xvii**

Acknowledgments . . . . **xix**

Introduction . . . . **xxi**

1 The Firefighting Mind-Set **1**
    The Last Alarm 1
    The Dominoes of Complacency 2
    The Firefighting Mind-Set 4
    The Full Service Fire Department 5
    American Fire Service History 6
    Dangers of Modern Firefighting 8
        High Btu fuels 8
        Early detection systems 9
        Full personal protective equipment (PPE) 9
        Lightweight construction 9
    Building the Proper Mind-Set 9
        First gear: Daydreaming mind-set 10
        Second gear: Conversational mind-set 10
        Third gear: Preparation mind-set 10
        Fourth gear: Life-or-death mind-set 10
    Preparation 10
        Mental fitness and preparedness 10
        Physical fitness and preparedness 11
        Continual training and development 12
        Check and maintain your equipment 14
        Listen to the radio 14
        Get prepared, stay prepared 15
    Summary 16
    References 16

2 The High-Rise Building **17**
    Introduction 17
    History of High-Rise Buildings 17
    High-Rise Defined 18
    Practical Definition of High-Rise 19
    Usable Aerial or Ground Ladder Length 19
    High-Rise Building Size 21
    High-Rise Building Fireground Strategy and Tactics 22
    High-Rise Building Construction 23
    Four Generations of High-Rise Construction 24
        First generation high-rise construction 24
        Second generation high-rise construction 25
        Third generation high-rise construction 25
        Fourth generation high-rise construction 26

High-Rise Building Features 27
    Specific high-rise construction concerns 27
    Fire detection and protection systems. 28
    Water supply 29
    Stairs 29
    Elevators 29
    Heating ventilation and air conditioning (HVAC) systems 30
    Utilities 30
Fire Extension and Smoke Spread. 30
    Poke-thru construction 31
    Curtain wall gap 31
    Auto exposure 32
Summary 33
References 33

3 Low-Rise and Other Standpipe-Equipped Buildings 35
Low-Rise Buildings 35
    Low-rise multiple dwellings. 36
    Shopping malls 37
    Warehouses. 38
    Sports venues. 39
    Parking garages 41
    Educational facilities 42
    Medical facilities 42
Other Standpipe-Equipped Structures. 43
    Bridges 43
    Tunnels 44
    Water vessels 45
Summary 45

4 The Standpipe System 47
Water Supply Priority 47
Lesson Learned 48
Standpipe System Defined 49
Classes of Standpipe Systems 50
    Class I standpipe system 50
    Class II standpipe system 51
    Class III standpipe system. 51
    Class IV standpipe system. 52
Types of Standpipe Systems 52
    Automatic dry standpipe system 52
    Automatic wet standpipe system 52
    Semiautomatic dry standpipe system .52
    Manual dry standpipe system. 52
    Manual wet standpipe system. 53
Standpipe System Components 53
    Fire department connection (FDC) 53
    Pipe and tube. 54

    Standpipe system valves. 55
    Hose connection valves 55
  Pressure-Regulating Devices 56
    One Meridian Plaza 57
    Pressure-restricting devices. 58
    Pressure-reducing valves (PRVs) 60
    Characteristics of PRVs 60
    Terminology 61
  Pre-Fire Planning 66
    Giacomini PRV 66
    New Elkhart Brass PRV 68
  Identifying Pressure-Regulating Devices 68
  False Information 69
  Fire Department Operations 70
  Summary 71
  References 71

5 The Problem Defined 73
  Logistics 74
    Elevators 76
    Stairs 76
    Reflex time 77
    Standpipe system. 77
  Exposures 78
  Summary 78
  References 79

6 Elevator Operations. 81
  Elevators 81
  Elevator History 81
  Electric Traction Elevators 82
    Elevator machine room 83
    Hoist-way 85
    Elevator car. 87
  Hydraulic Elevators 87
  Elevator Banks. 88
  Elevator Operations During Fires 88
    Fire service recall and control 90
    Recommended operational procedures for elevator use. 90
    Additional size-up considerations (for elevator use) 94
  Elevator Rescue Operations 94
    Rescue or removal 95
    Removal procedures. 95
  Methods of Removal and Rescue. 97
    Resetting the system. 97
    Fire Service recall 97
    Hoist-way door. 97
    Poling down 100
    Poling across 100

        Poling tools. 101
        Roof hatchway door 102
        Side emergency exit door 103
    Summary 104
    References 104

7 Stairway Operations **105**
    Types of Stairs 106
        Return stairs 106
        Scissor stairs 108
        Access stairs 111
    Tactical use of Stairs 113
        Attack stair 113
        Evacuation stair 115
    Stairwell Pressurization. 115
    Climbing Stairs. 116
        Heat buildup 117
        Carrying equipment 117
        Teamwork. 118
    Additional Logistical Techniques. 119
    Stairway Support Unit. 119
    Summary 121
    References 121

8 Hoseline Selection **123**
    Hoseline Size. 123
    The War Years 124
    Proper Weapon Selection 124
    GPM versus Btu 124
    1¾-Inch Handline 125
        Realistic flow 125
        Guidelines for using 1¾-inch handline 126
        Hydraulic calculations for 1¾-inch hoseline. 127
    2-Inch Handline 128
        Hydraulic calculations for 2-inch hoseline. 128
        Advantages of 2-inch handline 129
    2½-Inch Handline 130
        ADULTS acronym 130
        Hydraulic calculations 2½-inch hoseline. 131
        Reaction force formula for smooth-bore nozzles:. 132
        Reaction force formula for combination fog nozzles: 132
    Seven Keys to Success with the Big Line 133
        User-friendly apparatus and equipment 133
        User-friendly standpipe hose packs 133
        Properly stretching the attack line 135
        Proper operating pressures 136
        Low-pressure nozzles 136
        Unglamorous operating positions 137
        Training, training, and more training. 138

The High-Rise/Standpipe Hose Pack........139
Overwhelm the Enemy with Disproportionate Force........141
Summary........141
References........142

9 Nozzle Selection........143
　The Great Debate........143
　My Personal Experience........144
　　The mentoring begins........146
　　Andy Fredericks........148
　Smoothbore versus Combination Fog........148
　Protection Myth........149
　The Line-of-Duty Death of Firefighter James Heenan........151
　Treat the Disease, Not Just the Symptom........151
　Fog Application in a Center Core High-Rise........151
　The Effect of Wind........152
　Lloyd Layman........153
　　The insurance industry........153
　　The fire service........154
　　Water damage........154
　Proper Nozzle Selection........155
　　Hands-on training (HOT)........155
　　Flow tests at FDIC West, 2005........159
　　Steam and disruption of thermal layering........159
　　Standpipe system debris........159
　　Debris passing: fog versus smoothbore........160
　　Nozzle weight........160
　　Cost, durability, and maintenance........161
　Summary........162
　References........163

10 The Standpipe Equipment Kit........165
　Standpipe In-Line Pressure Gauge........166
　　Testing the SPG........167
　　Labeling the SPG........167
　　Static versus flow pressure........168
　60-Degree Lightweight Angle Elbows........169
　Spare Nozzle........170
　1½-Inch to 2½-Inch Increaser........170
　　Extending a hoseline........170
　　Creating a second 2½-inch outlet........174
　Other Adapters........175
　Hand Tools........176
　　Spanner wrenches........176
　　Pipe wrench........176
　　Spare hand wheel........176
　　Hand wheel wrench........177
　　Wire brush........177
　　Wooden wedges (door chocks)........177

   Door strap/search markers 178
   Colored markers, grease pencils, and chalk. 179
  Summary 179

11 Engine Company Operations 181
  Priority One: Stopping the Fire. 181
  Engine Company Functions 182
  Water supply 182
   Proper weapon selection 182
   Two engine companies, one handline 183
   Stretching the handline 184
   Advancing the handline 191
   The nozzle team 193
  Dangerous Procedures 194
  The First Handline, First 194
  The Backup Line. 195
  Using Master Stream Appliances Offensively 197
  Overall Strategic Plan 200
  Summary 202

12 Water Supply 203
  Types of Standpipe Systems 203
   Automatic wet systems 203
   Non-automatic dry systems 204
  Fireground Operations 204
   The engineer/pump operator 204
   The engineer's/pump operator's standpipe kit. 205
  The Fire Department Connection (FDC) 205
   Apparatus placement 209
   Securing the supply lines 210
  General Pumping Considerations 211
   Second due engine company 211
   Third and fourth due engine companies 212
   Bypassing the FDC. 213
   Pumping into a standpipe system. 214
   Hydraulic calculations. 215
   Protecting supply lines and engineers/pump operators 215
  Summary 217

13 Truck Company Operations 219
  Basic Truck Company Functions. 220
   Laddering 221
   Overhaul 222
   Ventilation 224
   Forcible entry/exit 224
   Search and rescue 226
   Salvage 226
   Utilities 227
  Summary 228
  References 228

14 Search and Rescue Operations 229
    Identifying Who Needs to be Rescued 229
        Protect in place 230
        The three floor rule 230
        Evacuation 231
        Removal 231
        Rescue 231
    Prioritizing Search Areas 231
    Fire Floor 232
        Residential occupancy 232
        Commercial occupancy 233
    The Three-Phase Search Operation 234
        Primary search 234
        Secondary search 236
        Final search 236
    Equipment for Search Operations 236
        Search markers (door straps) 236
        Alternative search marking system 238
        Improvised methods 239
        Thermal imaging cameras (TIC) 240
    Search Rope Procedures 241
    Summary 243

15 Air Movement and Ventilation 245
    Natural Air Movement 245
        Stack effect 246
        Neutral pressure plane 247
        The effects of wind 248
        Stratification of smoke 248
        Mushrooming of smoke 249
    Mechanical Air Movement (HVAC) 250
        The processing system 250
        The supply system 250
        The return system 250
    Operational Air Movement 251
        Built-in smoke management systems 251
        If in doubt, shut it down 252
        Manual control 252
        Fire blanket 253
        Breaking windows 253
        Elevators 254
        Fire/smoke towers 254
        Natural vertical ventilation 255
        Positive pressure fans 255
        Operational guidelines 257
    Summary 258
    References 258

16 Rapid Intervention Team (RIT) Operations **259**
    RIT History. 259
    Using RIT at Larger Buildings 260
    Operations at Low-Rise Buildings 260
    RIT RECON 260
    High-Rise RIT 262
        Fire attack RIT 262
        Upper search and evacuation (USE) RIT 263
        Roof RIT (air support). 264
        Mutual and automatic aid agreements 264
    RIT Basics 265
        Long-duration SCBA 265
        RIT tools 266
        Resources needed for RIT operations 266
        Door strap/search marker 267
        Extreme operations 267
        Improvised carrying devices 268
        Superman carry 269
    General RIT Strategies 269
        Commercial high-rise operations. 269
        Residential high-rise operations 271
        Low-rise and other standpipe-equipped buildings 272
    Summary 273

17 Command and Control **275**
    ICS/IMS 276
    The Initial Incident Commander 277
    Formal Command 277
    Incident Commander 278
    Location of the Command Post 278
        Interior command post 278
        Exterior command post 278
    Lobby Control Unit 279
        Systems control unit 281
        Stairway Support Unit 282
    Fire Attack Group 283
    Upper Search and Evacuation Group 284
    Staging Area Manager 286
    BASE   286
    Other ICS Operational Components 289
        Rehabilitation Support Unit (REHAB) 289
        Medical Unit 290
        Ventilation Group 290
        Utilities Control Unit. 290
        Salvage Group 291
    ICS Branches 292
    ICS Organization Sections 292
        Operations Section 293

   Logistics Section 293
   Planning Section 294
   Finance/Administration Section 294
  Summary 294

18 Air Support Operations 295
  Outside Resources 295
   Police department aircraft/helicopter 296
   Medical aircraft/helicopter (air ambulance) 296
   News media aircraft/helicopter 297
   Military aircraft/helicopter 298
   Forest service aircraft 299
  Operational Plan 299
  Training 300
  Helicopter High-Rise Teams (HHRT) 300
   Size-up 301
   Fire department operations 302
   HHRT equipment 303
   En route 303
   On arrival 303
   HHRT deployment onto the roof 304
   HHRT fireground operations 306
   Extended roof operations 309
  Rooftop Evacuation of Victims By Helicopter 310
  Summary 311
  References 311

19 Communications 313
  Fire Department Radios 314
   Non-repeated radio channels 314
   Mobile repeaters 314
  Internal Building Communication Systems 314
   Phones 315
   Cell phones 316
   Sound powered phones 316
   Public address (PA) system 317
   Reverse 9-1-1 318
   Runners 319
  Summary 319

20 Preparing for Battle 321
  Preparing Your Firefighters for the Worst 321
  The Denver Fire Department Hands-On Training Program 322
   Large-scale academy recruit drill 322
   Fire-attack scenario 323
   Initial assignment 323
   At the twenty-seventh-floor staging level 325
   The fire attack 326
   Truck company assignment 327

    Water supply evolution 328
    Models of excellence 328
  Quarterly District/Battalion Training 330
  Quarterly Air Support Operations Drills. 331
  Do We Really Need to Do That Every Time? 331
    A motivational factor 332
    Prediction versus preparation 333
    Lessons learned from American history 333
    History lessons within the fire service 334
  The American Fire Service's "Dream Team" 335
  Carpe Diem: Seize the Day 336
  Index 339

Foreword

One of our objectives as firefighters is to meet the challenges of a dynamic fire environment brought about by ever-changing techniques of building construction, building materials, and new designs and technology, by applying proven tactical methods of fire attack as well as newer methods.

Chief McGrail's book is a comprehensive and fresh look at how to deal with a fire in high-rise, low-rise, and other standpipe-equipped buildings. It is a book long overdue and one that promises to guide us well into the 21st century.

Having worked for a fire department with a large number of high-rise buildings in its jurisdiction, where standpipe operations are a daily event, I found that seeking information that might enhance our own procedures was a never-ending task. Dave McGrail first came to my attention in 1992 when I read his article, "Denver's Polo Club Condo Fire: Atrium Turns High-Rise Chimney," in the March issue of *Fire Engineering* magazine. It was well written, insightful, and contained a number of valuable lessons learned, ones that would apply to my own department. I knew then that I had just tapped into a gold mine of a resource, and his subsequent articles proved this.

I had the opportunity to meet Dave almost 10 years ago while serving on the Educational Advisory Committee for FDIC West, and over the years we have become good friends. This friendship has given me the chance to watch his professional development as one of the leading authorities on high-rise and standpipe operations in the country. As an international lecturer and instructor, Dave McGrail's knowledge has proven invaluable to thousands of firefighters. A number of fire departments, including my own, have sought his advice and expertise when modernizing their operations.

It should be noted that Dave had a distinct advantage in his intellectual growth as a firefighter. His father, a fire service icon in his own right, retired Division Chief of Operations for the Denver Fire Department Pat McGrail, was no doubt influential in formulating Dave's fire service philosophy and deserves his own accolades.

This book is a unique contemporary resource written by a man with firefighting in his blood and written from the perspective of someone in the field who isn't afraid to get down and dirty with the troops. It is a book that reflects years of experience, research, networking, common sense, and the integrity and character that define Chief Dave McGrail.

Ted Corporandy, Battalion Chief (Retired)
San Francisco Fire Department

Acknowledgments

Preparing this book has been a very long and arduous task. I would have never completed it had it not been for the constant encouragement and help from a long list of top shelf Brothers, friends, and associates. Most of the help came from my many friends and Brothers throughout the American fire service. I was also fortunate to have been given assistance from many good people who are not directly associated with the fire service. Ultimately, there are countless individuals who deserve recognition for their assistance, but after this long journey, I'm afraid I might forget someone. If I do, please forgive me, and please understand that I truly appreciate your assistance.

First, let me thank the members of the Denver Fire Department (DFD), especially the countless fire companies and individual firefighters who have always been willing to assist with various evolutions, training drills, and allowing me to photograph them doing the job, especially DFD Engine Co. 3 and Rescue Co. 1.

Special thanks to Pat McGrail, division chief (retired); Sean Roeper, firefighter, Tower Co. 4; Dave Frank, district chief; Don Burkhardt, captain, Tower Co. 1; Mike Berlin, lieutenant, Fire Prevention Bureau; Joe Gonzales, division chief, Fire Prevention and Investigation Division; Carl Johnson, captain, Truck Co. 16; Phil Miller, lieutenant, Engine Co. 3; Vern Scott, captain, Truck Co. 8; Mike Shepherd, lieutenant, Engine Co. 11; Andy Singer, lieutenant, Dist. #4; Pete VanderMiller, lieutenant, Engine Co. 24; Dave Borelli, captain, Engine Co. 24; Mike Young, firefighter, Rescue Co. 1, Eric Zuick, firefighter, Truck Co. 8; Zach Bousman, firefighter, Engine Co. 23; C. J. Haberkorn, captain, Engine Co. 6; Thor Hansen, captain, Engine Co. 13, and all the dedicated members of the DFD, the nation's best fire department.

Special thanks to my many Brothers across the American fire service:

Tracy Raynor, deputy chief, Boise Fire Department; Rick Payne, captain, Boise Fire Department; Jay Comella, captain, Oakland Fire Department; Daryl Liggins, lieutenant, Engine Co. 13, Oakland Fire Department; Ted Corporandy, battalion chief (retired), San Francisco Fire Department; Dave Franklin, chief of special operations, San Francisco Fire Department; Tom Murray, battalion chief, San Mateo Fire Department; Paul Schuller, battalion chief, San Jose Fire Department; Vincent Dunn, deputy chief (retired) FDNY; John Norman, deputy chief, FDNY; Jerry Tracy, battalion chief, FDNY; John Ceriello, lieutenant, FDNY; Harry Lee Davis, firefighter (retired) FDNY; Mark Wesseldine, firefighter (retired) FDNY; Jeff Shupe, firefighter, Engine Co. 11, Cleveland Fire Department; Dave Karn, firefighter, Rescue Co. 2, Columbus Fire Department; Jim Sewnig, firefighter, Engine Co. 46, Chicago Fire Department; Bob Hoff, deputy commissioner of operations, Chicago Fire Department; Ed Enright, battalion chief (retired) Chicago Fire Department; Bob Athanas, firefighter, Rescue Co. 3, FDNY; Scott Eckels, lieutenant, Castle Rock Fire Department; George Szczerba, firefighter, Rockingham County Virginia Fire Rescue; Roger Gillis, battalion chief (retired), Los Angeles City Fire Department; Gary Seidel, chief, Hillsboro Fire Department; John Nowell, battalion chief, Los Angeles City Fire Department; Rick Kolomay, lieutenant, Rescue Co. 1, Schaumburg (IL) FD; Steve Redick, senior dispatcher, Chicago Fire Department; Dennis Pattie, captain, Engine Co. 5, Ontario Fire Department; Tim Adams, captain, Sacramento Fire Department; Jim McCormick, lieutenant, Indianapolis Fire Department; Larry Collins, captain, Los Angeles County Fire Department; John "Skip" Coleman, deputy chief, Toledo Fire Department; Dave Traiforos, chief, Franklin Park (IL) Fire Department; Nate DeMarse, firefighter, Squad Co. 61, FDNY; Peter McBride; Dave Fornell;

Jim Reagan; George Goldbach, lieutenant (retired) FDNY and chief (retired) West Metro Fire Rescue.

The late Andy Fredericks, lieutenant, FDNY Squad Co. 18

    The good people at Elkhart Brass and Mr. Paul Albinger;

    The good people at Otis Elevator and Mr. Stephen Showers;

    The good people at PennWell Publishing and Fire Engineering Books and Videos: Mr. Jerry Naylis, Mr. Tony Quinn, Ms. Marla Patterson, Ms. Julie Simmons, Mr. Steve Hill, and Ms. Francie Halcomb.

    The good people at Zurn Wilkins, Mr. Pete Chapman and Mr. Mike Littlejohn;

    Ms. Heather Hays, KIDDE Fire.

    Mr. Ted Beck III, Biomarine, NTRON, Inc.

    The Denver Police Department Air Support Unit.

    Colorado Army National Guard.

    Professor Glenn Corbett; Mr. Jeffery M. Shapiro; The late Frank Brannigan; The late William Clark;

    Mr. Rick Bobka; Ms. Suzi Latona;

    Mr. Shawn Murphy; Mr. Mark Bennett;

    Mr. Jeff Navarro, Navarro Studios, Art and Illustrations.

Thank you all very much.

Introduction

Serious fires in high-rise and standpipe-equipped buildings are not an everyday occurrence. As history has proven time and again, however, both minor and serious fires, as well as numerous other emergency situations have occurred and do occur in high-rise and standpipe-equipped buildings. Most importantly, the potential for a wide range of fire or various emergency events in these types of buildings is ever present, not only in the large cities, but also many of the numerous small communities across the country. Operations in these buildings may be low-frequency events, but they are also very high-risk events.

Long before the events of September 11, 2001, tragic evidence of this high risk was graphically apparent when one took into account the numerous injuries and firefighter fatalities that occurred in high-rise building fires. These tragic fires occurred from the East to the West Coast, once again, in both large cities and small communities. Many of the fire buildings where these tragic fires occurred were very similar to the types of high-rise and standpipe-equipped buildings that can be found in almost any city. If you have these types of buildings in your city or fire district, you ultimately must be prepared for battle. This book is designed to help you do just that: prepare for battle.

I spent several years contemplating writing this book. In fact, I wasted a lot of time just thinking about it—valuable time that I wish had been spent writing. My reluctance revolved around my concern that I am certainly not an expert in high-rise firefighting. In fact, what I do consider myself to be is a lifelong student, not just of high-rise firefighting, but of structural firefighting in general. As a student, I spend lots of time analyzing my past fireground operations, in an attempt to continually refine how I do business. In addition, I dedicated countless hours listening to my mentors and those fire service icons who I consider to be the true experts, in an attempt to learn from them and hone my skills as a professional firefighter.

During my long period of contemplation, I was actually expecting that someone else would write a new "high-rise firefighting" book. In fact, I contacted many of those fire service icons, including Deputy Chief Vincent Dunn, FDNY (retired), and inquired as to whether or not he was contemplating such a project. I certainly didn't want to step on any toes, but especially not those of someone of Chief Dunn's extremely high caliber. Chief Dunn is a true American fire service icon, with many years of high-rise experience, and I wanted to ask for his advice. Chief Dunn gave me a green light and encouraged me to take the necessary steps to move forward with the project. After that conversation with Chief Dunn, I decided to formally get started with this monumental task. Many of the basic fundamentals and principles of this book are based on what I have learned from many remarkable fire service leaders and experts in the field, including Chief Dunn. These experts have all made a tremendous contribution to the development of this book.

Although I wouldn't consider myself an expert, I am very fortunate to have been given the opportunity to work for a very busy fire department in a large city, and have been assigned to some of the busiest companies and fire districts in the city. Denver has a very diverse fire problem, with a wide range of different buildings and occupancies, including over 350 high-rise buildings, and thousands more low-rise, standpipe-equipped buildings. During my 25 years of fire service experience, I have responded to and operated at numerous working fires in both high-rise and various standpipe-equipped buildings. Each of those events was loaded with opportunity to gain experience, and with each one came mistakes and lessons learned.

A large portion of my high-rise/standpipe firefighting experience was gained through *vicarious* experience, that is, learning from the experience of other firefighters. From the very beginning of my career, I have dedicated hours to the vicarious experience by reading many fire service periodicals, trade journals, and books, as well as attending the countless seminars and classes presented by those whom I consider to be the true experts. For me, the greatest expert of all is my father, retired Denver Fire Department (DFD) Division Chief, Pat McGrail. He spent nearly 42 years on the DFD, and I still count on him to this day to provide expert advice on a wide range of topics, including high-rise firefighting.

My affiliations with *Fire Engineering* magazine and the Fire Department Instructors Conference (FDIC) have been some of the most significant events of my career. I wrote my first article about my personal experience and the collective experience of my fire department at a serious high-rise fire that occurred in 1991. While instructing at the FDIC, I have met and become friends with many of the greatest minds the American fire service has even seen, among those the late Andy Fredericks (Lieutenant Andrew A. Fredericks, FDNY, Squad Co. 18, made the supreme sacrifice on September 11, 2001). I have developed close friendships with many fire service brothers from fire departments across this great country. They are the co-authors of this book—true experts who continue to guide me through my career. Brothers, thank you all, very much.

For whatever reason, since the beginning of my career I have been particularly interested in high-rise firefighting. I recall designing and building a new type of high-rise hose pack as a young firefighter over 20 years ago. I spent hours on the apparatus floor building, rebuilding, deploying, and practicing with that hose pack. Eventually I presented the new hose pack to my bosses, and it was quickly adopted by the fire department.

My first good working high-rise fire became a career changing experience, as I realized just how complex, dangerous, and overwhelming the entire high-rise operation can be. Over the past 25 years, my experiences, research, and thoughts have grown to represent something I want to share with you, my fire service family of brothers and sisters.

It is my fervent belief that firefighting operations should be the number one priority for a fire department. Unfortunately, during the 1980s and 1990s, much of the American fire service was hypnotized by the words, "customer service." I am not anti-customer service, but I believe that during this bizarre attempt to reshape the fire service, we truly lost sight of our original mission. In an attempt to provide better customer service, I believe we actually forgot how to provide customer service period; that is, we forgot how to provide good "emergency service." A frenzied quest for new fancy logos and unique mission statements replaced the critical, yet basic tenet of any fire service organization: to protect life and property from the ravages of uncontrolled nature—*fire*.

While many individuals in fire service leadership positions preached that the occurrences of fires were down and that we must expand our roles into other areas in order to survive, fires, in fact, continued, as did the death toll for both civilians and firefighters alike. In fact, as we left the 20th century, it was on another sad and tragic note: we lost 15 more of our brother firefighters in December of 1999 alone. The number of firefighter deaths significantly increased from the previous eight years, and firefighter injuries had also increased. Here we are, nearly eight years since the turn of the century, and we continue to see an average of 100 firefighter line-of-duty deaths each year. Many of those occurred at structure fires.

I believe that the families of those firefighters killed in high-rise and other building fires, as well as the countless families of the numerous civilians killed or injured in fires, would not subscribe to the theories of most of those individuals who make decisions behind desks in the comfort of air-conditioned buildings. We firefighters must take back control of the fire service, which has been out of control and on a downward spiral for too long. The priority is and must always be centered on our ability to provide excellent emergency service, especially that of fire suppression activities.

As I compile the countless mounds of material for this book, like many other brother firefighters across the country, I still remained shocked and saddened by the events of 9/11. I plead with the fire service professionals who have taken the time to read this book to keep in mind that our job is not getting easier and safer, but harder and much more dangerous. We must not fall victim to the widespread complacency that is the underlying foundation and literally the trademark of some individuals within the fire service, including some in leadership positions. Once they're finally gone, we will be left to clean up the mess!

As firefighters, we must always be prepared, both mentally and physically, for any type of emergency operation. The high-rise building fire is an especially demanding operation that requires a tremendous amount of preparation and training, as well as the appropriate equipment. There is a widespread belief that most high-rise buildings are fully protected by built-in fire protection systems, namely, fully automatic sprinkler systems. This may, in fact, be true in some newer cities and jurisdictions, however, in most jurisdictions and many older cities, numerous high-rise buildings are not fully sprinklered.

Of course, most fire codes require that all new high-rise construction must include the installation of full sprinkler protection throughout the building. A large number of high-rise buildings in this country, however, were constructed prior to the full sprinkler requirement. Manual firefighting, therefore, performed by a well-trained force of mentally and physically prepared firefighters, will be required in the event of a fire in many of these buildings. Also, we must not overlook the countless buildings in most jurisdictions that have standpipes, but are not considered high-rise buildings.

As you read through this book, I would ask you to keep in mind the following: this book was not written by an architect, or a fire protection engineer, or a mathematician, or an extremely analytical research scientist. It was written by a fireman from Denver. This is not a comprehensive dissertation about the intricacies of various high-rise building components. I don't refer to mathematical coefficients and various factors to explain the effects of wind during a high-rise fire. There is very little discussion about sprinkler systems and fire prevention activities related to high-rise buildings.

The high-rise firefighting tactics and standpipe operations described in this book are a compilation of information intended to help firefighters, company officers, chief officers, and fire departments to safely and successfully complete the high-rise and standpipe operations that they encounter. The heart of this book revolves around the basic fundamentals of firefighting, and the basic equipment associated with those fundamentals. Simple concepts, like good elevator discipline and selecting the appropriate weapons to combat the high-rise fire, which are not the same ones used to fight a car fire, represent the meat of this book. This book was written for firefighters by a firefighter.

The central theme of this book revolves around the most important component of any fireground operation, especially those in high-rise and standpipe-equipped buildings, and that is, *preparedness*. Fire service professionals must take the time to train and develop the skills associated with the use of basic standpipe tools and equipment. A regular training program, even if it simply involves a short firehouse drill, can make a tremendous difference. Please take the time to conduct quality training drills!

CHAPTER 1

The Firefighting Mind-Set

The primary mission of a fire department is to save human life and extinguish fire.
Everything else is, and must be, secondary.

David M. McGrail

The Last Alarm

At 0203 hours, it has already been a busy shift for Engine Company 18. Members of this company are trying to get a few minutes of uninterrupted sleep between calls when the lights in the firehouse come on and alert tones break the silence. The voice of a dispatcher announces that a private fire alarm system has been activated at a high-rise multiple dwelling located a few blocks from the firehouse. Eighteen is first due with Ladder 11 and the chief of the Third Battalion.

The men from 18 are moving slowly. They're exhausted from the myriad of activities that have consumed the first 19 hours of their shift. They attended a mandatory sensitivity training class that morning, followed by CPR recertification that afternoon. They answered seven EMS calls, three false fire alarms, two car accidents, a couple of dumpster fires, and a pretty good car fire. One of their three false alarms was at the building they are now responding to, again!

As Engine 18 pulls out of the firehouse on their 16th run of the shift, human imperfections start to manifest in the way of complacency. Over their communication system headsets, crew members, including the officer, start to air complaints about the continued nuisance alarms at the same building. Between all three platoons, Engine 18 has been there numerous times over the past several months. Prior to 18's arrival, all four of the members have managed to convince themselves that this too, will be another false alarm recorded in the company's journal.

As Engine 18 pulls into the front driveway of the building, the company officer gives an initial radio report and size-up information, having not actually even looked up at the building that he's seen several hundred times before, assuming nothing will be different this time. The men from 18 make their way into the lobby and are quickly joined by the members of Ladder 11. They have some, but not all of the equipment necessary to mount an effective attack against a working fire. Self-contained breathing apparatus (SCBA) packs are hanging off shoulders in a haphazard manner, and waist belts are not secured. The company officers take a quick look at the fire alarm panel in the lobby and note that a red light is illuminated, indicating an smoke detector for the ninth floor.

Seven firefighters pack into a passenger elevator—an elevator that is rated to carry a load much less than their collective weight and one that is not equipped with Fire Service Recall and control features. The officer from Engine 18 pushes the button on the interior panel requesting the ninth floor, and turns to the men, "Well, let's go see who burned their food this time, guys."

The above scenario is not fiction! It's real life, and plays out all too many times, everyday across the American fire service. It happens, because the greatest people on earth, firefighters, are still human beings. As human beings, we are imperfect, fallible creatures that, even with the best intentions, sometimes make mistakes. At the root of many mistakes lie the human weaknesses of complacency and laziness.

At the vast majority of calls to this building, the members of Engine 18 and Ladder 11 will more than likely find something other than a working fire or bona fide emergency condition when they arrive in response to an alarm. In fact, their past experience is loaded with statistical probability that this call is going to be just another nuisance alarm in the form of burnt food, a system malfunction, a maliciously activated pull station, steam from a shower, or perhaps a discharged fire extinguisher in the public hallway. It's certainly a gamble, assuming this alarm is false, but their assumptions have proven correct many times before.

Unfortunately, their luck has run out. Just before they reach the ninth floor, the Engineer/Pump Operator of Engine 18 makes a frantic communication over his portable radio, "18 Bravo to 18 Alpha. You guys got a fire! There's fire and smoke venting out a couple of upper-floor windows on the D-Delta side of the building. Do you copy?"

In the elevator, panic sets in as the captain from Engine 18 tries in vain to push the floor eight button; and just as he thinks to push the emergency stop, the elevator arrives at the ninth floor and the doors open. The firefighters from 18 and 11 are greeted by a very heavy smoke and heat condition that quickly floods the elevator car. The sounds of tools being dropped and SCBA air cylinders being turned on fill the air. In their haste to quickly don their face pieces, several men lose precious air supply—it can be heard free flowing due to open valves. All seven men are now fighting for their lives. Some of them end up out on the ninth floor in the elevator vestibule, and some are still inside the elevator car as the doors close. One of the men inside the elevator tries to block the elevator doors open with his leg, but fails in his attempt. The elevator now descends to the lobby without the other members. Inside the elevator car is the one and only standpipe hose pack brought into the building. The men on the fire floor are without the critical tools necessary to protect themselves and the occupants of the building. Two of the men from Engine 18 will eventually run out of air and be overcome by smoke when they are separated from the others in the ninth floor hallway of this multiple dwelling.

In the course of the next week, having succumbed to their injuries, those two brother firefighters from Engine 18 will be laid to rest. The sound of bagpipes will fill the air, two wives are now widows, and several children become fatherless. The names of these two brave men will be chiseled permanently into a hard granite stone. After all the pomp and circumstance, white gloves and salutes, a long and extensive investigation will take place. The investigation will yield nothing new at all. The human errors and the long list of complacent decisions that led to this tragedy have been listed before under *lessons learned*. Most disheartening is that the same issues will probably be listed again at the conclusion of another investigation!

The Dominoes of Complacency

When firefighters are injured or killed in the line of duty, the events leading up to these tragedies are usually based on numerous factors. I like to compare it to a line of dominoes falling (fig. 1–1). We've all been there and seen the fireground dominoes fall, one after another. Often, a good leader steps in and makes the necessary decisions to recover from a bad start and ultimately stop the dominoes. Sometimes it's just plain fire service luck that leads to a tragedy-free conclusion. Unfortunately, if your operations are based on luck, eventually, your luck is going to run out.

We see it time and again, with statements like "suddenly, and without warning, the building collapsed." Chief Vincent Dunn (retired deputy chief, FDNY), and the late Francis Brannigan (author of *Building Construction for the Fire Service*), in their attempt to educate the fire service, have told us time and again that the collapse of a burning building, especially one with a lightweight truss, should certainly not come as a surprise to any of us.

Arriving at the floor of alarm, opening the elevator doors, and encountering smoke and or fire would certainly be a surprise to a couple of guys from the parks department, the city librarian, or even a couple of cops. Firefighters, however, work for the *fire* department. Our primary mission is to respond to and extinguish fires. The company officer who takes his crew members directly to the floor of alarm demonstrates a very weak mind-set and eventually will get someone injured or possibly even killed. Those members might like their officer in the short run because he makes their life easier by not having them stop the elevator two floors below and walk up. But believe me, they have little or no respect for him at all, simply based on the fact that he's not protecting them. Surprised is the last thing firefighters should be upon encountering smoke or fire at the floor of alarm.

The primary responsibility of a company officer is to keep the members under his command as safe as possible. This is not an easy proposition, based on the fact that firefighters operate in very unsafe conditions. After all, we're talking about going after a fire on the upper floors of a high-rise building. The firefighter safety and survival premise is based on the principles of developing, maintaining, and practicing good habits, plain and simple. Furthermore, those good habits are not just applied once in awhile, but *every time*, and at every operation. I firmly believe that the actions firefighters usually take, good habits or bad, are the same actions they will take when it really counts. Those who normally take the elevator directly up to the floor of alarm will most likely do it the night that there's really a fire on the other side of those elevator doors (fig. 1–2).

Fig. 1–1. Once the dominoes of complacency begin to fall, they are hard to stop, and can result in firefighter injuries and deaths.

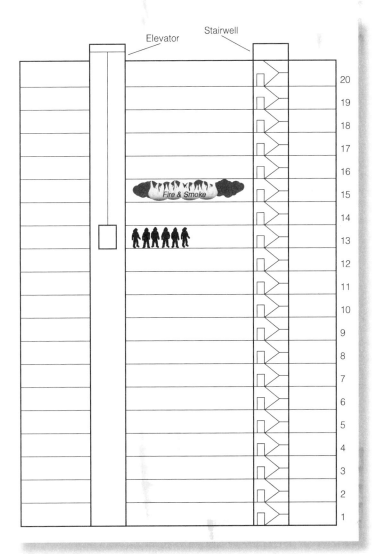

Fig. 1–2. Taking an elevator no closer than two floors below a reported fire floor or floor of alarm is a good habit, which equates to firefighter safety and survival.

The Firefighting Mind-Set

The primary subject of this book is directly related to fireground operations at high-rise as well as other low-rise, standpipe-equipped buildings. The central theme, however, that all the topics in this book will relate back to is that of a proper firefighting mind-set. Simply put, the most important characteristic that a firefighter possesses is his *mind-set*. A firefighter's mind-set is based on a compilation of factors, including his attitude, preparedness, and an overall belief that a serious event *will* happen. Maybe not today, certainly not every day, but eventually, someday, *it* will happen.

What is *it*? Well, *it* can be almost anything from a serious house fire, to a large commercial building fire. It might be the major collapse of a large, occupied auditorium or maybe even a plane crash. It can also be a serious fire in a high-rise or standpipe-equipped building. A truly dedicated firefighter is always prepared for the worst, and must be prepared for almost anything (fig. 1–3).

Over the years, I have found that most firefighters are familiar with the basics of what has to be, and what should be, done at a given event. As human beings, however, we sometimes cut out some of the elements of a comprehensive operation due to complacency and laziness. For example, even the newest firefighter who reads this book won't be surprised that I recommend exercising serious discipline when using elevators. One of the basic rules of high-rise operations is to never take an elevator directly to a reported fire floor or *floor of alarm*. This is nothing new, just good, basic, common-sense procedures. Have you, or has any one you know, ever witnessed firefighters bending this rule, or totally disregarding it? The example of Engine 18, and most of the material contained in this book, relate back to this chapter, under the umbrella heading of "The Firefighting Mind-Set."

What if I were to say that many fire departments in the United States currently are equipped to deliver less water at a fire on the 20th floor of a high-rise building than would be acceptable

Fig. 1–3. The possibility of a serious fire in a high-rise or standpipe-equipped building exists in many large cities and small communities.

Fig. 1–4. Far too many fire departments are improperly equipped to fight high-rise fires with weapons more appropriately used to combat a vehicle fire.

at a modern vehicle fire? (fig. 1–4). It's absolutely true. Far too many fire departments have a high-rise/standpipe hose pack equipped with $1^{3}/_{4}$-inch hose (and, in some very frightening cases, $1^{1}/_{2}$-inch hose) with an automatic combination fog nozzle. This complement of weapons, supplied from a low-pressure standpipe system, especially one with built in, automatic pressure reducing valves (PRVs), and a flow of less than 100 GPM, possibly much less, is likely a reality. A fire department that sends its firefighters into high-rise buildings, especially large commercial high-rise buildings, with this equipment is on a collision course with disaster and tragedy (fig. 1–5).

The scope of the problem is much greater than the obvious tactical and equipment error—it also encompasses an overall, widespread organizational complacency and lack of proper mind-set from the top down. When was the last time your department flow-tested its high-rise/standpipe nozzles at the low discharge pressures typical of most standpipe systems? Many nozzles, but specifically the automatic combination type, can truly be deceiving and may be a deadly choice for standpipe operations. It all starts at the top and the priority must be firefighting.

Fig. 1–5. A one-size-fits-all weapons package with 1^3/$_4$-inch or 1^1/$_2$-inch hose and a combination nozzle, often an automatic combination nozzle, is not designed for, and is certainly not safe to use, for high-rise firefighting.

The Full Service Fire Department

In today's fire service, most fire departments have expanded their role to reflect a so-called, "full service fire department," which includes all of those new services from EMS to Customer and Community Service. There is nothing wrong with expanding our services. After all, we are in the best position to provide such services. The problem has been the increased focus on all these new services at the expense of our primary service, firefighting.

Many new firefighters coming on the job today, and certainly over the past decade, have been inundated with an extremely damaging message. In fact, some academy recruits have even been told that, "If you're here to fight fires, go home, we don't do that anymore." While some politically-appointed people in leadership positions make these absurd statements, firefighter line-of-duty deaths continue to occur. Brave firefighters are killed while operating at a wide range of emergency incidents, including some of those allegedly rare events called structure fires.

A bizarre attempt to reshape the fire service has included visions of the fire department becoming more of a civic organization rather than an emergency services provider. All the while, a massive deterioration of the fire service has taken place. In some cases, fire departments have hired individuals who would make outstanding social workers, and trained them to be everything but firefighters. All of these destructive elements eventually add up and create a recipe for disaster.

We must be able to provide excellent emergency service, including fire suppression. In order to achieve that all-important goal, we must hire those physically and mentally prepared individuals capable of this excellence, and we must continually train them for their survival. All the other stuff is just that: other stuff.

For the past several years, many fire departments, and especially new firefighters on those departments, have been exposed to the widespread disease of complacency. At the heart of this disease are individuals, many unfortunately in leadership positions, that profess that "fires are down," and that if fire departments are to survive, they would have to change their names and provide other services. All the while, fires still continue to occur, and firefighters continue to be injured and killed, many times at these rare incidents called *fires*. Organizational complacency starts at the top and is one of the most deadly aspects of the modern fire service. Shame on those individuals who have led so many young and impressionable firefighters down the path of *killer complacency;* their conduct is irresponsible and completely reprehensible!

American Fire Service History

In order for today's firefighters to truly chart an appropriate course for their future, they must first fully understand their past. American fire service history is loaded with countless lessons that, if carefully studied and taken seriously, could save the lives of many future firefighters. At the center of our history is the question: Are fires really down, and if so, based on what measurement?

It wasn't until 1977 that factual statistics were first collected in the American fire service, through the National Fire Incident Reporting System (NFIRS). That's less than 30 years ago, and certainly doesn't give enough factual information to conduct a fair analysis of our entire past, which includes a couple hundred years of history. If we were to simply compare today's fire service with that of the 1960s and 1970s, we would obviously see a reduction in the total number of working fire incidents. But therein lies the problem: those who profess that fires are down are not taking into account all of the factors associated with that issue.

We must look closely and carefully at the entire history of the American fire service. Without factual statistics prior to 1977, we must seek out those individuals who actually lived through and experienced some of that fire service history themselves. Whenever I want a history lesson, I simply pick up the phone or sit down over lunch with my father, Pat McGrail (Division Chief, Retired, Denver Fire Department). During his 42 years on the front lines of the DFD, he lived much of the history that I am referring to and was also mentored by men who lived and experienced firsthand the many years of firefighting history before him.

In fact, my father tells me of a senior firefighter who showed him the ropes when he first came on the job. That fireman, George Girard, was one of my father's first mentors, and came on the DFD in 1913. This is a great testament to the importance of simply sitting down at the kitchen table of a firehouse and taking the time to listen to the senior man. The vicarious experience gained over a cup of coffee is at the soul of our very existence.

My career will hopefully take me past the year 2013. In that time, I will actively seek out and absorb as much information as possible from my many mentors, both within and outside of the DFD. I will also work hard to provide the critical guidance and mentoring for countless enthusiastic and dedicated firefighters who represent the future leadership of our beloved fire service. By making a commitment to pass on your experience, you can quickly see the connection between several generations of firefighters. The knowledge gained can span well over 100 years, and these historic lessons can become part of our long-term safety and survival.

In many cases, accurate statistics do not exist regarding the numbers of fires from throughout our history, but information can be garnered from individuals who lived and worked during the periods in question. The turbulent decades of the 1960s and 1970s are known to many in the fire service as the *war years*. In many cities, it sometimes seemed like everything was on fire at once. This was actually somewhat of an anomaly. Society was in flux in many ways during those turbulent years. Civil unrest, a very unpopular war in Vietnam, and other various phenomena of the times combined to result in an unusual and heavy workload for the American fire service. Some of the most senior members of today's fire service got in on the tail end of that tremendous workload. Therefore, having lived it, they sometimes base today's work on that period alone. But that simply doesn't tell the full story.

My father tells me that when he came on the DFD in 1955, a busy fire company was doing perhaps 500 to 750 calls per year (similar to the numbers of most large cities of that era). Many of those calls were, in fact, fire related or were actual working fires, but not all of them. Many firefighters of that era felt that a company would really be stretched thin if they got any busier than that. Today, a company that does less than 1,000 calls per year is referred to by many as a "retirement home." In fact, the busiest engine companies on the DFD, and in most large city fire departments across the country, are doing over 5,000 calls per year, and that includes many working fires.

Before the incredible peaks in fire activity that occurred in the turbulent '60s and '70s, the entire American fire service was significantly slower than it is today. Ironically, for most fire departments, those early years, with a much lower overall call volume, included a greater number of fire companies, and manpower levels that in most cases were at least twice what they are today.

Most fire departments are literally doing much more work, with fewer resources today, than ever before.

As we all know, today we have expanded our role significantly. Emergency Medical Services (EMS), Technical Rescue, Hazardous Materials, and of course, many fire departments use the term *Customer or Community Service* to describe some of the other things that they do. And guess what—we also respond to and operate at working fires. In fact, we respond to significantly more fires today than the vast majority of our forefathers, with, of course, the exception of those who worked during a period of approximately two decades in the latter part of the 20th century. The graph in figure 1–6 shows fire activity over the past 100 years. The figures before 1977 are estimated, while the statistics from 1977 up to the present are factual. Long-term projections in to the 21st century show an increase in fire activity.

Today, with the significant expansion of fire department–based EMS, statistics indicate that EMS represents the vast majority of our total call volume. In most cases, 50% or more of our total calls are EMS, whereas fire calls represent just a small fraction of that. However, when you truly examine the facts and figures, you quickly find that the statistics fall far short of an accurate representation of most fire departments, overall emergency response activity and workload. To begin with, the statistics include all of the EMS responses, but not all of the *fire-related* responses.

A large percentage of the total number of EMS responses are certainly not true emergencies. Anyone associated with the urban fire service can attest to the fact that many of these responses are the result of the widespread abuse of the 911 system. On the other hand, the statistics on fire responses are usually limited to those incidents that meet the definition of a true *working fire*. For the sake of statistical consistency, the fire service should count only the EMS workers or conversely, count all of the times that firefighters don their fire suppression equipment to investigate a possible fire or extinguish a small fire that had tremendous potential, and the numbers will be more realistic.

Furthermore, if the statistics for total work time were to be included, the comparison between fire duty and EMS would be much less lopsided. For example, a working fire incident necessitates a much greater overall work time than does a typical EMS working incident. Also, a much larger contingent of resources is required to bring a working fire incident to a safe and successful conclusion than the typical resources necessary for a working EMS incident. Both emergency medical services and fire suppression are important components to an

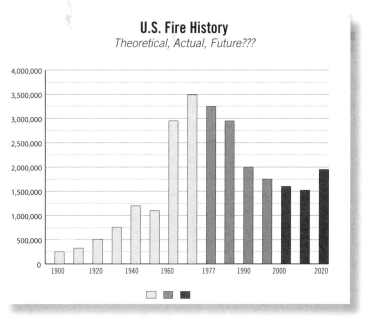

Fig. 1–6. Overall fire activity in the American fire service over the past 100 years

all-service organization. However, no organization can safely and successfully sacrifice the firefighting component for EMS.

I am not an anti-EMS person, or anti any emergency service for that matter. In fact, I consider good quality EMS to be an essential service provided by all modern fire departments. But one must keep in mind, for organizational success, it must still be the *Fire Department* with a *Division of EMS*. The landscape of the American fire service is littered with countless fire departments that have lost their ability to provide excellent fire suppression service, while significantly, and in some cases, solely focused on honing only their EMS and Customer or Community Service skills. When you ask the customers what they want, most will place at the top of their list the expectation of the fire department to put out their fire, save as much of their property as possible, and most importantly, rescue their loved ones from harm. Fire departments who barely manage to confine fires to the zip code of origin are not providing good customer service.

Even with solid statistical data indicating that the occurrences of fires today are down from that of the 1960s and 1970s, the fact is that there simply has never been a more dangerous time to be a firefighter than today. As the frequency of structure fires has decreased since the war years, the annual number of firefighter line-of-duty deaths (LODD) has remained relatively consistent at approximately 100 firefighters per year. Hence, the rate of LODD's per fire has actually increased. Even with all the fires that occurred in the '60s and '70s, one must keep in mind that some of the most serious fires in American history have occurred in the 21st century. Obviously, the most notable of all is the series of high-rise fires that occurred in the lower Manhattan section of New York City on September 11, 2001. That's right, those were high-rise fires. The collapse of the twin towers, both high-rise buildings, took place under fire conditions. The smaller, World Trade Center Building Seven, at 47 stories, would have been considered a very large high-rise building, probably the tallest in most cities. It is rarely talked about, but it, too, collapsed under extreme fire conditions. Nothing even close to this tragic event ever occurred in the war years, or at any other time prior, in the history of the American fire service.

Dangers of Modern Firefighting

Even if the events of 9/11 had never happened (and we all wish that were the case), there are still numerous significant factors that one must consider regarding the dangers associated with fighting fires in the 21st century. At the top of the long list of critical factors is, of course, the dreadful issue of terrorism. Terrorism is alive and well, and none of us can rest assured that we are immune from another attack. Whatever type of attack occurs, the soldiers on the frontlines of our homeland defense will be firefighters. Indeed, there has never been a more dangerous time to be a firefighter than today.

Serious danger lurks at even the most benign-looking apartment fire in a residential high-rise. Just a little over a month after 9/11, Captain Jay Jahnke, a 20-year veteran of the Houston Fire Department, died from injuries sustained while operating at a residential high-rise fire. His was not the first name placed on the granite stone after a tragic high-rise fire, and, unfortunately, won't be the last.

Several critical factors contribute to making this the most dangerous time in history to be a firefighter. Of those factors, some of the most notable are high Btu fuels, early detection systems, full personal protective equipment, and lightweight construction.

High Btu fuels

The significant role that plastics, synthetics, and other various high Btu fuels play in the modern fire environment cannot be underestimated. Today's fire environment has been described by many as containing *solid gasoline* as the primary fuel. Brother firefighter Mark Wesseldine (Firefighter Retired, FDNY Ladder Co. 58), makes a very interesting comparison as he mentors firefighters across the country. He asks them how many TVs they had in their homes growing up, at least the guys born before Generation X. Most, including Mark, attest to the fact that there was typically only one, if any TVs at all, in American homes in the 1950s, '60s, and '70s. He goes on to ask how many TVs we all own today, other than the one in the family room of modern homes.

His point is simple; most single-family dwellings of today are loaded with countless TV's, computer monitors, and other various Class B combustibles, the so-called *solid gasoline* of the modern fire environment. These types of combustibles are the primary factor in today's fires being significantly hotter, faster burning, and more difficult to extinguish, especially with small-diameter, low-flow handlines and nozzles.

How many TVs and computer monitors are on the 20th floor of most high-rise buildings? A small, 10-story residential high-rise, with perhaps only 10 apartments per floor could have close to 50 such devices on each floor. Between TV's and computer monitors, a large commercial high-rise might have several hundred on just one floor. That's a lot of fuel!

Early detection systems

I first heard this concept from our brother, Chief Vincent Dunn. He speaks of the dangers associated with today's smoke detectors and other various early detection devices. All of these components have been designed to, and have fortunately, saved many lives. As Chief Dunn says, these systems are also responsible for frequently placing firefighters at the scene sooner than ever before. Truly a double-edged sword, however, these early detection systems mean we can stop fires in their early stages, but they also increase a firefighter's chances of encountering a deadly flashover if the fire is caught in a growth phase.

Full personal protective equipment (PPE)

Full personal protective equipment is an incredible benefit for firefighters in terms of protection and prevention of injuries. The inexperienced or overconfident firefighter, however, may overestimate the protection and underestimate the heat conditions. This quickly leads to a false sense of security and can create a situation where firefighters literally place themselves in greater danger by getting in too deep without a proactive water application to cool the environment, or in some cases without the protection of a hoseline at all. PPE can sometimes mask the rapid development of high heat and potential flashover conditions.

As modern technology continues to advance the protective characteristics of our PPE, we must proactively educate and train our firefighters, especially the younger ones, to better understand fire behavior and the complex dangers of the modern fire environment.

Lightweight construction

Firefighters from any city, district, or jurisdiction need not look far to discover the sea of lightweight construction that is overtaking the landscape of many urban and suburban areas. This construction has already led to the deaths of too many firefighters. Our window of opportunity in these buildings is significantly shorter than that of our forefathers who operated in mostly ordinary and heavy timber-constructed buildings. And as we now know, the high-rise building, once thought to be very stable, with a long-term work environment under fire conditions is also home to lightweight construction. We all witnessed the collapse of two lightweight-constructed high-rise buildings that occurred on 9/11.

Building the Proper Mind-Set

Take a little closer look at this thing I call *mind-set*. You can't buy it at a store, and it isn't something that is easily acquired. A firefighting mind-set is created from a positive attitude and strong self-discipline. It is not a one-time-only achievement— it is true excellence, and, as we know, excellence is a continuous journey not just a one-time destination.

I like to use an analogy when referring to the firefighting mind-set. That analogy involves an automobile with a good old-fashioned standard transmission, nothing automatic about this. Picture yourself driving a vehicle with a standard transmission. You start out going up a street that has a slight incline. As you reach about 15 to 20 miles per hour, you shift from first gear, not to second, but fourth, and after letting the clutch out, the vehicle's engine labors to keep running. The car starts to bog down, the vehicle's speed slows, as well, and the engine RPM starts to slow down. With this slight incline, eventually the vehicle's engine stalls and the vehicle comes to a complete stop.

Only a very basic understanding of motor vehicles is necessary to understand what occurred in the previous scenario. Simply put, the automobile with a standard transmission is not designed to go directly from first gear to fourth gear. It's a process, starting with the lower gears and moving up to the higher gears, first, second, third, and then fourth. It all relates to vehicle speed and engine speed, and they must correlate.

I believe that we, as human beings, are similar to automobiles. We too have a process that we must go through in order to achieve top speed and performance. There is, in essence, a mental transmission with several gears, or steps toward top performance. We simply can't be expected to go from 0 to 60 by shifting from first gear to fourth. This mental transmission directly relates to ones mind-set. And as with the gears of a vehicle's transmission, the gears of mind-set each represent a process of slowly building up to top performance. These mental gears are each components of our overall preparation.

First gear: Daydreaming mind-set

✸ First gear is equal to a day-dreaming mind-set. This is not an acceptable mind-set for firefighters and represents complacency and laziness within the fire service. On-duty firefighters must be focused on providing emergency service and fire suppression, protecting themselves, and protecting the other members of their crew. The firefighter who is consumed with his day-off job or serious personal issues is not prepared for battle. His mind is not in the game and therefore, he should sit this one out until he's ready to go.

✸ The firefighter, or perhaps I should say employee, who answers his cell phone en route to a fire or other emergency is far from being in the proper mind-set. A fire department shouldn't need to establish rules regarding this common sense issue. The day-dreaming, or first gear mind-set, should be reserved for our off-duty hours.

Second gear: Conversational mind-set

Second gear is equal to a relaxed, conversation-type mind-set. You are alert, and awake, certainly concentrating on the task at hand, but that task might not necessarily be related to our daily operations as firefighters. Once again, this mind-set must be reserved for our off-duty hours. If it's real estate that they do on their day off, then they should do it on their day off, not at the firehouse, when their mind-set should be focused on firefighting and other emergency services.

Third gear: Preparation mind-set

Third gear is equal to a preparation-type mind-set. This is where you need to be while on duty. You are not just alert and concentrating, but specifically concentrating on items related to one's current responsibility. This level of mind-set includes all of the critical preparation components associated with fire suppression and other emergency services.

Fourth gear: Life-or-death mind-set

Fourth gear is equal to a life-or-death mind-set. This is where the rubber truly meets the road. You can only make a smooth transition into this mind-set, or fourth gear, if you have completed the requisite preparation and are already operating in a third gear, preparation mind-set.

Preparation

Ultimately, we are focusing most of our attention on the preparation, third gear mind-set. There are several, very important factors associated with a preparation mind-set. Each one is of critical importance, but by itself will not result in overall preparation. In a synergistic effect, each of these components collectively will lead to a third gear, preparation mind-set.

Mental fitness and preparedness

✸ This is by far, the most important component associated with the firefighting mind-set. Everything else associated with our mind-set is driven by our mental attitude. We've all heard the saying, "attitude is everything." Nothing could be more accurate than that. It's a positive, professional attitude that we are talking about, an attitude that serves as the foundation for everything we stand for as firefighters.

✸ At the front end of this mental attitude is your willingness and ability to develop and maintain an attitude of belief—a belief that today could be the day. That is, today could be the day that *it* happens. And as we said before, *it* could be almost anything, from a serious house fire with several kids trapped inside, to a fire on the 30th floor of a commercial high-rise building. You must develop and maintain the belief that *it* could happen and eventually *it* probably will. Most importantly, is the belief that we must be prepared for *it*, even though *it* certainly doesn't happen everyday.

✸ This critically important mental attitude is the exact opposite of what we must continually fight as human beings, that is, the deadly disease of complacency that is all around us. In addition to being deadly, complacency is also a very contagious disease. We have all probably suffered from it at one time or another, and may easily suffer from it occasionally in the future. That's where strict self-discipline comes into play.

✸ Complacency manifests itself in a wide variety of ways, but none more potent than the discounting of the potential of an emergency event before we even get on scene. An all-too-often heard sentiment is, "We get false alarms in that building all the time; it's probably nothing." That statement, especially when it is spoken by a company or chief officer is at the very heart of the deadly disease of complacency, and only serves to greatly exacerbate a much more serious organizational complacency. That attitude serves as the catalyst of what can often result in a tragic conclusion to a life-changing event. Go online and read the long list of fire service Line of Duty Death reports, and see for yourself, but remember, it's not always complacency on the part of the firefighter who gets injured or killed. More often, it's a broader organizational complacency that leads to such tragedies.

✱ Once again, as imperfect human beings, we have all suffered from complacency at one time or another, to a greater or lesser extent. But as firefighters, it is imperative that we develop and maintain a strong mental attitude to continually fight complacency and all of its associated baggage. This is not always easy to do, but it is step one toward a safe and successful conclusion to an emergency event.

✱ Start developing this positive mental attitude by avoiding being influenced by self-righteous, condescending, negative individuals who speak from ignorance every time they open their mouths. They don't believe that anything serious is ever going to happen, and will be the first to profess "it's probably a false alarm" long before the fire department is on scene. Their complacency will be the first domino to fall on a potentially tragic path. Once they realize that it is actually a fire or some other bona fide emergency event, they're surprised by it, which always leads to panic and poor decision making. Furthermore, these initial actions of panic and playing catch-up set in motion all of the elements for failure. Worst of all, a firefighter may be injured, or possibly killed, due to this fundamental lack of leadership. You must believe. Believe that *it* can happen; *it* has happened before, and *it* will happen again. Preparation is our only option. Remember, you control your own attitude!

Physical fitness and preparedness

This aspect of the firefighting mind-set requires that firefighters spend quality time in the gym. Not just a couple times a month or once a week, but a schedule that includes a fitness routine at least two or three times a week. This routine must include both cardiovascular fitness as well as muscular endurance.

It would seem obvious that firefighters should be in top physical condition. After all, our job is an extremely demanding occupation, both physically and mentally. But our own statistics relating to firefighter line of duty injuries and deaths indicates that we sometimes fall short of being physically fit. This is not to mean that you must spend every waking hour—morning, noon, and night—working out, taking vitamins and other various supplements. Like our fireground operations, our physical conditioning should be based on common sense and good balance.

I am talking about making an effort to maintain a well-balanced diet. This happens to be an area where I fall far too short. My own lack of self-discipline, especially when it comes to anything chocolate, requires additional stair-climbing every week, just to counter the effects of my favorite comfort foods. In addition to working on diet, walking, jogging, running, or perhaps climbing stairs two or three times a week, coupled with lifting some weights, will do the trick. Five or 10 pounds overweight is not critical, but it can quickly lead to being 25 or 50 pounds overweight.

✱ I know of several people employed by fire departments who would be categorized by most medical practitioners as morbidly obese. They are a danger to themselves and to others on the fireground. When they are either unwilling or unable to initiate a program of physical fitness, the organization must get involved, even if it is not the most politically correct action a fire administration can take. Ignoring the issue of grossly overweight or physically incapable employees will eventually lead to major problems, perhaps even tragedy. If these individuals do not step up to the department's physical fitness standards, they should be removed before tragedy occurs. It really doesn't matter how much time, effort, and money has been invested in them already, they are an unacceptable high risk to themselves and their fellow firefighters. Anyone who shows up for the first day of instruction at a fire academy in poor physical condition is either ignoring or not educated on the academy's physical standards and should be advised immediately on the conditions for acceptance. It's been my experience that those marginal recruits, rarely get any better, and usually become nightmare problems for the next 25 years.

All fire departments should have annual, mandatory physical fitness tests to evaluate every member. This evaluation should require the completion of a combat-style test in a reasonable time period. Those who can't complete such a test successfully need to be reassigned to a light-duty position, before they hurt themselves or someone else.

At a serious high-rise building fire, those who can't perform will probably become part of the problem. The obese individual employed by the fire department might make it up ten flights of stairs before the shortness of breath and chest pain become unbearable. At that point, it's not just one individual, but an entire crew or fire company that will no longer be part of the solution. That crew along with perhaps several others will be needed to carry the extremely heavy member down and out, and thus attempt to rescue him. Be part of the solution, not part of the problem. Get to the gym, and stay physically prepared for battle.

Continual training and development

In addition to believing that *it* can happen, part of the preparation involves additional brain development. This aspect of the firefighting mind-set requires that firefighters dedicate themselves to continual training and development throughout their career. For some individuals, the last time they picked up a book was in the fire academy. They did what was necessary to pass the exams, graduate from the academy, and complete their probationary period. Once they are off probation, that's it, no more studying, no more books, no more mental challenges, no more work, period!

I admire those fire service professionals who dedicate themselves to continued development. It might mean reading a periodical such as *Fire Engineering* magazine, logging onto a fire service Web site like FireNuggets.com, studying a new book, attending a college class, a two-day seminar, or perhaps even spending a week at the Fire Department Instructors Conference (FDIC). You see a lot of different faces at conferences like the FDIC. I admire most the gray-haired, senior members in attendance. They probably have the most experience, and yet they are still looking to develop and improve professionally.

Unfortunately, some individuals reach a certain level or rank and believe that they now have all the answers. Brother Firefighter, Tracy Raynor, Deputy Chief with the Boise Fire Department, refers to it as the "white shirt syndrome." Too many times we see individuals get their white shirt and gold badge, then they feel that's it, they know it all now. As Brother Raynor points out, these individuals are a disservice to the entire profession.

There is something new on this job everyday. One only needs to keep an open mind. And that's what this component of mind-set is all about, keeping an open mind and having a willingness and desire to learn, even after ten, twenty, or perhaps thirty years on the job. I believe that the learning curve in the fire service is actually a straight line upward (fig. 1–7)—it never really levels off, unless you let it. This component has a lot to do with spending some time in the library, so to speak. Read, study, listen, and learn: it is a continual process that never really ends.

Continual training and development runs the gamut from books to burn buildings. The critical importance of hands-on training cannot be overstated! It seems, however, that for some individuals in the fire service, the word "training" is like a dirty, four-letter word. Some officers have used it as a disciplinary tool, threatening to take the men out for training if they didn't straighten up. What a shame.

Training is not something to be ashamed of or used in a negative manner. It is one of the most important tools available to us. Firefighter safety and survival is directly related to the type and amount of training you receive. From the top of the chain of command, to the bottom, the responsibility for training falls on everyone. However, the company officer is by far in the best position to ensure that a healthy diet of training is established and maintained at the company level. Get out and train. Do it as often as possible, and make the training good, quality training, with a direct application to your operations. Train on the evolutions you perform frequently, and of course, train on the rare and unusual operations as well (fig. 1–8).

Be creative, and you can turn anything into a short training drill. After an EMS call in a high-rise multiple dwelling, stop at a standpipe outlet on a lower floor. Look for a pressure reducing valve (PRV). If one exists, take a close look, and conduct a short, but productive company drill. Keep the

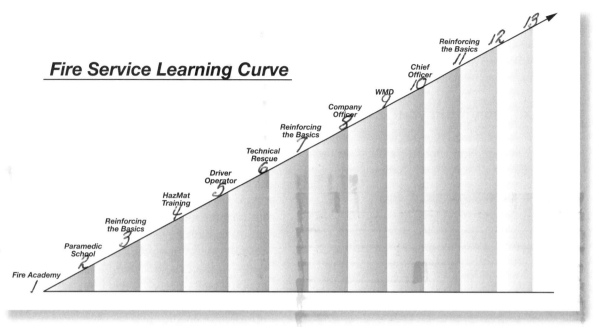

Fig. 1–7. The fire service learning curve is actually a straight line upward

Fig. 1–8 A good strong company officer provides the leadership to mentor and develop the members under his command, starting with daily company level training drills.

training positive and productive, and never use it as a tool for punishment or discipline. Train daily. As Firefighter Jim McCormick of the Indianapolis Fire Department often says, "Training saves lives, firefighters."

Check and maintain your equipment

This aspect of the firefighting mind-set requires that firefighters spend the time necessary at the beginning of every shift to thoroughly check and maintain all of their equipment. Start with your own personal protective equipment (PPE) with a special focus on the SCBA. For example, a strong firefighting mind-set is demonstrated by the firefighter who doesn't accept 4,000 psi as an indication of a full air cylinder. A full, high pressure air cylinder contains 4,500 psi of air. Changing out your air cylinder, even if it means just an additional three, or four hundred pounds of air, could make a big difference during your shift. It could save your life or the life of someone you are sworn to protect (fig. 1–9).

Full air cylinders on the SCBA harnesses and full spares in the compartment are basic, yet critically important, tenets of a strong mind-set. Hose loads and preconnected lines must be loaded properly and neatly, and ready to be stretched. High-rise/standpipe hose packs must be properly assembled, tight, compact, and ready to go. Nozzles, adapters, saws, forcible entry/exit tools, and so on, must all be thoroughly checked. Everything needed to do battle with the enemy must be in proper working order and ready to go. Your tools will take care of you if you take care of them.

Last but not least, take every opportunity to train with these tools. If it has been a while since your last fire, put your gear on, including the SCBA, go on air, and go to work. I prefer to climb stairs at a high-rise building with full gear, SCBA, and on air. It certainly is a good refresher course for living inside the bubble that is our SCBA face piece.

Just getting on a treadmill or simply climbing the stairs in the firehouse is very beneficial. Find out how long you can actually work in an SCBA with the air from a standard one-half hour rated air cylinder. I'm good for less than 20 minutes before I completely empty a one-half hour rated air cylinder. Translate that to real life, and you have quality information regarding how much time you or an average firefighter can operate on the 20th floor of a high-rise building. Second alarm, third alarm, etc., there's a good reason behind calling for help early.

Listen to the radio

I can walk into any firehouse, in any city, and quickly determine if it is a good fire company or not. One of the things I look for is whether or not the firefighters have a fire radio or monitor in their main living area, kitchen, dayroom, living room, and so on. Is there a radio or monitor (even a

Fig. 1–9. Firefighters should ensure that they have a completely full air cylinder, even if it means taking five minutes to change it out at the beginning of their shift.

portable radio from the rig) that receives citywide radio transmissions? Is it turned on? Can the men hear it, and are they listening to it? The good fire companies have a radio or monitor, it's on, and they're listening.

No, that's not the only thing they're doing. Yes, they have the hockey game on TV, they're talking and joking with one another, and they are preparing the evening meal. But they all get quiet and listen up when they hear a fire coming in over the radio, and that's when they start to think, are we due on the second alarm? What is that building? What's the best response route this time of day? When the Incident Commander strikes the second alarm, these guys are on the rig and out the door. That's called initiative, and those who demonstrate it are to be commended, not disciplined (fig. 1–10).

Even if your company isn't due, and won't go on the second or even the third alarm, listening to the radio, and switching over to the tactical channel is an excellent vicarious experience. You can learn a lot listening to the other guys. The first thing I do at the start of every work shift is to grab my portable radio, turn it on, and switch to channel one, which receives all citywide radio traffic from the DFD. I also turn on the radio monitor at my desk in the office. Unless I am actively involved in an incident of my own, if a fire or emergency event takes place in the city of Denver during my work shift, I will know about it, and hear it transpire from start to finish. It's not going to come as a surprise to me watching the ten o'clock news that a fire occurred in a district on the other side of the city.

I have been listening to the radio throughout my career. Over the years, especially as a young firefighter, I have heard some people made sarcastic comments about this practice. One such statement from the negative complacent types is, "they'll call us if they need us." Well, they might, or they might not. What if there's a failure with the communication system? That does happen from time to time. Mr. Complacent will be surprised when his company is called, and will be playing catch-up from the word go.

The good fire company has been listening to the development of the incident and has been gathering critical size-up information since the incident began. They know well ahead of time who's there, what they have, where they're operating, and what progress is or isn't being made. Ultimately, when called upon, this company will certainly be part of the solution rather than part of the problem. Listen to the radio. It's very interesting at times.

Get prepared, stay prepared

In a nutshell, the firefighting mind-set ultimately boils down to getting prepared, staying prepared, and believing that all your hard work

Fig. 1–10. Firefighters with a strong mind-set listen to the radio and are aware of the emergency operations that are occurring, some of which they might be called upon to assist.

and preparation will pay off some night, perhaps on the 21st floor of an occupied high-rise building, at a time when it's desperately needed. It boils down to quality time in the gym, the library, on the drill ground, and on the apparatus floor. All of these mind-set components add up to preparation. There is a synergistic effect when one has done the requisite preparation and all of the components are brought together. Good habits, plain and simple, translate to good, effective, safe, and successful operations every time.

This practice is truly hard work. It's just not that easy maintaining this high level of preparedness. We are talking about being 100% committed to a mind-set, and truly to a way of life, that is designed to pay dividends for only a very small fraction of our overall emergency responses. It is the extremely high-risk, low-frequency events that this mind-set has us prepared for, and may appear to be unnecessary for all the other non-working events. Without a crystal ball, we simply can't predict when the real serious events will occur. So, as for mind-set, you either have it, or you don't.

Summary

Unfortunately, we simply cannot control all of the countless factors associated with the dangers of modern firefighting. However, there is a long list of factors that we truly do have control over. Maintaining good physical fitness, achieving proper mind-set, continual training and development, checking and maintaining equipment, exercising proper elevator discipline all represent good, basic habits.

Habit, as defined in *Merriam-Webster's New Collegiate Dictionary* is, "an acquired mode of behavior that has become nearly or completely involuntary."[1] Good habits are not something we do occasionally, but every time. Remember, it's in your hands.

References

1. *Merriam-Webster's New Collegiate Dictionary*, 11th Edition, s.v. "Habit."

CHAPTER 2

The High-Rise Building

Introduction

As the title implies, *high-rise firefighting and standpipe operations* is the central topic of this book. So to start, I briefly examine the history of high-rise buildings, and provide a practical definition and explanation of what it means for a building to be classified as a *high-rise*. Keep in mind that the strategy and tactics, as well as the actual operating procedures, recommended and designed specifically for high-rise buildings can and should also be applied to many low-rise buildings that are equipped with standpipes.

Throughout the book I make reference to *high-rise buildings*, which also include other standpipe-equipped buildings, from low-rise buildings, to the so-called "wide-rise" buildings. Many wide-rises, such as large warehouses, may only be one story in height, but still have as much square footage as many high-rise buildings. Wide-rise buildings are sometimes compared to a high-rise lying on its side.

History of High-Rise Buildings

Tall structures have been around for centuries, dating back to the great pyramids of Egypt. The largest of the pyramids, the Great Pyramid of Cheops was built around the year 2560 BC to serve as a tomb for the Egyptian pharaoh, Khufu, when he died. At 481 feet, it ranked as the tallest structure on earth for more than 43 centuries, and was only surpassed in height in the 19th century.[1] This pyramid is taller than most modern 40-story high-rise buildings. Throughout history numerous tall structures have been built, from the pyramids to gothic cathedrals with tall spires. The Ulm Cathedral in Ulm, Germany, rises 528 feet in the air and has the tallest church spire in the world.[2]

The first high-rise buildings were built around the middle of the nineteenth century. According to the Institute of Real Estate Management, "The modern office building was created in response to rapid population increases and industrialization that occurred during the late nineteenth century."[3] From a construction standpoint, the catalysts for the construction of the first high-rise buildings were two major developments—the invention of the elevator and the use of steel frames. In 1853, Elisha Graves Otis invented the first so-called *safety lift*, known today as an elevator.[4] In the 1870s, the use of steel became available to form the framework of a high-rise structure. This replaced a much weaker combination of cast iron and wood. Buildings

Fig. 2–1 The very strong steel framework of the modern high-rise building was one of the advancements that made high-rise construction possible.

High-Rise Defined

In the National Fire Protection Association (NFPA) Fire Investigation Report of the 1993 World Trade Center bombing, a *mega-high-rise* is described as "a large, tall (greater than 50 stories), densely populated structure where emergency evacuation is difficult or impractical. They are further characterized in that the ordinary fuels which they contain may result in rapid fire growth, development, and spread because of their geometric arrangement, and in extensive smoke spread throughout the structure which threatens occupants in remote areas from the fire origin. Further, the time required for firefighters to establish effective firefighting operations can be extensive, because of the vertical arrangement of the structure."[6]

The actual definition of a high-rise can be based on one's own interpretation; in other words, "high" is in the eye of the beholder. Remember, the first high-rise buildings, which were constructed over 100 years ago, were thought of by many back then to be skyscrapers, even at heights as low as 10 stories. However, for our purposes, the eye of the beholder belongs to the fire service and a multistory building, of any size, can negatively affect our operations.

Specific definitions of high-rise may vary from city to city and depend on which building code was utilized by the authority having jurisdiction during the construction of a given building. The NFPA 101, *Life Safety Code*, which has been widely adopted across the American fire service, defines a high-rise building as "a building greater than 75 feet in height, where the building height is measured from the lowest level of fire department vehicle access to the floor of the highest occupiable story."[7] In addition, the new, widely adopted *International Building Code* uses the same measurement to define a high-rise. The 75-feet criterion, seems to be the most commonly used definition for a high-rise building. So, for all practical purposes, at approximately 9 to 15 feet per floor, generally a building can be classified as a high-rise if it is roughly five to eight stories or more in height.

could now be built to much greater heights due to the strength of the steel structural framework and the ability to safely transport occupants to the upper floors via elevators (fig. 2–1 and fig. 2–2).

The term *skyscraper* has been used over the years to describe buildings of varying sizes and heights. "Originally, a building of ten stories was considered a skyscraper, but today the word is seldom used to describe a building of less than fifty stories."[5] The vast majority of high-rise buildings in most cities are less than 50 stories tall.

However, strictly from a fire service point of view, whether a building is described as a high-rise has a lot to do with the capabilities of the fire department charged with protecting that building. Ultimately, any building, where all or even just portions of the building are beyond the reach of ground-based firefighting equipment, specifically aerial ladders, could and should be considered a high-rise.

Practical Definition of High-Rise

A simple definition, which can certainly apply to most fire departments, is that a high-rise building is any building in which the highest level, or portions thereof, cannot be reached by fire department ladders. In most medium to large cities, the longest aerial ladder is typically somewhere in the 100-foot range. For example, in Denver, we operate with both tower ladders and straight aerial ladders. Our longest aerial ladders are heavy duty, 110-foot rear mount apparatus, with the newest two aerial ladders at 105 feet, and our newest four tower ladders at 100 feet. Two of the newest aerial apparatus on the DFD are Truck Co. 16, which is equipped with a 105-foot straight-stick aerial, and Tower Ladder Co. 22, which has a 100-foot ladder tower aerial (fig. 2–3). Both are new, state-of-the art aerial apparatus, but both have their limitations. Although longer aerial ladders do exist, especially in other countries, historically, 100-foot has been the most common and standard length of aerial ladder across the American fire service.

Many smaller fire departments, however, are not equipped with aerial ladders at all. A 65-foot telesquirt, or even a simple 35-foot extension ladder carried on the side of a pumper, might be the longest piece of equipment available to gain access to the upper levels of a multistory building from the exterior in many jurisdictions.

Usable Aerial or Ground Ladder Length

You can establish what the usable length of aerial ladder would be for various buildings in your response district. Residential high-rise

Fig. 2–2 The development of a safe elevator was a critical advancement that allowed for widespread high-rise construction starting in the late 1800's.

Fig. 2–3 DFD Truck Co. 16's 105-foot straight-stick aerial ladder

buildings are typically shorter per floor than commercial high-rise buildings. The average distance of a floor in a residential high-rise building is approximately 9 to 10 feet per floor, compared to the typical commercial high-rise building with floors from 12 to 15 feet or more per floor (fig. 2–4 and fig. 2–5).

In addition, numerous other considerations and factors are associated with the effective use of aerial and ground ladders at multistory buildings. Fire department access around the perimeter of the building is a significant factor. Items such as setbacks, landscaping, trees, utility poles, overhead electrical power and high tension wires, cars parked on the street and around the perimeter of the building, underground parking garages with structures not rated to carry the weight of a typical fire apparatus, and other buildings located nearby can all negatively effect the usable length of an aerial or ground ladder. Furthermore, the actual spotting of the aerial ladders must take into consideration the climbing angle and various safety factors associated with the operation of that specific type of apparatus.

Fig. 2–4 A typical residential high-rise building

Rear-mount aerial ladders are very common throughout the fire service. A good, well trained, and experienced aerial apparatus operator is a significant key to successful apparatus placement on the fireground. Taking the extra time to back a rear-mount aerial into a good position has many times been the difference between reaching the objective or not. Hastily nosing a rear-mount aerial into position in the heat of battle might be faster, but with most of these type apparatus, close to 50% of the usable ladder length will be lost before it is even brought out of the bed. In the long run, a lot more time will be required to rebed the ladder, retract the outriggers, and reposition the apparatus. A victim hanging out a seventh floor window might not have that much time.

Fig. 2–5 A typical commercial high-rise building

Generally speaking, a well-positioned 100-foot aerial or tower apparatus will be capable of reaching at least the seventh or eighth floor of a typical residential high-rise, and the fifth or sixth floor of a typical commercial high-rise. As previously addressed, there are obviously numerous other factors associated with this, and actual usable length will vary depending on the collective impact of all those factors. With good access around the perimeter of a building, and a properly placed and spotted aerial apparatus, a 100-foot aerial ladder might actually reach well beyond what is expected based on past experience. I have seen DFD Truck Co. 16 access the 11th floor of a residential high-rise with their 105-foot ladder. However, the flip side of that involves a building with severe access problems and poor apparatus placement, which will likely yield results far short of those anticipated and will negatively affect the operation.

Ultimately, fire departments, and the individual ladder companies within those departments must preplan their respective districts in order to determine exactly how far their aerial will reach at a specific multistory building. The key here is, don't wait for the screams of a frantic civilian or chief officer to find out that your aerial will not reach an objective. Take your rig out into the district and practice spotting actual buildings. Make an attempt to do it as often as possible, and work toward the goal of identifying your apparatus's capability at every multistory building in your first due response district.

This is a tall order in many cities and jurisdictions, especially those with a large number of densely located multistory and high-rise buildings. In midtown Manhattan for example, Ladder Companies 2 and 4 of the FDNY have tens of hundreds of multistory and high-rise buildings in their first due areas. For these extremely busy fire companies, it would literally take years to spot their apparatus at every single building in their district. However, practicing on as many buildings as possible will help the apparatus operator develop excellent skill that can be used to accurately judge distance and effectively place the aerial at most buildings. Many fire companies seize the opportunity to practice at the end of calls that do not turn into working incidents. Vigilantly practicing and developing skills is one of the things that truly differentiates a great truck company from those that are mediocre or poor at best.

High-Rise Building Size

In many jurisdictions, terms such as *low-rise*, *medium-rise*, and *high-rise* are used to narrow the definition of various buildings. However, in this book, we will simply use two terms to describe building height, *high-rise* (buildings taller than 75 feet), and *low-rise* (buildings shorter than 75 feet). In most cities the average high-rise building is generally less than 50 stories in height. Buildings over 50 stories, the so-called *mega-high-rise*, should be considered the very largest of the high-rise family. And of course, a few cities have those monster high-rise buildings that are over 100 stories tall. Examples include the Empire State Building in New York City, and the Sears Tower in Chicago (fig. 2–6).

Fig. 2–6 The Empire State Building in New York City is considered by many to be the "grandaddy" of all high-rise buildings.

High-Rise Building Fireground Strategy and Tactics

This discussion regarding size and height of high-rise buildings, and defining high-rise, is so important because there are specific strategies and tactics that are routinely used by most fire departments at low-rise building fires that cannot be utilized during true high-rise operations. Those strategies and tactics are logistical in nature, and are directly associated with our inability to reach above the level of our longest aerial device.

Specifically, those strategies and tactics involve rescue and fire suppression from the exterior of the building. Trapped occupants, above a fire, and out of the reach of our typical, longest aerial ladder (110-foot) have only a few, very disturbing options. Those options include, dying via smoke inhalation, burning to death, or jumping to a certain death.

Every one of us recalls the graphic video footage as the events of September 11, 2001, unfolded on national television. News commentators were expressing their disbelief at seeing countless people jump to their deaths from the upper floors of the World Trade Center Towers. Only firefighters who have experienced the violent atmosphere and high heat conditions of an interior structural fire could begin to understand the rational behind such a deadly decision. In fact, at a January 2005 fire in New York City, six firefighters were forced to jump from the fourth floor of a Bronx tenement in order to escape the heat of a rapidly developing fire. Two of the men were killed, and four others were severely injured.

When those in need of rescue have run out of options, as rescuers we too have only a few viable options. Fire department rescue operations at fires in multistory buildings include the following strategic and tactical rescue options, listed from best to worst:

1. Rescue via the interior stairs during primary search operations.
2. Rescue from the exterior via fire escape (if the building is so equipped).
3. Rescue from the exterior via tower ladder apparatus.
4. Rescue from the exterior via aerial ladder apparatus.
5. Rescue from the exterior via ground ladders.
6. Rescue from the exterior via rope.
7. Rescue from the exterior, roof, via aircraft, specifically helicopter.

Only options 1, 6, and 7 are possible during true high-rise operations. Because we can't reach these victims via ground, aerial, or tower ladders, we must attempt to rescue them via the interior of the building, by way of interior stairs. This, obviously, is much easier said than done, and in many cases downright impossible, especially above the fire, such as the circumstances that faced our FDNY brothers on 9/11. For our high-rise operations, we must attempt to establish and maintain an evacuation stair, but if it is destroyed by some sort of terrorist attack such as an explosion, or if it is heavily contaminated by high heat and heavy smoke conditions, that evacuation stair is no longer a safe and viable logistical tool for rescue purposes.

The implementation and use of rope for rescuing occupants trapped above a fire is also an alternative which must sometimes be implemented. However, this is a rescue that requires an extremely high level of technical expertise to complete. Furthermore, the time needed to effect such a rescue, is in most cases significant, and with a fire burning below, the logistics of such an operation are very complicated and extremely dangerous. Few fire departments have trained and equipped members to handle such a difficult and dangerous task. A proactive training program and equipment acquisition that addresses this type of rescue operation should be initiated by all fire departments with high-rise and multistory buildings in their communities.

The use of aircraft, specifically helicopters, to rescue trapped occupants from the roofs of high-rise buildings is also an option. This operation has been seen before and documented at numerous high-rise fires in the past. However, most fire departments, including some of the largest, do not have an internal air support resource. Furthermore, most fire departments have not identified such an external resource, nor have they established the necessary relationships needed to rely on such a resource at short notice.

Fire departments without an internal air support resource must look outside the organization for help. Some possible resources include police departments, medical air ambulance, news media, and the military. I specifically address this issue in chapter 18.

The strategy of implementing a defensive operating mode and applying water via master streams from the exterior of a building is also eliminated when the fire is burning above the effective reach of master streams. The level of effective master stream reach is slightly higher than that of the actual ladder length, when the aerial is equipped with a master stream device at the tip. Obviously, reaching and rescuing a trapped person beyond the actual length of the ladder will, in most cases, be impossible. However, because the effective reach of the stream from an aerial-mounted master stream device typically goes up to 100 feet or more (horizontally) beyond the tip of the ladder or tower ladder bucket, the use of exterior master streams gives us a slightly greater logistical advantage.

A great example of using the master stream strategy was the Philadelphia Fire Department's use of portable master stream appliances at the One Meridian Plaza fire. Philadelphia firefighters encountered almost every imaginable problem at this fire. As true problem solvers, one of their many tactical solutions involved accessing adjacent high-rise buildings, supplying the standpipe systems in those buildings, and using portable master streams to attack the fire from the upper floors of these neighboring high-rises. This is obviously a worst-case scenario, but a very creative alternative for attacking a high-rise fire beyond the reach of conventional aerial apparatus. This operation would only be possible in areas where adjacent high-rise buildings are within a reasonable distance to the fire building. It is specifically applicable in those very large cities, with numerous high-rise buildings located close together.

High-Rise Building Construction

The construction of most high-rise buildings has historically been considered to be a significant advantage during fireground operations. Unfortunately, the tragic events of 9/11 changed that forever. Never before had there been such a catastrophic collapse of a high-rise building under fire conditions. Extremely serious and destructive high-rise building fires in both Los Angeles and Philadelphia burned for several hours before being brought under control. Isolated collapses, the *spalling*, or breaking up, of concrete, and the twisting of steel structural components did occur in the Philadelphia fire, but there was never a complete collapse.

As for the First Interstate Bank Building fire in Los Angeles, the United States Fire Administration (USFA) report contained the following information.:"Analysis revealed that no significant damaged occurred to major structural elements. Part of this credit must go to the unusually good application of fire resisting materials on support members".[8] During the post-incident review of this fire, LAFD leadership believed that had the fireproofing on structural components been compromised, a collapse would likely have occurred. This position was also later supported by the National Institute of Standards and Technology (NIST), which emphasizes the importance of building and fire code inspectors conducting comprehensive inspections, during and prior to completion of construction, to identify any area of fire protection deficiencies and order corrections immediately.

It is also important to note that the post-incident investigation revealed that the intensity of the fire at the First Interstate Bank building was so significant that heat passed through the steel decking on which the 4-inch concrete slab was poured. This actually caused additional vertical fire spread when the flammable adhesive mastic for the carpet began to vaporize. Carpet on the floors above, where the fire was finally stopped, showed many areas of significant charring from this internal, vertical fire spread.

With regard to the successful outcome of the First Interstate Bank Building fire, I would also add that the extremely aggressive attack by the LAFD was a significant factor in limiting the structural damage sustained by this building. However, one LAFD chief officer observed in a post-incident critique of the operation that their resources were so taxed, had the fire not been stopped when it was, it might have burned out of control, perhaps to the roof, which would have probably led to a major structural collapse.

Much discussion has occurred since 9/11 regarding fire department operations in high-rise buildings. We now have the knowledge that a high-rise building can, in fact, completely and catastrophically collapse, but we must keep in mind that this will hopefully be the exception rather than the rule. Understanding that the world we live in today presents the possibility that what occurred in New York City could occur in any city at anytime. A plane, missile, bomb, or any other massive assault on a high-rise building might yield the same devastating results. In those extreme situations, our procedures will have to be altered, limiting the operational time within the structure—no different than what we face on a daily basis with lightweight construction in dwellings and small commercial buildings.

A similar terrorist attack on a high-rise building in any city will certainly present major strategic and tactical challenges, and countless difficult decisions. Personally, I felt that the FDNY's operation on that terrible day was nothing short of spectacular. Obviously, FDNY or any other fire department will likely alter their future operations to include a specific operational timeline for these rare, but ever-possible, overwhelming fires in high-rise buildings. Once again, with regard to an operational timeline, our high-rise operation will not be unlike what we have been doing for years, especially most recently, at our more frequent fires in low-rise buildings assembled using lightweight structural components.

With that in mind, fireground commanders and chief officers must continually evaluate the structural integrity of the high-rise building. If a high probability of saving a human life exists, an offensive operating mode is mandatory. However, our window of opportunity to get in, search, evacuate, remove, and/or rescue any victims, and still potentially attack and extinguish the fire will be much narrower.

Even with the knowledge that a high-rise building can collapse under fire conditions, it is still generally one of the best buildings to fight a fire in—that is, in terms of the relative resistance to collapse when exposed to fire conditions. Once again, the focus here revolves around those high-rise fires that occur on a daily basis across the American fire service that we would consider controllable with one or two handlines.

Most high-rise buildings are built using protected noncombustible and fire-resistive components. The noncombustible steel components are protected and made to be fire resistive with the application of a sprayed-on, fire-resistive material. If quality materials and judicious application processes are used, the main structural components in these buildings can have a two- to four-hour fire-resistive rating under typical fire conditions. As for a plane, bomb, missile, or any other extremely brutal assault on a high-rise building, producing very high Btus, all bets are off. Remember, Tower One of World Trade Center, the second building to be attacked, collapsed in less than one hour.

Four Generations of High-Rise Construction

The construction of high-rise buildings has continually evolved over the years, from what was typically a very heavy and sturdy structure to today's lightweight superstructures. Most authorities agree that there are basically three different generations of high-rise construction. I believe that in our post-9/11 world, we are now seeing a fourth generation of high-rise construction emerge, one with an extremely strong emphasis on protecting the building and its occupants from the wide range of threats posed by terrorism.

First generation high-rise construction

The first high-rise buildings, constructed during the late nineteenth and early twentieth centuries, had extremely heavy load-bearing exterior walls, built of brick or stone. These exterior walls had to be exceptionally thick in order to carry the overall load of the entire building. A good example of this is the Monadnock Building in Chicago. Built from 1889–1891, it stands 16 stories tall and is considered the tallest load-bearing structure in the world. Its walls are six feet thick at the base in order to carry the entire load of the upper floors.

In this first generation of high-rise construction, most authorities agreed that a 12-inch thick wall was needed to support the first floor of a building, with an additional four inches added to the thickness of that load-bearing wall to support each additional story. Needless to say, this limited practical building height to about 10 stories. The Monadnock Building was a notable exception.[9]

Many of these first generation high-rise buildings were also characterized by cast iron facades. In addition, many buildings of this era, had unprotected cast iron columns, with wrought iron and steel beams. Floors were made of wood and were usually the weak link during a fire, which led to many a collapse.

Unprotected vertical openings were common to buildings built during this first generation of high-rise construction. It was not uncommon to see open stairways, light wells, and elevator shafts.[10] None of these features would be allowed in modern high-rise construction (fig. 2–7 and fig. 2–8).

Second generation high-rise construction

The second generation of high-rise buildings (often referred to as *pre-World War II* high-rise construction) gave rise to protected steel frame structures. These structures were characterized by their use of fire-resistive assemblies, compartmentalization, shaft enclosures, and the use of noncombustible materials. These buildings included masonry enclosures for all metal structural members, and vertical shafts were enclosed in masonry and tile. Floors were concrete on brick or hollow tile arches; floor areas were subdivided and combustible materials were limited.[11]

Fig. 2–7 The Equitable Building in downtown Denver, built in 1892, is considered a first-generation high-rise building.

Pre-World War II buildings are considered by most to be excellent buildings. Relatively small floor areas were common in these buildings, due to a need for close access to natural light and ventilation.[12] The invention of fluorescent lighting and the use of central heating, ventilation, and air conditioning (HVAC) systems led to the larger, more open-floor areas of modern high-rise buildings. The floors of these pre-World War II buildings were well-segregated fire areas. Probably the best example of this generation of high-rise building construction is the Empire State Building located in New York City.

Fig. 2–8 An example of an open stair in the Longworth House Office Building located in Washington, D.C.

Third generation high-rise construction

The third generation of high-rise buildings (often referred to as *post-World War II* high-rise construction) gave rise to a much lighter-weight construction: steel frame work with core type construction, typically a center core, is the norm. An exterior curtain wall of glass and or some sort of stone is common. The curtain wall is fastened to the steel frame structure in a manner that leaves a gap between the structural frame and the curtain wall. Fire-stopping material must be placed in this gap, or vertical fire extension will occur. This has been a significant factor at several major high-rise fires.

As for vertical fire extension via the curtain wall gap, even if this area has fire stopping, the method and type of fire stopping used is important. In many buildings, compromises in code requirements have allowed for a so-called *friction fit* of fire-stopping material, such as rock wool, to be placed in the curtain wall gap, but to be held in place only by friction. If this material becomes wet, for example from a water leak, or from condensation from perimeter air conditioning units, the material can quickly absorb the water, making it heavy enough to fall out of place. If this happens, the curtain wall gap once again is unprotected, thus creating an open path for vertical fire spread.

The use of a central heating, ventilation, and air conditioning (HVAC) system within the building is very significant, specifically with regard to air movement and extension of smoke and fire. Because of the common HVAC system, these third generation buildings are, in essence, sealed and can be defined as windowless. This fact tremendously exacerbates the *stack effect* within these buildings, a concept addressed in detail, in chapter 15.

A lack of fire towers in modern high-rise buildings and the evolution of less remote scissor stairs have contributed to egress problems for occupants. Most modern high-rise buildings have all of the stairwells located within the core area, which means they are no longer remote from one another. Because of their location within the core, these stairwells are much more likely to become contaminated with smoke and products of combustion during a serious fire.

Although many modern high-rises provide for a pressurization of the stairwells, this too has been known to fail. Stair pressurization can quickly be lost when multiple doors to the stairwell are being opened simultaneously during a mass evacuation. Furthermore, at least one stairwell, the *attack stairwell*, can become contaminated with smoke during fire department operations. Every effort should be made to ensure that occupants are not inside this stairwell, especially above the fire floor, prior to commencing attack operations. However, even with a thorough search of the attack stairwell above the fire floor prior to opening up and initiating attack, occupants might still enter the stairwell and can become trapped above the fire floor.

Many of the so-called "mega-high-rise" structures have unique construction features that had to be developed in order to reach such tremendous heights. Buildings such as the former World Trade Center Towers and the Sears Tower in Chicago use a tubular construction design. This feature places most of the load bearing on exterior walls in the form of numerous columns, and thus requires less structural support in the interior. In addition, truss construction, (lightweight trusses at the World Trade Center) is used to support the floor assemblies. The evolution of these construction features could be considered an additional generation of high-rise construction. For our purposes, we'll consider it a subcategory of the third generation, but, nevertheless, one that should be identified and closely examined.

Fourth generation high-rise construction

A fourth generation of high-rise building construction has begun to emerge. This will likely come to be known as *post-9/11* high-rise construction. With this type of construction, we will begin to see a resurrection of many of the features typical in the very heavy, well-built buildings of the second generation. Professor Glenn Corbett of John Jay College in New York City says that these newer high-rises will feature much more "robust" construction features[13] (fig. 2–9). Future high-rise buildings will likely not contain the lightweight bar joist trusses that were used throughout the World Trade Center Towers for the floor support assemblies. Furthermore, these newer high-rise buildings will probably be built with a means of egress, specifically a stairwell, and a means to assist fire departments with logistical operations, such as an elevator, both contained within a much more heavily fortified enclosure that is significantly resistant to fire, smoke, explosion, and collapse.

The implementation of such robust construction features will be costly, and thus likely to be met with resistance from many in the building construction community. However, the continued work of those dedicated to fire and life safety will hopefully win the war against those solely concerned about profit margins. In this post-9/11 era, we can no longer live with a complacent mind-set and hope that such an attack won't happen again.

High-Rise Building Features

High-rise buildings have numerous characteristics and features that make them unique from other buildings. The modern high-rise is a product of evolution and contains a myriad of sophisticated systems and components. Each of these various systems and components may play a role, either positive or negative, in the fire department's overall operation. Many systems and components are required by building and fire codes, and are designed to assist the fire department with operations in the event of a fire or other emergency. Furthermore, most are considered life safety systems, designed to protect the occupants.

Specific high-rise construction concerns

Understanding the construction features of a high-rise building is paramount to our success on the fireground. One of the first things that must be determined at a high-rise building fire or emergency is the total number of floors in the building, both above grade and below. The specifics of how these levels are identified and labeled in the building is also of critical importance to operating firefighters. Such items as whether there is a floor numbered 13; whether there any other intermediate levels such as mezzanines and or concourse levels; where the mechanical levels are located and if there are multiple mechanical levels. Is there a penthouse level, and is it an occupied tenant space or is it another mechanical level?

Immediately upon arrival at a high-rise fire or other emergency event, there are going to be a whole lot more questions than answers regarding the high-rise building and its construction features. Therefore, the Incident Commander (IC) must assign various teams with group leaders and division supervisors to conduct an ongoing reconnaissance. On the Denver Fire Department (DFD), we utilize a position called the *systems officer* (or, simply, *systems*), who, as the event evolves, and grows in complexity, can and should become a *systems group*. The systems officer is usually positioned inside the fire command center (FCC), and is responsible for monitoring the various building systems, including the fire alarm panel, the elevators, and the HVAC system. The systems officer is an excellent resource for the IC to gather critical building data and information.

The IC must also attempt to quickly determine the age of the high-rise and thus which generation of high-rise construction it falls into. Of particular importance is whether the building construction features any lightweight components, particularly any truss assemblies. This information will be a

Fig. 2–9 The new World Trade Center Building Seven, built with much more robust construction features, including a very thick application of fire-resistive material protecting steel structural components.

significant benefit throughout the operation, and will help guide the IC as to how long firefighters can operate inside the building with reasonable safety.

In addition, identifying whether the high-rise building in question is of core type construction, and if so, what type of core it uses. Is it a center core or some other type? Does the building have a central HVAC system, or not? Much of this information can be gathered ahead of time via a quality pre-fire plan. However, many large cities have numerous high-rise buildings; time constraints and a lack of necessary resources make collecting accurate information and assembling a quality pre-fire plan for each building, nearly impossible. There are private companies that develop quality pre-fire plans; however, these can be very expensive, and unfortunately, only a small fraction of high-rise buildings have them.

Another issue is that very few high-rise buildings have features that remain static. The modern high-rise building is dynamic, and many features and systems inside the building are changed and upgraded over time. A pre-fire plan is only as good as the accuracy of the information contained within and how up-to-date it is. Because of constant changes, the need to establish and maintain a comprehensive ongoing size-up of conditions at all operations becomes imperative. However, if a pre-fire plan is available, it may be a valuable resource if all the information is verified by firefighters during fire department operations.

If the building is of core-type construction, firefighters need to know what structural components make up that core, concrete or steel. What are the structural components of the building's frame: concrete, steel, or a combination of both? What are the structural components of the floors? Typically, in many modern high-rise buildings, the floors are made up of concrete poured in place over a metal decking. Most important is, what structural components are carrying the load of the floors? Is it protected steel I-beams and other large dimension structural steel components? Or is it being held up by a truss assembly, and is that truss assembly lightweight, such as the construction at the World Trade Center Towers? What type of fire proofing material has been used to protect the steel structural components?

The roof construction is also a significant concern. What are the structural components of the roof? What type of equipment is on the roof, and what kind of load is being carried by these structural components? Of particular concern with regard to the roof is, can it hold the weight of a helicopter? Is there enough room on the roof to land a helicopter or is there a dedicated helipad? Are there any other obstructions on the roof, such as antennas, wires, and other elevated equipment that could interfere with air support operations?

Is there a parapet wall on the roof, and how high is it? A higher wall will provide some additional safety for firefighters operating on the roof, and civilians who may potentially be evacuated to this location. Roofs with shorter parapet walls provide little or no protection, and firefighters or civilians could easily fall from the roof, especially if visibility is compromised due to smoke or during nighttime operations. Wind conditions at the roof level are also significant—unpredictable, and dangerous winds can literally blow an unsuspecting person off the roof.

Fire detection and protection systems

The types of fire detection and protection systems within the building must be identified. What types of devices are utilized—smoke detectors, heat detectors, rate of rise, manual pull stations, and so on. Most importantly, where are the alarm(s) located in the building? An early indication of multiple alarms, from different locations means there is a higher probability that there is an actual fire. Any time there are multiple locations of alarm, the lowest level of alarm should be investigated first.

The fire protection systems and equipment used to combat a fire must be identified. Is there a properly working standpipe? Is there only one standpipe riser in one stairwell, or are there multiple standpipe risers, one in each stairwell? In some cities, the standpipe is not in the stairwell at all, and hose valve connections are located out on the floors. What is the type and class of the standpipe system(s)? Are there any components associated with the standpipe system that might negatively effect our operations? Items such as pressure regulating devices are of critical concern and must be identified as early in the operation as possible.

Is the building equipped with a sprinkler system? What type of sprinkler system is it? Is there full sprinkler protection throughout the building, only partial sprinkler protection, or no sprinkler protection at all? If there is a sprinkler system, is it operating properly? Perhaps the fire has occurred during a maintenance period, and the sprinkler system is temporarily out of service, which leads to the next questions, can the sprinkler system be placed back in service and how long will that take?

The good news is that most fire and building codes require that any new high-rise construction include full sprinkler protection throughout the building, as well as an automatic wet, Class I standpipe system (which is defined in chapter 4). However, as I have previously stated, we must be prepared for potential fires in the countless buildings that were built prior to the stricter fire and building codes now in existence. Anticipating and preparing for the failure of the fire protection systems in any building, including modern high-rises, will keep us from being caught off guard when it happens; and it will.

3 Water supply

Having a comprehensive knowledge and understanding of the built-in fire protection systems within a high-rise building is essential during a fire or emergency. However, one must also take into consideration the water supply necessary to support such systems. Specifically, how is water supplied to the fire protection systems? Is this water supply designed to support the flow necessary to control a large fire? Water supply equipment can include gravity tanks, city water mains, fire pumps, and various other components.

The DFD, like most fire departments, will always set up and prepare to supply the built-in fire protection systems via the fire department connection. This is true, even if the system is a fully automatic, wet system. Hooking up to the fire department connection (FDC), and preparing to supply water to the building's fire protection system, serves as a backup water supply in the event that the primary water supply or any other components within the building system, such as fire pumps, fail to operate properly. The subject of fireground water supply at high-rise, and standpipe-equipped buildings is discussed further in chapter 12.

4 Stairs

The type and number of stairs within a high-rise building must be identified early. Typically, most high-rise buildings have at least two stairwells that run the entire vertical distance of the building, although generally, only one of the two provides roof access. In newer, modern high-rise buildings, these primary stairwells are usually pressurized when specific zones of the fire alarm system are activated. In some high-rise buildings, primarily older buildings, one of the stairwells might be a fire/smoke tower.

Whether the stair configuration is of the *return* or *scissor* type is also of critical importance. What areas of the building can be accessed by the stairs and the location of the fire in relation to the stairs are all important pieces of information. With this information, fire departments can determine which stairwell to use for the purpose of attack and which one should be used for evacuation. An attack stair and an evacuation stair must be designated early in the event, preferably by the first companies and/or fire attack group to arrive.

In addition, many high-rise buildings also have access stairs within some of the tenant spaces of the building. These stairs provide convenience for tenants traveling between the floors. The existence of tenant stairs is a critical concern that must be identified and communicated to all operating units early in the operation.. A comprehensive discussion of stairs and all the associated issues and problems is given in chapter 7.

5 Elevators

All high-rise buildings will have elevators. A small, residential high-rise might have only two elevators serving the entire building. However, a large, commercial high-rise building will have numerous elevators contained in multiple banks, with each bank of elevators designed to serve a specific area of the building. For example, the mega-high-rise Sears Tower in Chicago contains 104 elevators.

As stated early in this chapter, it was the development of the *safety lift*, now known as the elevator, that played a significant role in the rapid increase in the construction of high-rise buildings. Today, elevators are a fact of modern life. Millions of people use elevators every day, coming to and from work. Firefighters use them as well. In fact, they are a very valuable tool for us. However, their use must be based on strong discipline.

Identifying the number and type of elevators within a high-rise building is critical for operational success. Their location, and whether they have been recalled, or if they can be recalled, as well as if there are possibly any stalled elevators with trapped occupants, must be determined at the outset of an emergency event in a high-rise building.

A thorough and comprehensive discussion of elevators and their use during fireground operations is in chapter 6.

6 Heating ventilation and air conditioning (HVAC) systems

Most modern high-rise buildings are equipped with very sophisticated ventilation systems for the purposes of heating and air conditioning. In addition, it is typical to find many buildings equipped with ventilation systems designed to control the movement of air, and specifically smoke, in the event of a fire within the modern high-rise building.

Both the first and second generation of high-rise buildings relied on horizontal ventilation via windows that could be opened. Because of that, these older high-rise buildings typically presented less complicated problems with regard to removing smoke during fires.

The post-World War II, third generation high-rise is in essence a *sealed* building, sometimes even being referred to as a *windowless* building, not because of a lack of windows, but because the windows are generally not designed to be opened or easily broken out. When a fire occurs in a third generation high-rise building, smoke can travel throughout the building, often being found in areas very remote from the fire's origin. A building's HVAC system can have a positive effect on the control of smoke, or it can have a very negative effect with potentially tragic consequences.

The subject of ventilation in high-rise buildings is extremely complicated and loaded with many gray areas. It has been the smoke, not the fire, that was the culprit responsible for countless injuries and deaths at most high-rise fires. Directly related to that is the issue of ventilation, and how smoke and air travel within a typical high-rise. The entire subject is covered in more detail in chapter 15.

7 Utilities

In a typical high-rise building, utilities such as electricity, water, steam, and sometimes natural gas all play a significant role in the daily operations of a given building. Each of these can also play a significant role during a fire. Controlling utilities at a typical fire in a single-family dwelling or small commercial building is often a simple matter, and frequently completed with little effort. In a high-rise, the sheer size of the building and the complicated distribution and network of the various necessary utilities make control a very difficult operation that must be completed with finesse.

It isn't just a simple matter of cutting the gas and electric, as is typically done at our more common operations in smaller buildings. In a high-rise building, those utilities cannot simply be eliminated, because we rely on many of the building's systems to assist us during our operations. Electrical service is essential in order to power critical building features such as elevators, fire pumps, and lighting. Loss of any of these can exacerbate the long list of problems already facing firefighting forces at a serious high-rise fire.

Salvage operations on the fireground are typically considered to be post-control issues. However, at a serious high-rise fire, resources may be needed early in the event to set up protection for critical building systems and utilities. Most notably would be the electricity, as fire companies might need to be assigned to use salvage covers and other equipment to keep water from reaching electrical rooms and vaults.

Fire Extension and Smoke Spread

The vertical design of a high-rise building makes it particularly susceptible to rapid extension of fire and smoke throughout the building. Firefighters must fully understand and be prepared to deal with extension of fire and smoke spread to many areas, including those remote from the point of origin.

Fire extension in a high-rise building can occur in a variety of ways. However, the most common avenues of fire spread are via:

1. Poke-through construction
2. Curtain wall gap
3. Auto exposure

1. Poke-through construction

Examples of poke-through construction can be found in all multistory buildings, from a two-story house to a mega-high-rise. Simply put, all of the utilities that provide the necessary services to make multistory buildings inhabitable have to somehow reach all of these areas. That means floors from one level to the next must be penetrated in order to run plumbing, electrical and telephone wiring and conduits, HVAC supply and return air ducts, elevator hoist-ways and stairwells, and of course, in high-rise and standpipe-equipped buildings, the fire protection systems, specifically the standpipe (fig. 2–10).

When these various components penetrate a floor, there is usually a much bigger opening than is needed to run the utility. That's okay, but after the utility has been run, a gap exists—sometimes a large gap—that allows for vertical fire and smoke spread. This gap must be filled in with a fire-resistive material, typically fire-resistive putty for the smaller gaps. This is another critical area that diligent building and fire code inspectors must carefully scrutinize in order to identify and eliminate potential problems before they become an issue.

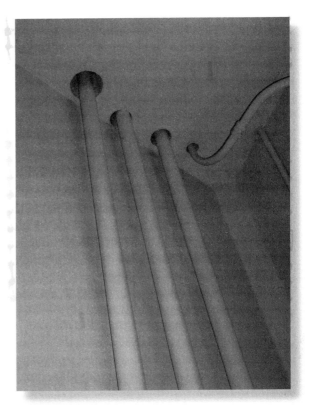

Fig. 2–10 Gaps left after poke-through construction methods become avenues of vertical fire and smoke spread.

Keep in mind, this vertical fire extension is not limited to small, utility poke-throughs. It could be a large, open stairwell, typical of older generation high-rise buildings. A significant fire that spreads out onto the fire floor, will seek this path of least resistance, sometimes driven by a strong stack effect, which could spread fire and smoke not just to one floor above, but several, if not all the floors, above the floor of origin.

2. Curtain wall gap

Many modern high-rise buildings utilize a core-type construction, with an exterior curtain wall of glass, aluminum, and or stone, or a combination of these. In some cases, as previously discussed in this chapter, there is a gap left between the main structural frame and the exterior curtain wall that can be an avenue of vertical fire spread. Of particular concern is whether there is a gap between the structure and the curtain wall, and if that gap has been properly filled in with fire-resistive material (fig. 2–11).

An unprotected curtain wall gap can lead to rapid fire extension to the floor above, creating a situation where operating forces might not ever get ahead of it. This has been a primary source of fire spread at many serious high-rise building fires. Identifying if there is, in fact, a curtain wall gap, and whether it is properly fire stopped, should be high on the list of size-up information. A reconnaissance team could be assigned to check

Fig. 2–11 The gap left between the building structure and the curtain wall must be filled with a fire-resistive material to prevent vertical fire spread.

one or more lower floors to examine the perimeter, in an attempt to visually inspect for a curtain wall gap and determine what type of fire stopping, if any, has been used.

Auto exposure

In a serious high-rise fire, once the fire has broken exterior windows and is allowed to vent freely to the exterior, rapid fire extension upward can now occur via *auto exposure*. The generic term for auto exposure is *lapping*. The fire, once again, is taking the path of least resistance, and moving in a natural direction, upward. Due to significant radiant heat and or direct flame contact, window glass, even strong and thick window glass typical of most high-rise buildings, will eventually fail. When it does, there will be an immediate and rapid extension of fire onto the floor above.

Of particular concern to operating forces, is the *spandrel distance* on the exterior of a building. That is the distance between floors that, in most cases, affords some protection from auto exposure. Of course, the greater the spandrel distance, the greater the protection from fire extension via auto exposure. For example, a high-rise building with large windows that span from close to the floor up to the ceiling is an indication that there will be a very small spandrel distance on the exterior curtain wall between floors. On the other hand, smaller windows, with some sort of noncombustible structural component forming the curtain wall between the windows, will buy us much more time and provide a greater protection from auto exposure.

In many buildings, especially high-rise multiple dwellings, exterior balconies provide significant protection from fire venting out windows from below (fig. 2–12). Balconies create somewhat of a vertical and horizontal spandrel distance, and significantly reduce the threat of vertical fire extension via auto exposure. That's not to say that it is no longer a threat—there is still going to be tremendous heat impinging on the underside of the balcony above, and fire traveling up the open areas between balconies, which will eventually lead to vertical fire extension. However, this type of construction has been to our advantage at numerous high-rise multiple dwelling fires in the city of Denver. Whenever I see balconies at a high-rise multiple dwelling fire, I am relieved, because this will generally slow down the vertical exterior fire spread, and in some situations, it also provides a reasonably safe (temporary) refuge area for some occupants, although certainly not directly above the fire.

Smoke and fire can also travel via open shaft-ways, such as open stairs, elevator hoist-ways, and other vertical arteries in the building. The network of supply and return air ducts used to move air throughout the building can also contribute to deadly smoke extension, often remote from the fire area. However, properly operating fire dampers are designed to stop fire and smoke extension within these HVAC ducts.

Regardless of the actual size of a building, a significant fire in any high-rise building can be a considerable challenge for any fire department. However, the larger the building the more complex the operation will likely become. The two most critical factors associated with operations in multistory buildings are logistics and exposures.

Fires that occur on lower floors can present some logistical advantages, in that we can typically get to the fire quicker and hopefully stop it from extending upward. However, the exposure concern

Fig. 2–12 Exterior balconies, typical of many high-rise multiple dwellings, provide significant protection from vertical fire spread due to auto exposure.

is generally a more compounding problem. The fire that occurs on a lower floor will potentially involve one, or several exposures in the form of floors above. All of the areas above will have to be addressed as soon as possible. Searches for any victims and identifying the locations of fire and smoke extension must be completed quickly. If this happens to be 10, 20, 30, or more floors, the manpower required for such an operation will be significant.

With that said, it is generally considered easier to deal with a fire that occurs on an upper floor, even if there is a significant logistical problem, such as an inability to use elevators. It will definitely take longer to get there and to get into operation, but the exposure concern is lowered, and thus it lessens the overall long-term problems.

Summary

Ultimately, it's the size that makes the high-rise building a much more formidable opponent than most other buildings. The bigger and taller the building, the more exposures there will be that must be searched and protected. And the bigger and taller the building, the more logistical problems we face.

These very large and complicated buildings require an extremely large commitment of resources from the outset if control of a given fire is going to be achieved. It is not unrealistic to prepare for, and anticipate using, a minimum of 100 firefighters at even a small fire in a high-rise building. That figure would apply to confined fires, in compartmentalized buildings, such as an apartment fire in a high-rise, multiple dwelling. Once the fire involves the public hallway, or fires that occur in the large, uncompartmentalized areas of a commercial high-rise building, the amount of resources needed to control such events grows exponentially.

In fact, the numbers will be far beyond what most fire departments will be capable of providing. Only the very largest can quickly muster three, four, and five hundred firefighters to operate at a major campaign fire in a high-rise building. Most fire departments must proactively establish automatic and mutual aid agreements with surrounding communities before a serious high-rise fire occurs. Failure to do so is tantamount to defeat before the battle even begins.

References

1. *Encyclopedia Britannica*, 2006. "Pyramids of Giza."

2. Dupre, Judith, 1996. *Skyscrapers, A History of the World's Most Famous and Important Skyscrapers*. New York: Black Dog and Leventhal. pp. 18–19.

3. Craighead, Geoff. 2003. *High-Rise Security And life Safety* Second Edition. Woburn, MA: Elsevier Science. p. 4.

4. Ibid. p. 2.

5. Ibid. p. 2 and Sonder, Ben. 1999. *Skyscrapers*, p.II.

6. Craighead, p. 45, and Isner, MS and Klem TJ. 1993, NFPA fire report, p. 55. *High-Rise Security And life Safety*, Geoff Craighead, 2003.

7. Craighead (Sonders).

8. High-Rise Security And life Safety, Geoff Craighead, 2003 (NFPA Fire Report, Michael S. Isner and Thomas J. Klen, 1993).

9. NFPA, Life Safety Code, 101. (Section 3.3.25.6, 2000), 7.

10. United States Fire Administration (USFA) Report, First Interstate Bank Building Fire, Los Angeles, 1988.

11. Craighead, Geoff. *High-Rise Security And life Safety*. 2003 (Institute of Real Estate Management 1985), p. 3.

12. Brannigan, Francis L. *Building Construction for the Fire Service, Third Edition*. p. 454 .

13. High Rise | Fire & Life Safety, John T. O'Hagan, 1977, p. 145.

14. Brannigan p. 458-459.

15. Interview with Professor Glenn Corbett, John Jay College, New York City, 2004.

CHAPTER 3

Low-Rise and Other Standpipe-Equipped Buildings

The strategies and tactics outlined in this book are primarily related to our operations at fires in *high-rise* buildings. However, you must not lose sight of the fact that in most cities, there are countless other buildings and structures that might not meet the fire and building code criteria necessary to be classified as high-rise, but nevertheless may have many of the typical characteristics and components found in a high-rise building. Specifically, we are talking about those non-high-rise buildings and structures that are equipped with a standpipe system.

Low-Rise Buildings

It is important to keep in mind that countless large buildings in many jurisdictions reach just short of 75 feet, sometimes by a mere few inches (fig. 3–1). This is not a mistake, but in many cases, these buildings were specifically designed that way by architects and building engineers under pressure to save money by reducing construction costs. This frequently used loophole allows designers to

Fig. 3–1 This large multiple dwelling, just short of 75 feet tall, is classified as a low-rise by definition, rather than a high-rise.

build a relatively large structure, but avoid the stricter fire code requirements, which would have mandated the installation of a more comprehensive fire protection system, specifically, an automatic sprinkler system throughout the building.

In many jurisdictions today, most modern fire and building codes currently in use require full sprinkler protection throughout, even for most large buildings below 75 feet. Unfortunately, most jurisdictions have not passed laws requiring the retrofitting of buildings in this category. So once again, fire protection rests with firefighters and manual firefighting.

Ultimately, we also must not overlook the significant number of buildings in many cities that have standpipes, but are not considered high-rise buildings, and not just the large, multistory, 60- to 70-foot structures. For example, generally buildings four or more stories tall, along with large shopping malls, large warehouses, and many other occupancies, all require standpipes according to most modern fire codes.

Fireground operations in these types of buildings are not, in many cases significantly different than that of a true high-rise operation, specifically in the multistory buildings below 75 feet. Here the logistical and exposure problems are not as compounded as they are in a true high-rise, but nevertheless these problems do exist, and must be addressed. A fire on the fifth floor of a six-story multiple dwelling at two o'clock in the morning presents many of the same very serious life safety issues as a true high-rise would. For most fire departments, the strategies and tactics at this fire will include the use of an interior standpipe system.

Low-rise multiple dwellings

There are many low-rise, multiple dwellings in many jurisdictions, so the occurrence of fires in these buildings is not uncommon. Many of these are equipped with standpipe systems. For example, in Denver, the fire and building codes used in the past required that all such buildings that were four or more stories tall be equipped with a standpipe system.

As you might imagine, a loophole also exists with these buildings. There are countless large, "four-level," multiple dwelling buildings in my city where the lowest level was built as a garden level (fig. 3–2). This was done specifically to receive a classification of not a four-story building, but a

Fig. 3–2 This large, multistory, multiple dwelling, with four levels, is considered a three-story building due to the lowest level being a garden level, and, therefore, it does not have a standpipe.

three-story building; therefore, a standpipe would not be required. Many of these buildings require very long hoseline stretches for engine companies, and in most situations, a typical long preconnect (300 feet on the DFD) will not reach a remote fire in an apartment on the upper floors. Engine companies should train frequently to hone their skills on long hoseline stretches from a static hose bed.

In situations like this, it is imperative that engine companies resist the temptation to extend a 1¾-inch preconnect with additional 1¾-inch hose. Preconnected 1¾-inch attack lines that are too long, (anything over 300 feet) require excessive engine pressure in order to overcome the friction loss in the line and supply an adequate volume of water. A 300-foot stretch of 1¾-inch should be the maximum length. Engine companies must practice stretching 2½-inch hose from a main, static hose bed to supply working lengths of 1¾-inch attack line for operations in these low-rise, non-standpipe-equipped, multiple dwelling buildings that require unusually long hose stretches.

For those low-rise multiple dwellings equipped with standpipe systems, fire departments can consider using these systems to gain a tactical advantage. Obviously, speed is the primary advantage associated with using a standpipe system, but there are many things to consider.

First, is the system a dry or wet system? Dry systems require time to fill and may be hiding other serious issues, such as open hose valves in remote locations of the building that can only be detected after the system is charged with water. Even with a reliable wet system, hand stretching an attack line from the pumper apparatus may, in many situations, be a safer, faster, and more effective operation.

Our key to success is to preplan buildings whenever possible and flow test the standpipe system. If the system is deemed reliable, then by all means use it when appropriate. Strict adherence to the basics, including always hooking up on the floor below the fire, is essential. Some buildings, in fact most in my city, have standpipe hose valve connections in cabinets on the floors rather than in the stairwell, creating a tremendous temptation to hook up on the fire floor during the heat of battle. For safety purposes, the handline must come from the floor below the fire. If the fire is on the second floor, first floor, street level, or any level below grade, the standpipe system should be bypassed and a line stretched from the pumper apparatus outside the building. In these situations, the standpipe system should still be supplied with water as soon as possible, so that lines can be utilized for possible extension as necessary and appropriate, but the first handline should come from outside the building.

Shopping malls

Large shopping malls are often equipped with a standpipe system. These buildings are usually very large and hand stretching an attack line from a pumper located outside the building is often times very difficult. The standpipe system in these buildings can usually be described as a *horizontal standpipe*. Because of that, its use, although sometimes necessary, can also be very dangerous. This is due to the simple fact that we cannot usually approach and attack from below. Quite simply, there may not be a safe position in the interior of the building from which to begin the attack. Because it is safer, fires in malls may require a more time-consuming and labor-intensive handline stretch from the outside.

Company officers must weigh their options, and let circumstances dictate procedures. Every effort should be made to initiate the attack from an area that is remote from the fire, clear of smoke and heat, and that has adequate visibility. In those situations where an interior standpipe hose valve on the same level as the fire must be used, fire companies should use some sort of method to identify an escape route for members operating on the attack line. Specifically, a rope tag line leading to the outside should be attached to the hose where it is connected to the standpipe hose valve. If conditions change, and visibility at the standpipe outlet is lost, members must have a secondary method in place for emergency egress. Many firefighters have been injured and killed when they attempted to exit a fire area by following a handline out, only to be taken to an interior hose valve location. Disoriented in the heavy smoke, these firefighters were unable to locate the actual exit point, and many have died only a few feet from the exit.

Graphic and tragic evidence of this was the line-of-duty death of Captain Jay Jahnke of the Houston Fire Department. Captain Jahnke died on the fire floor inside a high-rise multiple dwelling. The attack line he attempted to follow to safety was attached to a hose valve inside a cabinet located in the public hallway on the fire floor. Captain Jahnke was a very experienced fire officer from a large and busy fire department. This underscores the fact that under heavy smoke and heat conditions, an umbilical cord to safety is essential, even for the most experienced among us.

Ultimately, the decision to use a horizontal standpipe inside a large building, such as a shopping mall, must be taken very seriously, and made only after careful deliberation. Fire companies must establish methods, and frequently practice hand stretching hoselines from static hose beds long distances, so that in the event that the use of a horizontal standpipe is too dangerous, they will have a backup plan in place. Ideally, stretching from the pumper at fires in buildings with horizontal standpipes should become the primary method—it is the safest.

Warehouses

Numerous non-high-rise, but large-area buildings are found in many communities, which are often standpipe-equipped, even though they don't have multiple stories above grade. Large area warehouses are a good example. My friend and brother firefighter, Dennis Patti, captain with the Ontario, California, fire department, introduced me to the term *wide-rise*. Wide-rise is used by Ontario firefighters to describe the hundreds of warehouse buildings they protect. Referring to these as *wide rises*, implies that they are like a high-rise lying on its side. The logistical and exposure issues in these buildings are horizontal rather than vertical (fig 3–3). The inability to set up an interior staging area and hook up handlines to the standpipe from uninvolved areas below the fire, and then initiate attack from below make these operations particularly dangerous.

The standpipes in many of these buildings are usually Class II, designed specifically for occupant use (fig. 3–4). Some of these warehouse buildings are equipped with Class III systems, designed for use by both occupants and fire department personnel. When operating in buildings that are

Fig. 3–3 This fire occurred in a large warehouse, buildings that, due to their massive size, are often referred to a wide rises.

Fig. 3–4 A typical Class II standpipe system in a building such as this Home Depot is not designed to provide adequate fire flow, and should not be used by fire department personnel.

equipped with Class II standpipe systems, fire department personnel should not use these Class II systems, as they are not designed to deliver the fire flow necessary for safe and effective firefighting. Furthermore, it is not recommended to use the Class II systems either, as this requires connecting the attack line inside the fire building, which will not serve to lead firefighters to safety outside. Hand stretching attack lines from the outside, although logistically more difficult in most situations, provides for an umbilical cord to safety, guiding firefighters to the exterior in heavy smoke conditions. In addition, an appropriate fire flow can be achieved using fire department pumpers, hoselines, and nozzles, ensuring that firefighting forces have an effective fire stream to safely and effectively attack the fire.

Ultimately, success at these type operations rests heavily on the truck company personnel, who must locate the seat of the fire and guide engine companies into the area. Furthermore, the main entrance might not be the best access to the fire area. Taking the extra time to size-up the situation and determine the best access point will often yield a much more effective operation with a high probability for success and safety. For example, after the location of the fire has been determined, an engine company should be directed to stretch from a convenient location close to the fire area. A large, exterior dock door, perhaps very remote from the main entrance, might provide the shortest, fastest, best, and safest path to the fire. Yes, we want to protect the primary egress, however, at these extremely large buildings, quickly and efficiently stretching to the fire, by way of the most convenient means, may ultimately protect the greatest number of people.

Sports venues

Another good example of non-high-rise buildings that are often times equipped with standpipe systems are sports venues. Football stadiums, baseball parks, arenas, and coliseums are just a few examples (fig. 3–5). There are many unique aspects to these occupancies, not the least

Fig. 3–5 Large sports venues are another example of standpipe-equipped buildings.

of which is that when occupied, the life hazard can be extreme. Once again, the standpipe system may be a benefit to our operation, but its use must be given careful consideration.

Because these sports venues are frequently multistory structures, we usually have the ability to use the standpipe safely and successfully when we can hook up and initiate attack from below the fire. Furthermore, many of these venues are open-air structures, making it much easier to operate with the benefit of adequate ventilation. Thousands of plastic seats and countless other combustibles make for a very generous ("solid gasoline") fuel load, so firefighters must anticipate and prepare for large, fast-moving fires (fig. 3–6).

During my career, there were several fires at the old Mile High Stadium (which has since been torn down and replaced) where the Denver Broncos football team played their home games for years. Two of those fires were very significant with numerous operational problems for firefighters. I operated at one of those fires, and from the outset, we had major problems with low water pressure. Because of freezing temperatures during the winter months, this building, like many open air sports venues, is equipped with a dry standpipe system.

Our operational problems at such sports venues have been known to start with the inebriated sports fan who has no idea how severely he will affect our operation by opening a hose valve, and leaving it that way. What was intended to be a humorous activity, is only met with disappointment, when after turning the hand wheel on the hose valve several times, no water is seen discharging from the outlet. "Oh well...," the drunk sports fan just assumes it's broken.

Without any fanfare, his ignorant stunt ends up undetected by cops or security guards, and the consequences of this, along with many other hose valves that are left partially open from a variety of other causes, will only be seen when the system is charged with water. At that point, the problem usually manifests itself in the form of an inadequate fire stream, with very low pressure at the point of operation. Attempts to correct the low pressure at the pump panel don't seem to make matters much better. Then the reports of water flowing at numerous locations throughout the building are received, accompanied by massive amounts of water cascading down stairs, and collecting at the lower levels.

Needless to say, sports venues, and any other buildings equipped with manual dry standpipe systems are loaded with many potential problems.

Fig. 3–6 This fire at Mile High Stadium in Denver had a significant fuel load, including numerous plastic seats.

Routine inspections of these occupancies can help eliminate some of these problems, but ultimately, it is important to prepare for the worst, and have a Plan B in place.

Operational procedures at buildings with manual dry standpipe systems should include proactive procedures to quickly identify and correct the problem of open valves. Simply anticipating this and being aware of the potential problem is one step closer to success. A naïve approach, with a mind-set that results in firefighters being oblivious to the real reason behind a low-pressure situation will result in fireground failure.

A good engine company will be prepared to stretch a line off the pumper. Creative fireground procedures, including the use of rope bags to hoist an attack line up the outside of a building will result in a fast, efficient fire attack, oftentimes completed faster than stretching a line off the standpipe, but only by a well-trained, physically and mentally prepared engine company.

Parking garages

Multistory parking garages are another good example of a structure equipped with a standpipe system. In many areas, specifically those with cold winter climates subject to freezing temperatures, the system will be a manual dry standpipe. These parking structures present very serious fire problems with a significant fuel load of gasoline bombs, and perhaps some propane and natural gas bombs as well (fig. 3–7). Firefighters in most cities and jurisdictions will be no strangers to the occasional and sometimes frequent vehicle fires inside a parking garage.

These vehicle fires inside parking garages are very serious fires with a tremendous amount of potential. From a strategic standpoint, fire departments must understand that these must not be treated as "just a car fire," but, rather, as a structural fire with serious potential. Many of these parking structures are connected to other structures, with the most serious being the parking garage below an occupied, high-rise building. Although the potential for fire extension is not extreme, a serious petroleum-based fire, typical of a vehicle fire, will produce significant dark black smoke, which can easily travel into the occupied areas of the building.

Ultimately, it's been my experience that the manual dry standpipe systems, typically found in many parking garages, (at least in climates that are subject to freezing temperatures) are not very reliable, and should be avoided. This is once again primarily because of the hose valves left open by vandals (or just idiots with nothing else to do), but can also be due to much more serious problems, like broken pipes, that are hidden until the system is charged with water. A rope bag dropped down to the pump operator can be an efficient and fast way to quickly get water on a serious vehicle fire. The good engine company will practice this evolution and be prepared to use it, if not as an initial procedure, at least as an alternative if the standpipe system fails for whatever reason.

Fig. 3–7 This multistory parking garage is loaded with dangerous potential, and has a major life safety exposure above in the form of a high-rise office building.

6 Educational facilities

Another occupancy that will likely be equipped with a standpipe system is educational facilities. In my city, standpipes can be found in a wide range of educational facilities, from elementary schools to large, college classroom buildings. Only a few blocks from my firehouse is a very large, multistory public high school, with long hallways. It is equipped with a Class III, automatic wet standpipe system. Several standpipe hose valves are also equipped with automatic, factory preset, pressure reducing valves (PRVs). This public high school, like many educational facilities, has large, uncompartmentalized areas that are unsprinklered, and loaded with fire potential (fig. 3–8).

Obviously, these occupancies carry with them a huge life hazard problem during the daytime hours of the school year. Furthermore, numerous other problems are inherent at such occupancies, from malicious false alarms to frequent, small arson fires, usually trash set on fire in bathrooms. The processes inside many schools include the use of hazardous materials in chemistry and physics labs, along with the tools and hazardous procedures associated with wood, metal, and automotive shops.

I have also found that many of the elementary and middle schools in the city of Denver are equipped with Class II standpipe systems. The Class II standpipe system was not designed for fire department use and is likely to deliver less than 100 gallons per minute. In my city, a tremendous potential exists at these old public school buildings. The Class II standpipe system should be avoided, and attack lines should be stretched from the exterior via a static hose bed.

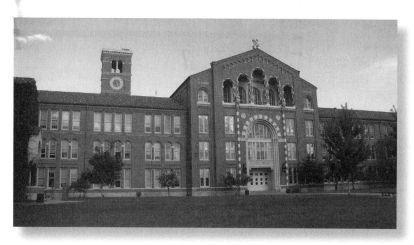

Fig. 3–8 A Denver public high school occupies this large, unsprinklered, multistory, standpipe-equipped building, wich is loaded with potential.

7 Medical facilities

The problems faced by fire departments at medical facilities are countless. Many of these buildings are equipped with standpipe systems, both vertical and horizontal. The large, multistory medical occupancy presents a similar operation to many high-rise buildings, with the added problem of an extremely serious life hazard issue (fig. 3–9).

Complacency can quickly take hold, as many of these type occupancies, at least in my city, are "frequent flyers," with many false alarms. The activity, processes, and complicated systems in these medical facilities all contribute

Fig. 3–9 A high-rise hospital building should be considered a major target hazard with a significant life safety problem.

to a long list of nuisance alarms from steam scares to burnt popcorn inside the microwave oven in the nurse's lounge. Nevertheless, we must maintain that strong mind-set and always be prepared for anything in these type occupancies that have countless hazards—everything from chemicals to radiation can be found in a typical medical facility. In my first due area there is a multistory hospital building. Near the entrance, which we use to access the building during alarms, is an NFPA 704 Hazardous Materials placard. The categories include a number 4 in the Blue and Red quadrants, and a number 2 in the Yellow quadrant (fig. 3–10). This is a real attention getter for me. It underscores the vast range of hidden hazards within this hospital complex, and reminds me each time we respond to this occupancy that nothing can be taken lightly.

The most obvious problem for firefighting forces in medical facilities, specifically hospitals, is the overwhelming life hazard that exists in these occupancies. Hundreds of nonambulatory medical patients, many tethered to thousands of pounds of life-sustaining equipment, will not be easily moved to a safe location during a serious fire. A protect-in-place strategy will, in most cases be best, however, there will be circumstances where this is will not be a viable option.

Fig. 3–10 This is the NFPA 704 Hazardous Material placard for a large, multistory hospital complex in south Denver.

Fig. 3–11 The Bay Bridge in San Francisco is another example of a standpipe-equipped structure.

Other Standpipe-Equipped Structures

In addition to those buildings and occupancies previously discussed, there are a wide range of numerous other structures and facilities that have a standpipe system as part of their built-in fire protection. Some of the most common are bridges, tunnels, and water vessels.

Bridges

There is a wide range of bridges in jurisdictions both large and small. Many of these bridges have a standpipe system in order to provide water remote from the source. Some bridges, such as the Bay Bridge in San Francisco span very long distances. The Bay Bridge connects the City of Oakland to San Francisco and is elevated high above the San Francisco Bay (fig. 3–11). Tens of thousands of passenger and commercial vehicles cross the Bay Bridge every day, and with that significant traffic volume comes numerous accidents, vehicle fires, and other emergencies. The Bay Bridge, like many very long bridges, has a standpipe system designed to provide a sustained water supply for fire department operations that take place a long distance from the closest municipal fire hydrant (fig. 3–12).

In addition to those more obvious standpipe-equipped bridges, there are also numerous smaller bridges with standpipes. A massive expansion project of Interstate 25 (I-25) through the city of Denver has included the rebuilding of numerous bridges that carry traffic on surface streets over the highway. Several of these bridges are equipped with a standpipe system, not designed to deliver water up to the bridge, but rather down to the interstate highway below (fig. 3–13).

Besides those bridges that carry motor vehicles, there are also bridges that carry trains. Large freight and passenger trains, along with public transportation trains, such as light rail and subway trains, typically utilize bridges of one form or another. Some of these bridges are equipped with standpipe systems.

The majority of standpipe systems found on bridges are dry systems. Like other structures with dry systems, climate is the primary reason for the utilization of this type. The potential operational problems previously addressed are also of concern. However, bridges provide one tactical advantage, in that most bridges are open-air structures, and smoke from a fire will likely dissipate into the open atmosphere, generally making operations less complicated and punishing.

Fig. 3–12 The fire department connection (FDC) at ground level for the Brooklyn Bridge in New York City

Fig. 3–13 This standpipe FDC is located on a bridge that spans an interstate highway in Denver and is designed to supply water to the highway below.

Tunnels

Another type of infrastructure that can be found in many cities are tunnels. Like bridges, the primary use of tunnels is that of transportation for vehicles and trains. Tunnels can be found in many forms, from short tunnels carrying vehicles a few hundred feet, to long, expansive tunnels that travel several miles underground. The famous "big dig" project in Boston is a good example. This massive project took an inner-city highway and placed it underground. Many large cities also have an extensive network of tunnels as part of a subway system.

Most of these tunnels are equipped with a standpipe system. Whether the system is wet or dry is a significant concern, but the primary concern is directly related to the more complicated and punishing firefighting operations typical of a below grade fire. Once again, extreme caution must be used whenever an attack line must be taken from a standpipe that is located below grade. If possible, and ideally, an attack line should be stretched from a safe position, preferably one that is completely outside of the fire area and tunnel.

Completing a longer hose stretch from a safe position is certainly much easier said than done, but is an issue that must be carefully considered by all fire officers involved with such an operation. The central concern is that firefighters operating deep inside a tunnel have a connection with the exterior. This is best established by having an attack line that comes from outside the tunnel, and one that firefighters can follow from the fire area to an area of refuge, preferably completely outside and in an open air atmosphere.

Water vessels

There is another location where standpipe systems can found that deserves special attention. That location is onboard water vessels of various types. Standpipe systems can be found on many different types of water vessels, from large cruise ships, to passenger ferries. Of course, most large military vessels are also equipped with standpipe systems for shipboard firefighting.

The occurrence of shipboard fires in the city of Denver is extremely rare, due primarily to the lack of any large bodies of water and ships. Therefore, my experience with standpipe operations on water vessels is limited to what I have learned from those who have done it, and the preplanning I have done at the beginning of the few cruises I have had the pleasure to take. Firefighters who work in locations with large bodies of water and various water vessels, should include shipboard firefighting among the long list of potential standpipe operations.

Summary

When firefighters speak of a standpipe operation, many will usually make the assumption that the operation took place in a high-rise building. The fact is that there are countless standpipe operations that take place in locations other than in a high-rise building. Firefighters must identify all standpipe installations in their response district, and train on the specifics associated with standpipe operations, regardless of where that standpipe system is located.

Every effort should be made to avoid using the so-called "horizontal standpipes," where the fire is on the same level and near the standpipe hose valve. Alternatives, such as hand stretching attack lines from the static hose bed of a pumper will provide for the ever-important umbilical cord to safety. In addition, for those multistory, low-rise buildings equipped with manual dry standpipe systems, engine companies should be prepared to implement a Plan B. Using rope bags to stretch handlines vertically up the outside of a structure, and thus bypassing a questionable standpipe system, is an excellent alternative in many situations.

CHAPTER 4

The Standpipe System

A building's standpipe system is one of the most important fire and life safety components in a high-rise, low-rise, or other standpipe-equipped building. Of course, there are countless other critical fire and life safety systems that serve a specific purpose and ultimately protect the buildings occupants. Specifically, a built-in, fully-automatic sprinkler system has been cited by many as being the most important, and an absolutely essential fire protection feature, especially for all high-rise buildings. A standpipe supplies the water for the sprinkler system, so in essence, these systems are truly interrelated components of an overall fire protection system. In fact, many new systems are considered *combination systems*, where the sprinkler system and the standpipe for manual firefighting are actually supplied by the same standpipe riser.

Water Supply Priority

Based on historical evidence, I would certainly agree that the sprinkler system is a most critical fire protection feature, especially for high-rise buildings. A properly designed and working sprinkler system will, in most cases, extinguish the fire or keep it small enough to be easily managed by firefighters upon their arrival.

My focus in this chapter, however, will be on the use of a standpipe system for manual firefighting in those rare situations where a sprinkler system fails, for whatever reason, to successfully extinguish a fire, or the countless situations where there is no sprinkler system to begin with in a building. These operations require the use of handlines from a standpipe, and manual firefighting by a well-trained force of mentally and physically prepared professionals.

A question that has been asked time and again is, when a building has a standpipe system for manual firefighting, and a separate sprinkler system, which fire department connection (FDC) should an engineer or pump operator supply first, the sprinkler or standpipe? In buildings where the standpipe and sprinkler system are interconnected (combination systems) the decision has already been made. When you supply one, you supply them both. However, for the countless buildings where the standpipe system for manual firefighting and the sprinkler system are truly independent of one another, a decision as to which one to supply first has to be made.

I believe the answer is a simple one, and it coincides with the theme of this chapter. A pump operator should always supply both, but the FDC for the standpipe should be supplied first. The reason behind this is simple: the fact that the fire department has been called to this building to begin with is a good indication that something is wrong. If it's an automatic alarm due to water flow, than perhaps the sprinkler system has already done what it is designed to do and our job will be made much easier.

However, if there is a working fire in the building, that indicates that possibly the sprinkler system may not have worked properly and manual firefighting will be necessary. Ultimately, when our members take a handline off the standpipe from the floor below the fire floor, we must ensure that they will have a sustained supply of water at the appropriate pressure, first, to protect themselves, and second, to give them the critical water supply they need to extinguish the fire and thus protect the occupants of the building.

Although a sprinkler system is a very dependable and lifesaving fire protection feature, it can fail, or even just one of the numerous components that make up a sprinkler system can fail. For example, while writing this chapter, my department, the Denver Fire Department (DFD), just concluded a week-long period of extremely high emergency activity related to the cold weather. Denver natives frequently say "If you don't like the weather here, just wait a few minutes and it will change." That's a bit of a stretch, but as I write this, the temperature outside, in late February, is close to 80 degrees. Ten days ago, it was 13 degrees below zero. Over the past week, thousands of frozen pipes, many supplying sprinkler systems, upon thawing out have burst, leaving countless buildings unprotected. My point is this; we must be prepared for sprinkler system or component failure. Furthermore, making sprinkler system water supply the priority might result in unknowingly supplying a broken water pipe, greatly exacerbating an already serious problem. All the while, operating members might not have sufficient water volume and pressure to safely initiate and successfully complete manual firefighting with handlines.

The FDC for the sprinkler system will also be supplied as soon as possible, but the priority, in my opinion and based on my experience, should be to supply the standpipe FDC first. In reality, a good engineer/pump operator will likely have charged supply lines to both, very quickly. Once again, in newer buildings these systems are typically interconnected, but a large number of buildings in many jurisdictions have separate FDCs and standpipe risers for the sprinkler system and the standpipe for manual firefighting.

Lesson Learned

Several years ago, I responded to and operated at a fire that occurred in a 15-story multiple dwelling. It was Christmas Eve, and on the day of the fire I was working as the lieutenant on Rescue Co. 1. The fire occurred in a third-floor apartment and heavy fire and dark black smoke were venting from the apartment windows on the 9th Avenue side of the building upon our arrival. We arrived just seconds behind the first due engine company, and prior to the arrival of the first due truck company. The chief assigned us to the fire floor to assist the engine company.

One of the first problems that we encountered was, you guessed it, low water pressure. The engine company was stretching an attack line off the standpipe as my men were preparing to force the apartment door. With a charged line in position, the door was forced, and dark black smoke pushed forcefully into the public hallway. This was a large, occupied, multiple dwelling, with four separate wings, all connected to a common public hallway that had no fire doors, so the need to hold this fire to the apartment of origin was critical.

With the heavy smoke conditions and high heat, the officer of the engine company ordered his nozzleman to open up the line into the overhead to cool the upper atmosphere for a safe advance into the apartment. Firefighters reading this will easily visualize these events, as this very scenario takes place time and again in cities across the country. This was a typical apartment fire in a multiple dwelling, which is usually brought to a safe and successful conclusion within minutes of fire department arrival.

As the engine company nozzleman cracked open the bail of the nozzle, a quick burst of water shot toward the ceiling, and than the line went limp, leaving a very weak, low-flow stream, far from effective for firefighting. The engine company immediately focused on solving their water supply problems. My crew entered the apartment in an attempt to conduct a quick primary search. One member was equipped with a *can* ($2^1/_2$-gallon water extinguisher) and was able to hold the main body of fire while the other men completed a limited search in the uninvolved portion of the apartment.

With the search completed and the can empty, we exited the apartment and closed the apartment door. There was a moderate smoke condition in the public hall by now, and many occupants of other apartments were starting to exit into the public hallway. We told those within ear shot to get back in their apartments, in an attempt to protect in place. Some opted to fight their way through the smoke to the exit stair, and some had to be assisted by other fire companies.

It seemed like an eternity, but the engine company finally got their water supply problem solved, and, with an effective fire stream, was able to quickly extinguish what was no longer a one-bedroom fire, but a well-involved apartment blaze. A thorough search came up negative, and pre-control overhaul verified that the fire did not extend to adjacent apartments. However, there was some minor extension to the apartment above from auto exposure that was quickly controlled with the handline on the floor above. The adjacent apartments were searched, as well as the apartment on the floor above, all were negative. A backup line was stretched onto the fire floor, but was not used. Ventilation via the attack stair quickly cleared the smoke from the building and the incident was under control.

A post-incident review of the fire revealed several items that led to the water supply problem. This was an older high-rise building with a partial sprinkler system in the basement level only. Two separate FDC's were located on the outside wall of the southeast wing of the building. This was also the location of the entrance to the attack stair. As companies made entry, the door was chocked open, which blocked a clear view of the standpipe FDC. The engine company engineer/pump operator mistakenly supplied the FDC for the sprinkler system. After receiving communication from the engine company officer regarding poor water pressure, the engineer/pump operator quickly identified the problem and supplied the proper FDC.

This was an honest mistake, but could have had disastrous results had it not been for several other factors, most notably the careful job of forcible entry by the Rescue 1 crew. Members were cautious to maintain the integrity of the door to the apartment as they forced entry. This allowed for the door to be closed during the period when they didn't have an effective fire stream. The heat and fire, as well as most of the smoke, were kept confined to the apartment of origin, allowing the majority of the other occupants to be protected in place.

Ultimately, engineers/pump operators must be particularly mindful of older buildings with systems that are likely not interconnected. When this situation exists, supplying the standpipe FDC first, if it can be identified, and then supplying the sprinkler FDC is the recommended procedure. If there are multiple FDC's, and the engineer/pump operator is not sure which FDC supplies what system due to a lack of or worn-out placards, then supply all of the FDCs. Also, engineers/pump operators must quickly determine what areas are going to be supplied by a given FDC. In the previous example, it was a building with partial sprinkler protection in the basement only. The fire was on the third floor. Pressurizing supply lines to the sprinkler system FDC could have led to additional problems, such as broken water pipes.

Standpipe System Defined

What exactly is a *standpipe system*? Firefighters who work in areas with high-rise and low-rise buildings equipped with standpipe systems must have a thorough understanding of the standpipe systems in these buildings in order to effectively operate at fires in them.

A standpipe system is defined in NFPA 14 as: "An arrangement of piping, valves, hose connections, and allied equipment installed in a building or structure, with the hose connections located in such a manner that water can be discharged in streams through attached hose and nozzles, for the purposes of extinguishing a fire, thereby protecting a building or structure and its contents in addition to protecting the occupants. This is accomplished by means of connections to water supply systems or by means of pump tanks and other equipment necessary to provide an adequate supply of water to the hose connections."[1]

For all practical purposes, a standpipe system is simply a water supply system used for fire suppression. It can be compared to that of a city water distribution system, with primary feeders, secondary feeders, and distributor lines, each supplying various components of the overall water delivery system. For a standpipe system, the primary feeder would be the piping used to deliver the water into the standpipe system from its supply, whether that happens to be a city main, gravity tank, pump, or a combination of those components. The secondary feeder could be the *riser* itself, the vertical or horizontal main used to deliver the water to all areas of the building. From there, the smaller piping, or distributors, delivers the water to all of the actual standpipe outlets or hose connections within the system.

Another definition provided by Shapiro in the *Fire Protection Handbook* reads: "Standpipe systems are fixed piping systems, with associated equipment, that transports water from a reliable water supply to designated areas of buildings, where hoses can be deployed for firefighting. Such systems are typically provided in tall and large-area buildings."[2]

Classes of Standpipe Systems

Standpipe systems have been broken down into three basic classes, depending on the designed use of a specific system. In other words, the standpipe system class is based on who will be using it.

Class I standpipe system

The Class I standpipe system is designed specifically for fire department use and for those trained in handling heavy fire streams. This class of standpipe has $2\frac{1}{2}$-inch discharge outlets strategically placed and designed to deliver high volume water when used with $2\frac{1}{2}$-inch hoselines. Requirements vary depending on the fire and building codes used by a particular jurisdiction. However, Class I systems are typically required for buildings three or more stories in height (in some jurisdictions, four or more stories in height) and buildings of large square footage, such as large, covered malls. Many older buildings, especially in older cities, may have several multistory buildings,

Fig. 4–1 The standpipe hose valve outlet of a typical Class I standpipe system

Fig. 4–2 The standpipe hose valve outlet of a typical Class II standpipe system

Fig. 4–3 A $1\frac{1}{2}$-inch × $2\frac{1}{2}$-inch increaser

typically low-rise, without a standpipe system. These buildings were built before the stricter, more comprehensive building and fire codes that are used by most jurisdictions today. The Class I standpipe system is required to deliver a rated flow of 500 gpm at the topmost, or hydraulically most remote, outlet at 65 psi residual pressure pre-1993, and 100 psi residual pressure post-1993 (fig. 4–1).

Class II standpipe system

The Class II standpipe system is designed specifically for occupant use. Its use is frequently designated for so-called "first-aid firefighting," by building occupants and/or fire brigades before the fire department arrives. This class of standpipe has $1^1/_2$-inch discharge outlets and, in some buildings, includes $1^1/_2$-inch hose attached to the outlet and stored on a rack. The hose is typically a single jacketed linen hose, 100-feet in length with a non controlling nozzle (fig. 4–2). These features were designed for persons with no actual fire suppression training. Like the Class I systems, requirements vary depending on the fire and building codes used by a particular jurisdiction. The use of Class II standpipe systems, however, is declining as many jurisdictions are actually removing the hose from the cabinet. The city of Denver began removing the hose from Class II standpipe systems several years ago. Denver, like many jurisdictions realized that firefighting should only be undertaken by a trained force of physically and mentally prepared professionals.

Class II standpipe systems are only required to deliver 100 gpm from the hydraulically most remote standpipe outlet. Because of this, fire departments should not utilize Class II systems for firefighting. However, some $1^1/_2$-inch outlets on Class III systems, depending on how they are supplied, can flow enough water to supply a $2^1/_2$-inch attack line. A $2^1/_2$-inch handline can be hooked up to a $1^1/_2$-inch standpipe outlet using a simple $1^1/_2$-inch × $2^1/_2$-inch increaser (fig. 4–3 and fig. 4–4). Before attempting to utilize a $1^1/_2$-inch standpipe outlet, the supply must be verified to ensure that it is capable of supplying enough water to produce and maintain an effective firefighting stream.

Class III standpipe system

The Class III standpipe system is a combination of the Class I and Class II systems. It can have both a $2^1/_2$-inch outlet and a $1^1/_2$-inch outlet, or simply a $2^1/_2$-inch outlet with a $2^1/_2$-inch to $1^1/_2$-inch reducer attached to the outlet (fig. 4–5 and fig. 4–6). This is designed for both fire department and occupant use. Once again, many jurisdictions are eliminating the occupant, or "first-aid," use of these systems, by simply removing the

Fig. 4–4 A $1^1/_2$-inch × $2^1/_2$-inch increaser threaded onto the $1^1/_2$-inch outlet of a Class III standpipe system

Fig. 4–5 The standpipe hose valve outlet of a typical Class III standpipe system with both 2½-inch and 1½-inch outlets

Fig. 4–6 The standpipe hose valve outlet of a typical Class III standpipe system with only one 2½-inch outlet, but with a 2½-inch to 1½-inch reducer attached

hose and related appliances. Therefore, if the hose has been removed and is not available for occupant use, the system is, for all practical purposes, a Class I system.

Once again, if both outlets are being supplied from the same riser, the 1½-inch outlet can potentially supply another 2½-inch handline (fig. 4–7). There will be some additional friction loss in the smaller diameter piping between the riser and the outlet; because of the short distance, however, this friction loss will have a minimal impact. A simple adapter, the 1½-inch × 2½-inch increaser, at a cost of about $10, is all that is necessary to complete this hookup.

Class IV standpipe system

In the *Fire Protection Handbook*, Shapiro also points our that "Some sources refer to a fourth class of system called a "combined" system; however, combined systems are simply Class I or Class III standpipe systems that also supply water to a sprinkler system."[3]

Types of Standpipe Systems

Whereas, the class of standpipe system refers to its intended use, standpipes are also classified by type, which further describes the characteristics of the standpipe system, most specifically whether or not it is a wet or dry system. There are five categories of standpipe system types listed in NFPA 14 and further described by Shapiro in the *Fire Protection Handbook*.

Automatic dry standpipe system

An *automatic-dry* standpipe system is a dry standpipe system that has "piping that is normally filled with pressurized air. These systems are arranged, through the use of devices such as a dry-pipe valve, to automatically admit water into system piping when a hose valve is opened. They are connected to an automatically available water supply capable of supplying the water demand necessary for firefighting."[4]

Fig. 4–7 When both outlets of a Class III standpipe system are supplied by a large diameter riser, the 1½-inch outlet (with an increaser) can be used for fire attack.

Automatic wet standpipe system

An *automatic-wet* standpipe system is a wet standpipe system, which has "piping that is filled with water at all times and [has] an automatically available water supply capable of supplying the water demand necessary for firefighting."[5]

Semiautomatic dry standpipe system

A *semiautomatic-dry* standpipe system is a dry standpipe system that has "piping that is normally filled with air that may or may not be pressurized. These systems are arranged through the use of devices, such as a deluge valve, to admit water into system piping when a remote actuation device located at a hose station, such as a pull station, is operated. They also have a preconnected water supply capable of supplying the water demand necessary for firefighting."[6]

Manual dry standpipe system

A *manual-dry* standpipe system is a dry standpipe system that has "piping that is normally filled with air, and these systems do not have a preconnected water supply. A fire department connection must be used to manually supply water for firefighting."[7]

Manual wet standpipe system

A *manual-wet* standpipe system is a dry standpipe system that has "piping that is normally filled with water for the purpose of allowing leaks to be detected. The water supply for these systems is typically provided by a small connection to domestic water piping, and it is not capable of supplying firefighting water demands. A fire department connection must be used to manually supply water for firefighting."[8]

Standpipe System Components

With a basic understanding of the different classes and types of standpipe systems a fire department may have to deal with, we can now look closely at the various components and hardware associated with a standpipe system. The first thing all firefighters must clearly understand is that the codes and standards used to design and build standpipe systems are minimum requirements. Rarely will a property owner, developer, investor, or any of the countless individuals and companies associated with a building construction project suggest or be willing to exceed the minimum requirements of the building and fire codes used by the authority having jurisdiction. To do so would cost additional money—these people are in the business of making money, not exceeding minimum requirements, even if it would be better for operating firefighters and the safety of building occupants. The codes and standards of the twenty-first century are very good, comprehensive requirements that builders must follow, but they are still *the minimum*.

Fire department connection (FDC)

Starting on the outside of a high-rise building or any low-rise standpipe-equipped building, one of the first standpipe system components that the fire department encounters and must deal with is the fire department standpipe connection, commonly referred to simply as the FDC. It is also sometimes referred to as the fire department siamese connection, because most are typically equipped with two identical inlets or hose connections, thus, the term *siamese*. For purposes of simplicity and standardization of terminology, I will use FDC in this book (fig. 4–8).

In most jurisdictions, typically, the most common size inlets or hose connections for FDCs are $2^{1}/_{2}$-inch. There are generally two $2^{1}/_{2}$-inch female swivel connections making up a standard FDC (fig. 4–9). Many fire departments supply the FDC with one or two $2^{1}/_{2}$-inch supply lines, but most use a larger supply line, typically a 3-inch supply line with $2^{1}/_{2}$-inch couplings or adapters, and in some cases, $3^{1}/_{2}$-inch supply lines are used. Some fire departments opt to use even larger supply lines to supply the FDC, up to and including 5-inch supply lines. In some cities, rather than seeing the typical siamese connection with two $2^{1}/_{2}$-inch connections, there is one large connection, typically a quarter turn, sexless, Storz coupling, for quickly connecting a 5-inch supply line (fig. 4–10).

Fig. 4–8 A typical standpipe system fire department connection (FDC).

Fig. 4–9 Most standpipe system FDCs are equipped with two 2½-inch inlets with female swivel connections.

Fig. 4–10 Some fire departments utilize a single 5-inch supply line to supply a standpipe FDC.

This is well suited for those fire departments that use 5-inch supply lines exclusively, and that have the ability to pump through it. However, because of the weight of this large diameter supply line, it will likely require some support just below the FDC in order to prevent kinking. Furthermore, most large-diameter hose is not designed to withstand the high pressures needed to supply the standpipe systems in tall high-rise buildings.

The FDC is also equipped with internal clapper valves, which are of particular concern to fire departments. The fire department member(s) assigned to hook up to the FDC, typically one member initially, usually the engineer/pump operator, must visually inspect the inside of the FDC including the clapper valves. Many FDCs are not properly protected against damage and vandalism; many are missing caps, which allows the FDC to become a catchall for trash and various debris (fig. 4–11). Firefighters in cities across the country have found everything from crushed pop cans to used hypodermic needles inside the FDC. Prior to connecting supply lines to the FDC, the engineer/pump operator must first ensure that all trash, debris, and any other loose items are removed from inside the FDC. I address this in depth, in chapter 12.

Pipe and tube

In a typical standpipe system, especially those in larger buildings, countless lengths, sometimes thousands of feet of pipe and tubes are used to transport the water from the supply to the standpipe outlets at various locations throughout the building. This system is a very good example of the axiom, "a chain is only as good as its weakest link." Particularly disconcerting is the fact that to supply the upper floors of high-rise buildings taller than 30 stories, pressures in excess of 250 psi, and 300 psi plus for buildings in the 40-story range are required.

One particular building in downtown Denver has a placard near the FDC with 415 psi listed as the required pressure to supply the high zone of the standpipe system. That's a lot of pressure for a standard pumper to handle, not to mention the fire hose which is tested annually only to 250 psi. We also have to consider the building components, piping, valves, and so on. According to NFPA 14, the pipe, fittings, and valves used for the standpipe system are rated for pressures of up to only 300 psi. NFPA 14 also recommends that the maximum pressure within the system at any point should not exceed 350 psi.

Standpipe system valves

Several valves are located throughout a typical standpipe system. Post indicator valves (PIVs), outside stem and yoke valves (OS&Y), check valves, and sectional control valves control the flow of water throughout the standpipe system.

The first of many valves is located just inside the building, usually only a few feet from the FDC. There are some exceptions where the FDC has to be placed remote from the building due to underground parking garages, landscaping, terrain, and various other barriers. In these situations, horizontal piping carries the water supply from the FDC to the building. Regardless, once the piping comes into the building, valves are used to control the water flow.

In many buildings, specifically high-rise and many of the newer low-rise buildings, a fire pump is included as part of the overall fire protection system. Water from the FDC is piped into the fire pump room, usually located at the lowest level of a building. A check valve in the piping from the FDC keeps the water in the system from flowing backward and out into the piping up to the FDC. Water from the FDC actually bypasses the fire pump in order to provide a reliable means of supplying the standpipe system in the event that the fire pump fails. There is also a wide range of piping and valves associated with the fire pump.

Piping connected to the water supply is equipped with an indicator-type valve and a check valve located close to the supply. That water supply may come from tanks, pumps, or direct connection from the municipal water supply. The FDC is not equipped with an isolation valve, so there will be no chance of a valve accidentally being closed and preventing the system from being supplied via the FDC.

Hose connection valves

The last valve in the standpipe system before water is actually discharged onto a fire is located at the hose connections. These connections are located on every floor within high-rise and in standpipe-equipped low-rise buildings. Most buildings have the hose connections located within an approved, fire-rated stairwell, at each landing. However, some jurisdictions allow for the hose connections to be located within standpipe cabinets, out on the floor. Regardless of the location within the building, the hose connections should be unobstructed, and are required to be located not less than 3 feet or more than 5 feet above the floor.

The most recent edition of NFPA 14 recommends placement of the hose connections at the half landing between floors, but this is only applicable to buildings that have return-type stairs. The idea is that the hose connection will be closer to the fire, allowing for a greater usable length of hose line. Unfortunately, hose management requires that much of the hose will need to be stretched out on the floor below the fire when the attack commences from the stairwell. Flaking hose out on stair landings and half landings is not a good practice, as it leads to severe kinking, especially with $2^1/_2$-inch handline. So, placing the hose connections on the half landings is good in theory, but in practice, it might not prove beneficial.

The type of valve used at a hose connection can vary, depending on the size of building and the location of the hose connection within the building. In many cases, the valve is a basic globe valve that operates by turning a hand

Fig. 4–11 Many FDCs become convenient receptacles for trash and other debris.

wheel located at the top of a threaded stem. With this hand wheel, a firefighter can manually control the flow and pressure at the hose connection. This can be accomplished with significant accuracy and finesse if a pressure gauge is used. Some newer systems come equipped with a pressure gauge, but the best procedure is to ensure that a standpipe inline pressure gauge is included as part of the equipment carried in every engine company's standpipe equipment kit. This is a standard piece of equipment for many large fire departments across the country including the DFD (fig. 4–12).

In larger systems, specifically those in large high-rise buildings, the head pressure on the lower levels of a standpipe riser can be significant. To add to that the pressure from a fire pump and excessive pressures can be encountered at many locations, but especially at the lower levels. Because of this, firefighters may need to reduce and regulate that pressure at standpipe hose connections. When the static pressure at a hose connection exceeds 175 psi, an approved pressure-regulating device should be provided to limit both static and residual pressures. This is another piece of the standpipe puzzle that can create some significant problems.

Pressure-Regulating Devices

For most in the American fire service, pressure-regulating devices, and specifically pressure reducing valves (also know as PRVs) were virtually unknown prior to 1991. It was only after the tragic One Meridian Plaza high-rise fire, which occurred in Philadelphia on February 23, 1991, that PRVs became an essential part of the high-rise standpipe operations training curriculum in the American fire service.

After One Meridian Plaza, a great deal of attention was paid to PRVs, and many in the fire service started to recognize these devices as a potentially serious problem. A National Fire Protection Association (NFPA) Alert Bulletin (91-3), "Pressure-Regulating Devices in Standpipe Systems,"[9] published in May 1991, highlighted the dangers posed to firefighters by improperly set PRVs, and was a much-needed wake-up call. In addition, an excellent article written by Captain Dave Fornell (author of the *Fire Stream Management Handbook*) published in the August 1991 issue of *Fire Engineering*, served as an excellent start toward educating the fire service.[10]

Unfortunately, like many fire service tragedies, we seem to forget the lessons learned all too soon, and as time passes by, the significance of those lessons tends to slowly diminish. This is readily apparent to me and the members of my FDIC Standpipe Hands-on Training (HOT) team as

Fig. 4–12 A standpipe inline pressure gauge should be standard equipment for fire department operations in high-rise and other standpipe-equipped buildings.

we tenaciously work to educate the American fire service on the critical components of standpipe operations.

It's been nearly 15 years since Philadelphia firefighters were confronted with improperly set PRVs at One Meridian Plaza. Yet, many fire departments, apparently having not been directly affected by that tragedy, have not taken proactive steps to effectively deal with PRVs during standpipe operations. As part of this chapter on standpipe systems, I will revisit the PRV issue and attempt to provide the necessary information for those interested in and dedicated to effecting change across the American fire service.

One Meridian Plaza

The fire at One Meridian Plaza was a very significant event in American fire service history. It will forever remain as one of the most important high-rise fire case studies, and the lessons learned from this event will carry on for years. It was in the 38-story Meridian Bank Building in Center City, Philadelphia. A fire started on the 22nd floor at about 8:40 p.m. and burned for the next 19 hours. The fire consumed eight full floors of the building and required the services of more than 300 Philadelphia firefighters. Three of Philadelphia's bravest lost their lives fighting this overwhelming fire.

Numerous factors contributed to the loss of the three members of Philadelphia Fire Department (PFD) Engine Co. 11, not the least of which was improperly set PRVs. The post-fire investigation revealed that when the PRVs were originally installed, the pressure settings were improperly adjusted. On the night of the fire, the PRVs reduced the discharge pressure to less than 60 psi, which was not sufficient either to overcome the friction loss in the $1^3/_4$-inch hose or to provide the proper operating nozzle pressure for the automatic combination fog nozzles used by the PFD at the time of this fire. This low pressure made it impossible to develop effective fire streams; thus, firefighters were unable to combat this large, fast-moving fire.

The type of PRVs in One Meridian Plaza were, in fact, field and fireground adjustable; however, it wasn't until several hours into the incident before a technician equipped with the proper tools and an understanding of the adjustment process was able to effectively adjust several of the PRVs. By that time, the fire had gained considerable headway and had control over several floors, making manual firefighting impossible. Ultimately, the improperly set PRVs made it impossible for Philadelphia firefighters to combat this fire when it was still a manageable size.

In most cities, standpipe systems, especially those in tall high-rise buildings, are typically equipped with pressure-regulating devices. These devices can generally be found in two locations: at zone/floor-control valves for combination standpipe/sprinkler systems or at standpipe hose-valve outlets.

In larger systems, specifically those in tall high-rise buildings, the head pressure on the lower levels of a standpipe riser can be significant. Add to that the pressure from the building's fire pump, and excessive pressures can be encountered at many locations, but especially at the lower levels. Because of this, firefighters need to reduce and regulate the pressure at standpipe hose connections.

Most fire code requirements for standpipe systems are based on recommendations from NFPA 14, *Standard for the Installation of Standpipe and Hose Systems*, and NFPA 13, *Standard for the Installation of Sprinkler Systems*, respectively. These requirements include the installation of an approved pressure-regulating device any time the static pressure within a standpipe system will exceed 175 psi. When the static pressure at a hose connection exceeds 175 psi, an approved pressure-regulating device shall be provided to limit both static and residual pressures. Furthermore, if flowing pressures exceed 100 psi, then NFPA 14 requires that an approved device be installed at the outlet to reduce pressures to a maximum of 100 psi.

Pressure-regulating devices, as defined in NFPA 14, "are designed for the purpose of reducing, regulating, controlling, or restricting water pressure in order to limit standpipe system outlet pressure so that firefighters can safely and effectively operate handlines for manual firefighting."[11] Pressure-regulating devices used in standpipe systems can be categorized into three basic types:

1. Pressure-restricting devices
2. Pressure-reducing valves
3. Pressure-control valves

Pressure-restricting devices

Pressure-restricting devices (PRDs), sometimes referred to as *pressure-restricting valves*, reduce pressures in flowing conditions only. These devices/valves operate in a manner similar to a fire department pump operator's gating down at the pump panel of a pumper. They do not compensate for changes in input pressure to maintain a constant discharge pressure, and they do not control pressure in static conditions. For these reasons, their use is limited. There are several different types of pressure-restricting devices/valves.

Orifice plate. According to Jeffrey M. Shapiro, "An orifice plate is a metal disk with a restricting orifice. Orifice plates may be inserted into outlet connections to standpipe systems to reduce residual pressure on the downstream side. Although relatively inexpensive, orifice plates are generally undesirable because they do not maintain a steady discharge pressure and because they can damage hoses, specifically the inner lining, if the pressure jet from an orifice plate strikes the inner wall of a hose. If used, orifice plates should always be installed at hose couplings to allow for easy removal."[12] (fig. 4–13).

An orifice plate basically looks like a very large metal washer. You have easy access to the orifice plate if it is installed at the standpipe outlet. Remove the orifice plate by prying it off with a small screwdriver. Simply insert the screwdriver between the hose valve outlet wall and the outside of the orifice plate; with a little leverage, pry the orifice plate off and out. A pair of channel locks can also be used: clamp down on the middle of the orifice plate and pull it out. The orifice plate is not field or fireground adjustable. It can usually be removed, however; thus it can be considered fireground removable, and firefighters can easily deal with it.

Mechanical PRD. This mechanical device has hose threads on both ends and resembles other types of fire department appliances. It is similar in size to certain adapters, such as a double female. This mechanical pressure-restricting device, also referred to as a *pressure-reducing device*, is a one-piece mechanical device designed to restrict/reduce outlet pressure (fig. 4–14).

Fig. 4–13 An orifice plate mounted inside a 2½-inch standpipe hose valve connection.

Fig. 4–14 An example of a mechanical PRD in place on a 1½-inch hose valve outlet inside a downtown Denver high-rise building. These types of mechanical PRDs are found mostly on older installations.

A mechanical PRD has a series of overlapping holes cut into the waterway through which the water flows. It is considered to be field adjustable and is fireground adjustable because the size opening can be manually adjusted by turning a knob on the outside of the device to fully open, fully close, or partially open the holes. Mechanical PRDs restrict, and thus reduce pressure, but only in residual or flow conditions. It does not compensate for excessive static pressures (fig 4–15).

For fire department operations, instead of adjusting the water flow by adjusting the device, it is recommended that the entire device be removed. This device, which is threaded onto the standpipe outlet, is easily removed, making it fireground removable, and thus, allowing an engine company to then attach its hoseline directly to the standpipe outlet, threads permitting. These types of PRDs are found mostly in older buildings, but can be found in some newer installations, specifically in older eastern cities. This type of device is particularly common in New York City.

Limiting PRD. A mechanical limiting device works by limiting the valve from being fully opened. This type of PRD is mounted on the valve handle itself. This limiting device, or *stop*, works to physically limit the distance the valve can be opened. Firefighters can easily deal with most PRDs of this type, as they are typically designed to allow for complete removal or they can be broken off without damaging the valve itself. Once the limiting device has been removed, the valve can then be operated normally and opened fully if necessary (fig. 4–16).

There is also another type of pressure-restricting device/valve that has a mechanical limiting device and works in the same manner as the previously described PRD. Once again, this type of device is mounted on the valve handle itself. This is not designed to be broken off; however this type of PRD can be removed by loosening two set screws with an Allen wrench. Once it's free, the restricting device can be completely removed from the valve handle. With the restricting device removed, the valve can then be fully opened (figs. 4–17, 4–18, and 4–19).

One more variation of a limiting PRD involves the use of a removable clip that holds a collar in position and is located on the outside of the valve stem. Once again, this device acts to limit the valve from being fully opened. This device is field and fireground adjustable/removable without any special tools. Simply remove the pin from around the collar by pulling outward; once removed, the valve can be opened fully (fig. 4–20 and fig. 4–21).

Fig. 4–15 The overlapping holes inside this mechanical PRD can be adjusted to restrict and reduce flow pressure only. However, when it is encountered, it is best simply to remove the device entirely and attach the hose directly to the hose valve. A threaded adapter may be required.

Fig. 4–16 This type of PRD operates by limiting how far the valve can be opened. It is considered field and fireground adjustable/removable.

Pressure-reducing valves (PRVs)

The preferred devices for regulating excessive pressures are *pressure-control* and *pressure-reducing valves* (PRVs). These valves are designed to actively regulate outlet pressures in both static and flowing conditions, and, based on NFPA 14 recommendations, most fire codes require them for outlets of systems where static pressure could exceed 175 psi. Pressure-control and pressure-reducing valves are preferred over pressure-restricting valves/devices because hoses and sprinkler systems connected to standpipes are exposed to both static and flowing conditions, and uncontrolled static pressures can cause equipment failure and pose a danger to firefighters.

Characteristics of PRVs

An increase in inlet pressure does not increase outlet pressure by the same amount on most PRVs. Some PRVs allow only a percentage of an increase; others have maximum outlet pressure. Pilot-operated PRVs are set for maximum outlet pressure; therefore, fire departments cannot pump additional pressure to an outlet. These types are normally found in line to low zone or to separate riser manifold.

Most PRVs also act as check valves; the direction of flow is marked with an arrow on the valve body. This is particularly important, as it eliminates the possibility of pumping into a lower-floor outlet. In situations where the outside fire department connection is damaged or not accessible—and when fire departments want to establish and maintain a secondary, redundant water supply for a standpipe system—an outlet, typically on the first floor, can be pumped into. However, if this outlet is equipped with a PRV, it will also act as a check valve and will not allow water to be back-fed into the riser.

Fig. 4–17 This PRD has a limiting device that can be removed.

Fig. 4–18 An Allen wrench is used to loosen two set screws, and then the limiting device can be removed.

Fig. 4–19 Once removed, the valve can be fully opened.

In one building in downtown Denver, a standpipe hose valve located at the lobby level is equipped with two separate valves. One valve is a factory preset PRV, and the other is a standard hose valve. A fire department engineer/pump operator could not pump water into the system by way of the PRV valve, but could pump into it using the standard hose valve.

Pressure-control and pressure-reducing valves employ a mechanism that compensates for variations in inlet pressures by balancing water pressure in an internal chamber or chambers, typically against a spring. Pressure-control valves, a type of pressure-reducing valve that is pilot operated, are considered by many to be the most reliable method of controlling pressure. Often, the outlet pressure is not field adjustable and valves are preset at the factory for use at a specific floor level in a building. In such cases, it is critical to ensure that valves are installed at the intended location.

✱ "Unfortunately, pressure-regulating devices, particularly pressure-reducing valves, have reportedly experienced high failure rates. Flow testing of pressure-regulating valves is probably the single most important aspect of testing standpipe systems that have these devices. The need for routine testing of these devices cannot be overemphasized."[13] As stated earlier, the NFPA issued an alert bulletin to this effect following the One Meridian Plaza fire. Also, "unpublished results of tests conducted by the Los Angeles City Fire Department (LAFD) between 1984 and 1986 further emphasize the point. In these tests, 413 pressure-regulating valves were tested in a dozen buildings. Test results indicated that more than 75% of the valves tested required recalibration, repair, or replacement."[14]

Fig. 4–20 This PRD uses a pin to hold a collar in place.

Terminology

The entire issue of standpipe system PRVs has been interesting and controversial for a number of years. My own experience has been very frustrating: every time I begin to think that I have found the answers to the myriad of problems and issues surrounding PRVs, I actually come up with more questions. To further complicate matters, when seeking out resources and conducting research to answer the many questions, you quickly realize that there is a wide range of different answers to the same questions and that, obviously, they can't all be correct. The terminology associated

Fig. 4–21 By removing the pin from this PRD, the valve can then be fully opened.

with PRVs is a very good example. The definitions of certain types of PRVs and PRDs provided by nationally recognized standards, such as NFPA 14, are sometimes different from those provided by the manufacturers of various valves. I am much more interested in simplifying the problems associated with our operations than getting caught up in semantics.

In an attempt to simplify what can be a very complicated subject, I prefer to break the whole PRV issue down into a couple of very basic, common sense categories. The first is whether the type of pressure-regulating device is a *valve* or a *device*. It can be defined as a *valve* when there are internal components associated with its pressure regulation that cannot be removed. It can be defined as a *device* when there are either external components, such as limiting devices, or internal components, such as an orifice plate, associated with its pressure regulation. These components are separate from the actual valve and can be removed, ultimately changing the valve from a PRD to a standard hose valve on removal.

Second, as for the true pressure-regulating and -reducing valves, there are critical concerns for firefighters on the fireground. It all has to do with whether or not firefighters can change and adjust the valve outlet pressure, specifically in the heat of battle during a fireground operation. For this purpose, we can categorize the pressure-regulating and reducing valves into four categories.

/ **Factory preset, not adjustable.** Just as the name implies, these types of valves have been preset at the factory during the manufacturing process. Most important to us, they cannot be adjusted and the pressure cannot be changed at anytime during fireground operations—what you see is what you get. This is not necessarily a bad thing if the valve has been properly designed, assembled, installed, and maintained. However, there is plenty of room for human error associated with these types of valves.

To begin with, a fire protection engineer designs the system and makes recommendations for the installation of PRVs based on his mathematical calculations. His math must be 100% accurate and based on accurate data provided by the building architect with regard to the specific characteristics of that building such as size, area, and the like. Erroneous data can affect one or several components within the entire fire protection system. Secondly, even if all of the fire protection engineer's calculations are correct, the valve must still be assembled properly for the factory-preset type or adjusted properly for the field-adjustable type. Once again, there's plenty room for human error.

A factory-preset valve assembled based on proper and accurate mathematical calculations can easily lead to a future fireground problem if it is not installed in the intended location. In many high-rise buildings, hundreds of PRVs could be installed and used throughout the building. Each and every one of them must be designed properly, assembled properly, installed properly, installed at the designed location, flow tested, and properly maintained to ensure safe and proper operation when needed. That's a very tall order.

Two examples of factory-preset PRVs are the Elkhart and Powhatan models. These devices have a similar appearance and the signs indicating

Fig. 4–22 A Powhatan factory-preset PRV. The ring at the top of the valve body provides a very subtle indication that this is a PRV.

that they are PRVs are rather subtle compared with some of the other larger PRVs. On both these models, there is the appearance of a large ring extending outward around the valve body near the top. Inside, these factory preset valves have a set of water chambers that work by internal hydraulic controls to regulate and reduce pressure under both flowing and static conditions. Once again, it is important to note that both of these types of PRVs are factory preset and are NOT field or fireground adjustable (figs. 4–22, 4–23, and 4–24).

Field adjustable. Several years ago, I was assigned as the captain of DFD Engine Co. 3. We had recently operated at a high-rise fire in which the building had built-in automatic pressure-reducing valves. We operated with a 2½-inch handline, the PRV operated properly, the fire was quickly controlled, and the operation was brought to a safe and successful conclusion. However, during our post-fire review, we realized that most of us didn't have a comprehensive understanding of the particular type of PRVs in this building; more importantly, we did not know how to deal with them during emergency conditions in the event of a malfunction.

After this incident, I immediately contacted a friend, who at the time was assigned to our Fire Prevention Bureau. Captain Thor Hansen was plenty sharp and was a tremendous resource with regard to fire-prevention and code-enforcement issues. A few shifts after that high-rise fire, we met with Captain Hansen at a building that had the same type of PRVs. He introduced my crew and me to the building engineer, who proceeded to educate us about this particular PRV.

The PRV in question was a Zurn valve. In this particular building, printed signs at the standpipe hose valve locations indicated that tools to adjust valve pressure were located in the Fire Command Center (FCC) and the fire pump room. This information would immediately give the unsuspecting firefighter his first clue that there was a PRV at this standpipe hose valve (fig. 4–25, and fig. 4–26).

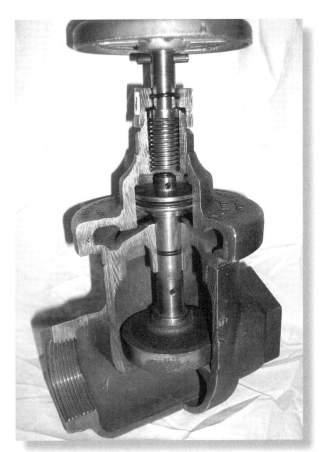

Fig. 4–23 A cut-away view of an Elkhart factory-preset PRV. Note the ring at the top of the valve body and the two separate internal water chambers.

Fig. 4–24 Note how small the waterway opening is on a Powhatan factory-preset PRV. This opening measures about 3/8 inch. This will still provide adequate volume, but at a low pressure.

Fig. 4–25 This printed sign on a standpipe cabinet in a downtown Denver high-rise clearly indicates the presence of a PRV and the location of tools to adjust valve pressure.

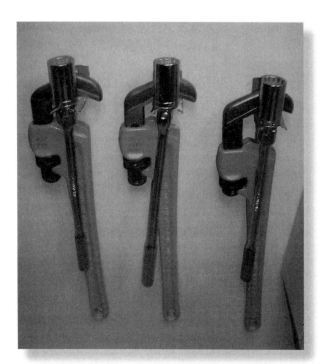

Fig. 4–26 These tools are kept in the Fire Command Center and the fire pump room and can be used to adjust the pressure on the Zurn brand PRV.

The building engineer showed us how we could use the large adjustable wrench to loosen and remove the bonnet portion of the valve. Once the bonnet was removed, we could use a separate tool, a wrench with a $1^{1}/_{16}$-inch deep well socket on a ratchet to adjust the valve pressure (figs. 4–27, 4–28, and 4–29).

The adjustment can be done by turning an adjustment nut clockwise to increase pressure and counterclockwise to decrease pressure. My crew and I were very excited with this new information. We felt as though we now had a significant advantage and could potentially solve low-pressure problems associated with these types of PRVs at any future high-rise fires in buildings equipped with these types of PRVs.

However, we also discovered that the term *field adjustable* can be somewhat misleading. This type of valve falls into the category of field adjustable. The term primarily applies to field adjustment during the installation process; in other words, a fire protection engineer, based on his calculations, could have his crew members make the necessary adjustments to these types of valves after installation at a particular building.

For example, let's say a 50-story high-rise building is being built. A two-zone standpipe system (high zone and low zone) was designed and installed in the building, using numerous PRVs—a total of 80—at specific locations. The manufacturer will design and build 80 identical valves and send them to the construction site. These valves will then be installed and field adjusted at the time of installation based on their respective vertical location within the standpipe system.

Although this type of valve is field adjustable, it could be a rather cumbersome process to make adjustments during fireground operations. First, the tools needed to complete the adjustment must be located and brought to the site where a valve needs to be adjusted. That location could be several floors away from where the tools are located, generally in a fire pump room located at one or more mechanical levels or in the FCC in the lobby. With this particular valve, the size and weight of the tools needed to make pressure adjustments make them impractical to carry in a fire department's high-rise standpipe equipment kit (tool bag).

The term *field adjustable* does not necessarily mean that it can be adjusted quickly, specifically in the heat of battle during fireground operations. Although this particular type of valve, the Zurn Valve, can be considered fireground adjustable, that is only true if firefighters receive comprehensive pre-fire training and if tools to make the adjustments can be quickly and easily accessed when needed.

3 Fireground adjustable. Because the term *field adjustable* can be a bit misleading, I felt it necessary to provide a couple of additional terms that can be applied to our operations. What is of critical importance to the fire service is whether or not a pressure-reducing or -regulating valve can be adjusted with speed during a high-rise building fire—in other words, is the PRV or PRD truly fireground adjustable? Keep in mind that a particular valve can be both field adjustable and fireground adjustable.

4 Fireground removable. One additional term that should be included is *fireground removable*. Many of the PRDs previously discussed fall into this category. If firefighters can quickly remove a restricting device and thus allow full pressure to be delivered from the hose valve outlet, that device is in essence, fireground adjustable, simply by removing the device. The adjustment occurred when the device was removed.

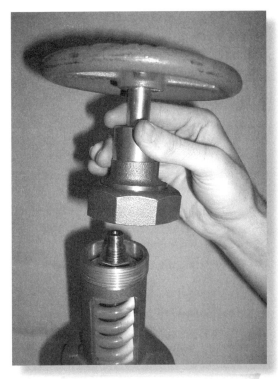

Fig. 4–28 Removal of the bonnet provides access to the adjustment nut

Fig. 4–27 Removing the bonnet of a Zurn PRV using a large adjustable wrench

Fig. 4–29 Using a 1¹¹/₁₆-inch deep well socket, the Zurn PRV can be adjusted. A clockwise rotation of the adjustment nut will increase pressure; a counterclockwise rotation will decrease pressure.

Pre-Fire Planning

In the process of solving fireground problems, many of the answers can come long before the fire, during *pre-fire planning*. Identifying that a particular building has pressure-regulating devices is the first step toward solving a potentially devastating fireground problem. Determining whether the particular valve or device can be adjusted or removed during a fire (is it fireground adjustable or removable?) is the next step. Last but not least, is the critical step of providing firefighters with the necessary tools and training for making the necessary adjustments.

The PRVs at the One Meridian Plaza fire were of the field/fireground adjustable type. Unfortunately, missing from the equation were the tools necessary to adjust the valves and prior training for the firefighters on how to complete such an adjustment. While preplanning many high-rise buildings in my first due response area, I have found a significant number of PRVs of the same type as those in One Meridian Plaza. Because of that fire, these types of PRVs had an ominous reputation within the American fire service. However, I have found, on taking a much closer look, that these types of PRVs are somewhat user friendly with regard to fireground adjustment.

Giacomini PRV

The type valve to which I am referring is a *Giacomini*, manufactured in Italy. Immediately upon seeing the valve, it is very obvious that it is a PRV type of valve. First, it is a very large valve, much larger than a standard standpipe hose valve. Second, it has several different labels affixed to the valve body, providing instructions on how to adjust the valve pressure (fig. 4–30 and fig, 4–31).

The method used to adjust the outlet pressure on this type of PRV is very clear and requires minimal training. Furthermore, there is only one tool needed to adjust this type of valve—a very small, compact, and short metal rod about $3/8$-inch in diameter and about 14 inches long.

The tools (valve adjustment rods) are typically kept in one of two locations. At one building in my district, there is small red cabinet in the FCC with a label that reads "Valve Adjustment Rods." A bundle of about 20 adjustment rods are stored inside. In another building, valve adjustment rods are located inside the standpipe cabinet at the hose valve (fig. 4–32 and fig. 4–33).

Fig. 4–30 A Giacomini PRV.

Fig. 4–31 Labels on the Giacomini valve give instructions for making pressure adjustments.

Fire departments should not count on finding these tools inside the building when they need them. Valve adjustment rods in the FCC or fire pump room may be several floors away from where they are immediately needed, and those that are located inside standpipe cabinets will eventually come up missing. Any number of things can happen to the tools between the time they are installed and the night of our fire. Ultimately, we must be prepared.

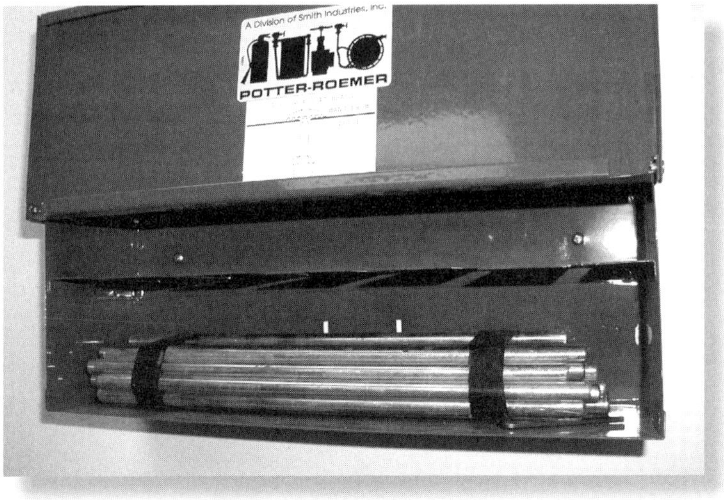

The captain of one of the engine companies in my district, Captain Dave Borelli of Engine Co. 24, suggested that we make our own tools and carry them inside our standpipe kits. He contacted our repair shop and requested that it fabricate several of these adjustment rods for the companies. We have added one adjustment rod to every engine company's standpipe equipment kit (tool bag). Because of the rod's small size and minimal weight, the addition to the equipment kit presents no problems. This was a very good, proactive step to ensure that we have tools available inside the building at the standpipe outlet if pressure needs to be adjusted on this type of valve.

The Giacomini PRV has four holes bored into a rotating barrel near the top of the valve. To adjust the

Fig. 4–32 This small red box, located in the Fire Command Center of a Denver high-rise, contains several adjustment rods that can be used to make pressure adjustments to the Giacomini PRV.

Fig. 4–33 A Giacomini PRV with the valve adjustment rod inserted into a hole on the rotating barrel. Note the hole in the rotating barrel. There are four holes on this barrel, one in each quadrant.

Fig. 4–34 The firefighter can adjust valve pressure when the valve is open and with water flowing by rotating the barrel clockwise.

pressure, a valve adjustment rod is inserted into the most accessible hole, and the barrel is rotated. In tight quarters, such as inside a standpipe hose valve cabinet, or close to a wall, the size of the valve adjustment rod will not allow for continuous rotation of the barrel. The rod will have to be removed and placed in the next hole to continue rotating the barrel. This process will continue until the pressure adjustment has been completed. Like the Zurn valve, pressure can be increased on the Giacomini valve by rotating clockwise, whereas a counterclockwise rotation will decrease the pressure (fig. 4–33 and fig. 4–34).

New Elkhart Brass PRV

Shortly before this book went to publication, my friend Paul Albinger from Elkhart Brass introduced me to a brand new fireground-adjustable PRV, designed and built by Elkhart Brass. This PRV has many similarities to the Giacomini valve, in that it can also be adjusted using a short metal rod to rotate an adjustment barrel. However, the torque required to rotate the barrel on the Elkhart Brass PRV, at 15 pounds, is significantly less than that of the Giacomini PRV. This new Elkhart Brass PRV is relatively small compared to some of the other PRVs, and has operating instructions printed on the outside, along with a shield to prevent vandalism. This new Elkhart Brass PRV appears to be very user friendly, and based on what I have seen, it will likely make our operations much easier when operating off standpipe systems equipped with this type of PRV (fig. 4–35).

Identifying Pressure-Regulating Devices

Ever since the One Meridian Plaza Fire, several members of the Denver Fire Department have been actively looking for and identifying the presence of pressure-regulating devices in buildings throughout our city. It has been determined that numerous buildings of various sizes and occupancies in our city do, in fact, have pressure-regulating devices. They have been found in a wide variety of buildings from large public high schools to the newest commercial and residential high-rise buildings. Firefighters have even identified the presence of PRVs inside a five-story assisted-living center. Pre-fire planning by firefighters and companies has determined that there are currently at least eight variations of pressure-regulating devices used in high-rise and standpipe-equipped buildings within the city of Denver.

Fig. 4–35 This new Elkhart PRV is well designed and extremely user friendly.

Ultimately, the primary concern associated with pressure-regulating devices has to do with discharge pressure during fireground operations. Research and flow testing have determined that most PRVs in the city of Denver are set to provide a discharge pressure of from 70 to 90 psi. That means very little pressure is available to overcome friction loss and achieve proper nozzle pressure. The use of low-pressure, high-volume weapons for firefighting operations is critical, specifically 2½-inch attack lines, with low-pressure smoothbore nozzles.

There are some specific indications that a pressure-regulating device is present. On pressure-restricting devices, the restricting device is typically located on the outside stem of the valve, making it easy to identify and, in most cases, easy to adjust or remove; these types of valves, therefore, are fireground adjustable/removable.

Pressure-regulating valves, such as the Zurn and Giacomini, are very large valves, which is a good initial indication that they are, in fact, PRVs. Also, the Giacomini and the New Elkhart Brass valve have written labels with instructions on how to adjust the pressure, clearly indicating the presence of a PRV.

The factory preset models, such as the old style Elkhart and Powhatan, are closer to the size of a standard hose valve and have very subtle differences, making them difficult to identify as PRVs. However, on both of these valves, the valve body is larger at the top, with a large ring near the valve stem, which indicates the presence of the internal water chambers and, thus, the presence of a PRV.

If all else fails, a dependable method you can use to determine the presence of a PRV is to simply remove the cap from the outlet and look inside. A smooth operating stem is typically an indication of a pressure-reducing valve. A threaded operating stem is typically an indication of a standard hose valve, but remember that restricting devices could still be on the outside (fig. 4–36).

It is important to note that even if there is no pressure-regulating device at the standpipe hose valve, PRVs could still be used in the system. Buildings that have a combined standpipe/sprinkler system using the same riser to supply both the floor hose valves and the sprinkler system frequently are equipped with PRVs. In these combined systems, the PRVs are often hidden and are sometimes located in the plenum space above a drop ceiling tile (fig. 4–37). In these situations, fire companies could have a low-pressure problem at the hose valve caused by an improperly set PRV that is hidden from view.

Fig. 4–36 Looking inside the valve, a threaded stem (left) indicates a standard hose valve. A smooth stem (right) indicates a PRV.

Fig. 4–37 This factory preset PRV is located at the ceiling level and is reducing pressure before water is delivered to a hose valve and the sprinkler system. Not seeing a PRV at the hose valve is not an assurance that one doesn't exist.

False Information

It has been said that false information is worse that no information at all. I would caution all readers to be particularly mindful of the fact that you will often times receive false, inaccurate information as it pertains to built-in fire protection systems, and specifically pressure regulating devices. This is usually not intentional, but rather based on plain ignorance.

A very good example of this was shared with me by a good friend. Firefighter Sean Roeper, DFD, has been a valuable asset to me in the long term quest to identify pressure regulating devices in buildings throughout our district, and the entire city. Sean relayed to me his experience of finding a building in his first due response district that did, in fact, have pressure regulating devices at the hose valve outlets. However, a large sign was placed just behind the valve with the words, "No Pressure Reducing Valves Used on Fire Protection System" (fig. 4–38). False information, on a large, printed sign!

This serves as an excellent example of just how ridiculous false information can be. Firefighters must be tenacious, and very cognizant of what is fact and what is fiction. Don't trust the information, unless you have verified it, and are 100% sure of its accuracy, particularly with regard to building information, and preplanning.

Fire Department Operations

The city of Denver can be used as an example of the many cities and jurisdictions across the fire service that are responsible for the fire protection of countless high-rise and other standpipe-equipped buildings. In Denver, more than 350 buildings are classified as high-rises (75 feet or taller in height) and of those, more than half are not fully protected by a sprinkler system. Hundreds more low-rise and large-area buildings in Denver are equipped with standpipe systems. Within all these buildings, there are thousands of pressure-regulating devices. There is simply no way to tell what the factual discharge pressure and flow of a specific valve is without actually flow testing the valve. This would be virtually impossible to accomplish based on the time required and the lack of resources to complete such a task.

Therefore, firefighters must operate within a strong mind-set and prepare for the failure of a PRV or anticipate that a PRV might be improperly set. We can't control the myriad of issues associated with PRVs and their performance, but we can select the best weapons to use for operations in buildings equipped with standpipes, specifically standpipe pressure-regulating devices.

The current edition of NFPA 14 recommends, as previous editions also have recommended, that fire departments use 2½-hose with a 1$\frac{1}{8}$-inch smooth bore nozzle for standpipe operations. Fire departments equipped with only 1¾-inch hose and combination nozzles, specifically the automatic type, need to take proactive steps and change their operational policy with regard to equipment used for standpipe operations. If for no other reason, this must be done for firefighter safety and survival.

Fig. 4–38 Although this sign located inside a fire department hose valve cabinet indicates no PRVs, the valve itself is actually a Zurn PRV.

Summary

All fire departments that respond to and operate at high-rise and standpipe-equipped buildings should be equipped with 2½-inch handlines and smooth bore nozzles. This complement of weapons is critically important for operations in buildings that use pressure-regulating devices, especially pressure-regulating valves that are non-fireground adjustable.

Furthermore, fire departments must keep in mind that this is not just a low-pressure issue. High-rise buildings, especially those that are commercial occupancies, require a very proactive and aggressive approach to weapons selection from the outset. Our window of opportunity to stop a serious fire in a commercial high-rise building is very narrow. These are high risk, low frequency events, and the 1¾-inch preconnect with tank water is not an option. The reflex time at these fires is significant.

Training conducted by DFD District #3 in 2004 indicated that even with firefighters who are very well prepared and extremely strong physically and mentally, water delivery to a fire on the upper floors of a high-rise building could take 20 or more minutes from the time of arrival. Because a fire can easily double in size every 30 seconds and because of the tremendous potential for rapid fire growth and spread within the large uncompartmentalized areas typical of a commercial high-rise building, proactively selecting and using a 2½-inch attack line from the beginning provides the greatest margin for success and firefighter safety.

This arsenal of weapons should always be carried into the high-rise building, especially the commercial high-rise. This should be the standard procedure even for automatic-alarm investigations, as we usually tend to fall back on our daily habits, especially when the heat is on.

Statements like, "The 2½-inch is too big," "too hard to maneuver," "takes too much time," "we don't have enough manpower," and the most pathetic two words that can be spoken, "we can't," must be eliminated from the fire service vocabulary. A strong commitment to mental and physical preparedness, along with a consistent and regular diet of training on the use of this most valuable tool is what differentiates the exceptional engine company from those companies that simply get by and sometimes don't get the job done at all.

The bottom line is this; we have little control over the design, installation, and maintenance of fire protection systems in the buildings we protect. Good code enforcement and preplanning will help, but when all else fails, the equipment we choose to operate with can either protect us, or harm us. Choose your weapons wisely.

References

1. National Fire Protection Association (NFPA), 2003, NFPA 14, Standard for the Installation of Standpipe and Hose Systems, 3.2.23 Standpipe Systems, 2003 Edition, Quincy, MA, p. 14-6.

2. Shapiro, Jeffery M., 2003, *Fire Protection Handbook*, Nineteenth Edition, Volume II, Section 10, Chapter 18, Standpipe and Hose Systems, p. 10-351.

3. Ibid, p. 10-351.

4. Ibid, p. 10-352.

5. Ibid, p. 10-352.

6. Ibid, p. 10-352.

7. Ibid, p. 10-352.

8. Ibid, p. 10-352.

9. National Fire Protection Association (NFPA) Alert Bulletin (91-3), 1991, "Pressure-Regulating Devices in Standpipe Systems."

10. Fornell, David P., 1991, *Fire Engineering*, "Pressure Regulating Devices," August, 1991.

11. National Fire Protection Association (NFPA), 2003, NFPA 14, Standard for the Installation of Standpipe and Hose Systems, 3.3.17 Standpipe Systems, 2003 Edition, Quincy, MA, p. 14-6.

12. Shapiro, Jeffery M., 2003, *Fire Protection Handbook*, Nineteenth Edition, Volume II, Section 10, Chapter 18, Standpipe and Hose Systems, p. 10-360.

13. Ibid, p. 10-360.

14. Ibid, p. 10-363.

CHAPTER 5

The Problem Defined

From the very first time that a fire occurred in a large multistory or tall high-rise building, firefighting operations in these buildings earned the well-deserved reputation of being particularly difficult, dangerous, and demanding. Numerous characteristics of multistory and high-rise fireground operations set them apart from the other

Fig. 5–1 A typical single family dwelling, approximately 2,500 square feet.

Fig. 5–2 A typical commercial high-rise building, approximately 1.5 million square feet.

"routine" daily operations. The obvious, is that of sheer size, as a typical high-rise building is a much larger and more complicated than structures such as a two-story, single-family house (fig. 5–1 and fig. 5–2).

The single family house is the place where most structure fires occur, and where firefighters fight the vast majority of their fires. These often-called routine operations are often comprised of a fast attack, using a short, 200-foot, preconnected, 1¾-inch attack line to quickly and efficiently control the most frequently encountered fire problem in the American fire service—the room and

Fig. 5–3 The single-family dwelling fire occurs everyday in big cities and small towns.

contents dwelling fire that occurs several times everyday, in big cities and small towns (fig. 5–3). On the other hand, a high-rise fire is a much more, rare event (fig. 5–4).

Two major concerns make operations in large, horizontal, standpipe-equipped, and large multistory low- and high-rise buildings extremely challenging—strategic and tactical problems. Those two items are the very complicated logistical aspect of these operations, coupled with the overwhelming exposure concern. That's what sets specifically the high-rise operation apart from all our other operations: logistics and exposures. Let me first address the issue of logistics.

Logistics

Merriam-Webster's New Collegiate Dictionary defines *logistics* as, "The aspect of military science dealing with the procurement, maintenance, and transportation of military material, facilities, and personnel."[1] Fire departments are considered to be paramilitary organizations, and thus, the military-based definition of logistics can be applied to fire department operations as well. Logistics can also be defined simply as the art or science of moving manpower and equipment from point A to point B. In this case, point A being the lobby, or lowest accessible level of a large standpipe-equipped, low-rise or high-rise building, and point B could be the staging location, which might be located several floors above the lobby level (fig. 5–5).

When a fire or other emergency condition exists on the upper floors of a tall high-rise building, logistics becomes an important component that can quickly affect the overall operation. Our preparedness, or lack thereof, will ultimately lead to either success or failure.

Whenever I present to other firefighters on the topic of high-rise operations, I always ask them a question: given a serious fire in a large, very tall high-rise building, would you prefer to have the fire in the lobby, on the top floor, or somewhere in between? Most veteran firefighters will answer, the top floor, whereas the majority of the less experienced members and laypersons with no experience will say the lobby. Some of the folks, usually those concerned about being politically correct, will choose a location somewhere in between as their answer, so as not to hurt anyone's feelings.

For fires in high-rise buildings, even if the location of the fire is on the lower floors, there will still be a significant logistical impact on the operation, because we must access all of the floors, stairwells, and elevators above the fire floor(s) to search for any occupants, and to identify any areas of potential smoke or fire extension.

For fireground operations in multistory and high-rise buildings, we typically have four specific logistical options:

1) Access from the exterior via fire department ladders, specifically at low-rise buildings, or if the fire area or portions of the fire area are within reach of our fire department ladders at high-rise buildings.

2) Access from the interior via elevators, if the elevators are safe to use.

3) Access from the interior via enclosed stairways.

4) Access from the exterior roof area, via aircraft, specifically helicopter.

When a fire exists in a high-rise building, above the exterior reach of fire department ladders, logistics becomes an extremely significant factor. When the problem is above the reach of our aerial ladders, we must then rely on the internal logistical equipment inside the building.

Fig. 5–4 Fires in large high-rise buildings are rare events, and present numerous problems not typical of our more frequent operations.

Fig. 5–5 There is a significant logistical problem at any large high-rise fire.

Elevators

The primary tool used to move people to the upper floors of a high-rise building is the elevator. The elevator has been described as a valuable tool that can also quickly become a dangerous and deadly trap. We cannot live with a mind-set that depends solely on elevators for internal logistical support. We must establish and maintain logistical alternatives, including the use of stairs, and in extreme cases, the use of a helicopter to insert firefighting teams onto the roof of a high-rise building, if that's possible.

Stairs

Professional, proactive, and well-disciplined fire departments and individual fire companies use elevators with great discretion. Furthermore, the firefighters are prepared for the worst by maintaining a strong mindset and a superior level of physical fitness, which is needed to climb several flights of stairs in full personal protective equipment (PPE) while carrying, in most cases, 50 to 100 extra pounds of heavy firefighting equipment. Mindset is an overall attitude and life style, and superior physical fitness is only achieved through an ongoing program of physical training (see chapter 1).

Fire departments should continually evaluate and adjust their operating procedures to include the best, most effective and efficient methods to transport firefighters and equipment to the upper floors of a high-rise building. Simply put, there are countless options, from using lighter-weight equipment, to employing specific techniques and methods to carry equipment in the most efficient manner. However, nothing beats good, old-fashioned physical training, specifically that which mirrors the type of physical activity that will take place at a high-rise event—climbing stairs in full PPE with equipment (fig. 5–6).

Strong physical and mental preparation is essential to achieve logistical success at high-rise operations, specifically when stairs must be used to access upper floors. However, even with the very best preparation, this method of ascent will still be difficult and cumbersome, primarily due to the fact that numerous occupants will likely be evacuating down the stairs while we are attempting to go up.

The typical 44-inch wide stairwell quickly becomes a congested, slow, and frustrating path for firefighters working their way up to a fire floor. Loaded with SCBA, spare air cylinders, hose packs, and countless other pieces of bulky firefighting equipment, there is simply not enough room to climb up while civilians are coming down (fig. 5–7). Evacuating occupants, who are reluctant to stop, must do so and turn, placing their back against the wall to minimize their profile as we go by. Firefighters must also adjust their body by turning to create as small a profile as possible in order to effectively pass the retreating occupants.

Continually compensating for passing occupants is much more physically exhausting and time consuming. If possible, every attempt must be made to direct occupants down a separate stairway from the one firefighters will use to access the upper floors. Establishing a separate attack stairway and evacuation stairway early in the event will greatly enhance the logistical operation. Using a public address system to continually direct

Fig. 5–6 Firefighters should regularly practice climbing stairs in full gear, and developing the best methods to carry firefighting equipment.

Fig. 5–7 There is very little room in a typical high-rise stairwell for firefighters to efficiently ascend while occupants are hastily coming down.

evacuating occupants to the evacuation stairway is critical. In addition, proactively training building management and security personnel, floor wardens, and occupants in the tactical use of stairways and the need to follow specific evacuation instructions will be beneficial.

Reflex time

The logistical problems of moving firefighters and equipment to the upper floors of a high-rise building, in addition to placing necessary firefighting equipment into operation once the fire floor is reached, result in an overall reflex time. This reflex time can be very significant, and is the end result of the significant and ongoing logistical problem at high-rise fires. A very good engine company might take perhaps only one or two minutes to stretch an attack line and apply water to a typical room and contents fire at a single-family dwelling, but that same action might take 10, 15, 20, or perhaps more than 30 minutes at a serious high-rise fire.

Standpipe system

Keep in mind, the logistical problem, although primarily associated with the actual movement of firefighters and equipment, also includes numerous other components that support the overall firefighting operation. For example, fire protection systems, specifically the standpipe system are logistical in nature. This is the system that moves water for fire extinguishment. As discussed in chapter 4, the standpipe system can fail for a variety of reasons. This can become a logistical nightmare that may require significant manpower to overcome. Such was the case at the Philadelphia One Meridian Plaza fire. Because of improperly set PRVs, and a severe low pressure problem, firefighters had to hand stretch, three separate 5–inch supply lines, up several flights, in three separate stairwells, to assemble a portable standpipe system. This is an example of a very severe logistic problem (fig 5–8).

From the elevators to the standpipe, and the HVAC smoke control to the communications system, each of these are logistical in nature. While typically not found in buildings and occupancies that comprise our most frequent operations, we must be aware of these various logistical components, and be prepared with backup plans in the

Fig. 5–8 If a standpipe system fails, having to hand stretch a supply line up several flights of stairs is a very difficult and physically demanding logistical operation.

event there is a failure with any one of them. Logistics is and always will be one of the most significant problems facing firefighters from the outset of an emergency event, especially a fire, in any large multistory building, but especially a tall high-rise.

Exposures

The other major problem that is encountered and must be dealt with at most high-rise events, is the exposure concern. All firefighters must understand that every single fire is a six-sided, multidimensional problem. All of the by-products associated with a fire and the process of combustion, including smoke, heat, and flame, are dangerous and destructive as they travel in all six directions; up, all four sides, and down away from the fire.

Clearly, a fire's natural movement in an upward direction makes the geometric configuration of any multistory or high-rise building particularly susceptible to rapid fire growth and extension upward. Therefore, once again, the lower a fire is the greater the problem, specifically from an exposure standpoint. Countless fire service case studies reinforce this concern. The 1980 fire at the MGM Grand Hotel and Casino in Las Vegas, Nevada serves as an excellent example. Although the fire occurred at the ground level, the vast majority of the 85 fatalities occurred on the upper floors, and most of those deaths were due to smoke inhalation. The MGM Grand was a 26-story, high-rise building. Had the fire occurred on one of the upper floors, most of the occupants would have been unaffected, and likely could have been protected in place.

Although it presents some significant logistical challenges to the firefighting forces, generally, the higher up a fire is in a high-rise building, the better off everyone is, from the occupants to the firefighters. The events of 9/11 can also be used as an example. The first tower to collapse was the second one to be attacked. That attack, however, was at a lower level, creating a situation with more exposures above, a greater number of exposures to burn, a greater number of occupants trapped above, and a more significant load (weight) that resulted in an earlier collapse.

Summary

There are a myriad of strategic and tactical problems that firefighters will face and must deal with at a high-rise event. The interrelated problems of logistics and exposures will likely present the greatest problems to firefighters. Fire departments and firefighters will have little or no control over many of the problems related to exposures at a typical high-rise event. We simply don't control where the fire occurs. On the other hand, there are numerous things that we can do, proactively, in order to better cope with the problems we encounter related to logistics. We must start by proactively preparing for the worst.

Fire departments, and specifically individual fire companies, should constantly be proactively working to develop new methods to quickly and efficiently transport tools and equipment to the upper floors of a high-rise building. There are numerous things that can make the very exhausting and arduous task of climbing stairs loaded down with heavy equipment much more efficient. Firefighters should practice a variety of different methods, with the ultimate goal of having a solid plan in place before a 30-story climb. Don't wait until the day of a serious fire in the lobby of the fire building, when you're forced to quickly implement an unproven method, to carry a spare air cylinder. Some specifics of efficiently transporting personal protective and firefighting equipment up several flights of stairs is addressed in chapter 7.

Complicated logistics and the extreme exposure problems are by far the two most significant problems that make firefighting operations in large, standpipe-equipped, low rise buildings, and especially tall high-rise buildings, much more difficult. Recognizing these problems and proactively addressing them will help firefighters achieve a safe and successful operation.

We have little control over the exposure problem, but we can and must address it by assigning companies to search the countless exposed areas. As for the logistics, it too is very much beyond our control, but once again, we must deal with it by proactively developing good solid operational procedures well ahead of time. Most importantly, our mindset must never underestimate the overwhelming problems associated with the logistical and exposure problems at these large, complicated buildings.

Reference

1. *Webster's New Collegiate Dictionary*, 11th Edition, s.v. "Logistics."

CHAPTER 6

Elevator Operations

In chapter 5, I listed *logistics* as one of the two primary and most significant problems firefighters and fire departments face with regard to firefighting operations in high-rise and other multistory, low-rise buildings. Access to the upper floors of a high-rise or multi-story building via an elevator is one of our logistical options. In this chapter, we will specifically address the logistical tool of elevators for fire department use during fires and other emergency operations in both high-rise and low-rise multistory buildings.

Elevators

Elevators are a standard component of all high-rise buildings and several other buildings of various types and sizes. In fact, with the implementation of the Americans with Disabilities Act (ADA) several years ago, just about any new public building taller than one story will likely have at least one elevator.

Countless people depend upon elevators to transport them throughout multistory buildings, specifically commercial high-rise buildings, during every business day. In addition, the numerous occupants of high-rise multiple dwellings rely on elevators to transport them to and from their homes each day. It would be hard to imagine life without elevators.

For firefighters, and specifically fire department operations in high-rise buildings, the elevator can also be an extremely valuable tool. However, firefighters know all too well that during fireground operations, an elevator can quickly become a very dangerous and potentially deadly trap. Therefore, a great deal of attention and discussion must be focused on elevators, elevator operations, and the proper utilization of elevators during fireground operations in multistory and high-rise buildings.

Elevator History

In chapter 2, I discussed high-rise buildings, including a brief history of their development. The rapid increase in the development of high-rise buildings during the late 19th and early 20th centuries can be directly attributed to specific advancements in building construction and the invention of a safe elevator.[1]

In 1853, Elisha Graves Otis was credited with having invented the first safe elevator, which was originally referred to as a *safety lift*. Prior to Otis's invention, a number of different types of "hoists" had been developed during a period of over 100 years. Those hoists were prone to failure, however, and countless injuries and deaths resulted when the lifting cable(s) used to hoist the occupants failed time and again, causing the hoist to plunge to the ground. Otis solved the problem with a safety lock, which is still in use today.[2]

Electric Traction Elevators

There are two basic types of elevators: *electric traction* and *hydraulic*. The electric traction elevator is most commonly found in taller buildings, generally those over six stories tall, and certainly in most of those that are considered true

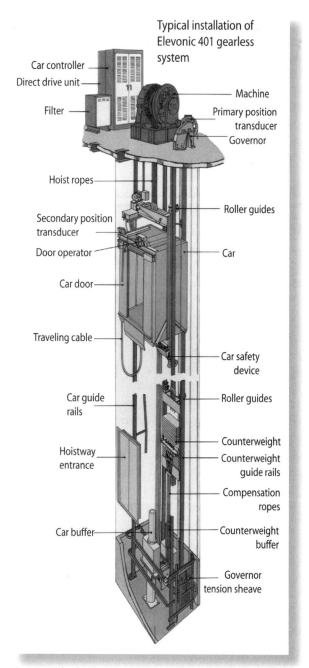
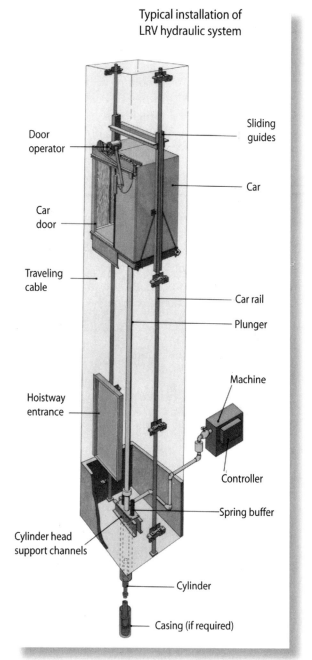

Fig. 6–1 The basic components of a typical electric traction elevator (left), and a hydraulic elevator (right)

high-rise buildings at 75 feet or more tall. Electric traction elevators operate by way of an electric motor, cables, and counterweights (fig. 6–1).

According to the Otis Elevator Company, a typical gearless traction elevator operates in the following manner:

"Six to eight lengths of wire cable, or hoisting ropes as they are known in the industry, are attached to the top of the elevator and wrapped around the drive sheave in special grooves. The other ends of the cables are attached to a counterweight that slides up and down in the shaft-way (hoist-way) on its own guide rails.

The result of this arrangement is that with the weight of the elevator car on one end of the hoisting ropes, and the total mass of the counterweight on the other, it presses the cables down on the grooves of the drive sheave. When the motor turns the sheave, it moves the cables with almost no slippage. Actually, the electric hoisting motor does not have to lift the full weight of the elevator and its passengers. The weight of the car and about half its passenger load is balanced out by the counterweight, which is sliding down as the car is going up. In the process, it supplies the necessary traction between the hoist ropes and the drive sheave."[3]

With six to eight wire hoisting ropes used to lift the elevator, this could be interpreted as an indication of a significant safety redundancy. However, the primary reason behind the multiple hoisting ropes is to provide increased traction, although the additional hoisting ropes do also provide a significant margin of safety.[4]

The primary components of a typical electric traction elevator installation can be divided into three general areas:

1) The elevator machine room
2) The hoist-way (or shaft-way)
3) The elevator car

Elevator machine room

The elevator machine room for an electric traction elevator is usually, but not always, found above the hoist-way (also referred to as the elevator shaft or shaft-way) at the top-most point that the elevator services (fig. 6–2). The weight of the machinery and all of the various components require a very substantial and strong structure at the top of the hoist-way. It is critically important for fire departments, and specifically individual fire companies, to preplan the location of elevator machine rooms, as access to these areas becomes essential, especially during elevator emergencies, in order to control the electrical power to the affected elevator. Within the elevator machine room are several very important components including the main electrical disconnect, the traction hoisting machine, the controller, the motor generator, and the speed governor.

Main disconnect. In most installations, there is a main electrical disconnect for all elevators. It might be in the form of a large fused knife-type switch or a large circuit breaker. In buildings with multiple elevators, and specifically where multiple elevators are in the same bank and controlled by

Fig. 6–2 A typical elevator machine room for a large commercial high-rise building

the equipment located in the same elevator machine room, there are multiple electrical disconnects. Each disconnect is typically labeled with a number corresponding to a specific hoisting machine and the elevator to which it provides power (fig. 6–3).

✳ It is important to note that the actual lighting and ventilation controls within a specific elevator are typically powered from a source that is separate from the main disconnect. This maintains a minimum level of comfort for occupants trapped in a stalled elevator. However, it can be confusing for firefighters during elevator rescue operations, because they may not be certain that the main power has, in fact, been shut off to that elevator. Once again, lights and ventilation fans typically remain on and running even after the main electrical power has been shut off to the affected elevator.

2 **Hoisting machine.** The traction hoisting machine will be located inside the elevator machine room. This is the main piece of equipment that moves the elevator up and down. The machine consists of a traction sheave, which is a large, grooved wheel that moves the hoisting ropes to raise and lower the elevator (fig. 6–4). The driving motor is a large, powerful electric motor. The hoisting machine is also equipped with motor brakes. These brakes actually do not stop the elevator during normal operations. They merely operate to hold the car in place and keep it from moving by retarding the motor after the car has made a normal stop.

3 **Controller.** The controller is the actual brains behind the elevator operation. Each individual elevator has its own controller. Signals calling the elevator to various locations are received by the controller. The controller determines where to send the elevator, and then dispatches it to the proper location.

4 **Motor generator.** On electric traction elevator installations, the motor generator converts the building's alternating current to direct current. Because direct current can be controlled more easily than alternating current, the elevator can be operated in a much smoother manner.

5 **Speed governor.** The first attempts to build elevators failed because they lacked the necessary safety features to keep the elevator from crashing down if the hoisting ropes failed. The speed governor is the device that activates the "elevator

Fig. 6–3 The main disconnect (electrical shutoff) inside elevator machine room

safeties" that prevent the modern elevator car from falling and crashing down into the pit if the hoisting ropes fail. The speed governor is located over the hoist-way in the elevator machine room or at the top of the hoist-way if the machine room is not located at the top of the hoist-way.

6 Hoist-way

Hoist-ways (also referred to as *shaft-ways* or simply as the *elevator shaft*), are vertical shafts in which the elevator or elevators travel and are typically located in the core area, separated from the rest of the building. The elevator machine room is typically located at the top of the hoist-way and there is an elevator pit at the bottom. The hoist-ways in modern buildings are totally enclosed to reduce the possibility of vertical fire spread.

Furthermore, the hoist-way walls are constructed of fire-resistive materials and the hoist-way doors are also fire-rated doors.

Within the hoist-way, the elevator car and the counterweights travel on vertical rails that are attached to the hoist-way walls. Terminal switches are located on the rails or on the elevator car itself near the end of its travel points to ensure that the car slows down and stops at the appropriate locations. If these fail, devices called buffers are located at the bottom of the hoist-way, which are designed to provide a cushion should the elevator fall.

Hoist-ways can be divided into three different basic types:

1. The single hoist-way
2. Multiple hoist-ways
3. Blind hoist-ways

Fig. 6–4 A typical elevator hoisting machine

7. **Single hoist-ways.** Single hoist-ways contain only one operating elevator. This type of hoist-way is usually found in smaller, low-rise buildings (fig. 6–5).

8. **Multiple hoist-ways.** As the name implies, a multiple hoist-way contains multiple elevators. There is typically no separation between elevators within a multiple hoist-way. Multiple hoist-ways are very common, especially in larger, high-rise buildings (fig. 6–6).

9. **Blind hoist-ways.** Blind hoist-ways do not have normal openings, except typically at the very bottom and top levels that the elevators serve. It is basically like traveling in a vertical tunnel (fig. 6–7). The idea behind the blind hoist-way has to do with efficiency. This type of hoist-way is common in very large high-rise buildings, where there are express elevators which travel from the main lobby at the ground level to an upper lobby, called a *sky lobby*. At this location, occupants can exit the express elevator and board a different elevator which provides local service to upper areas of the building, for example, the top 20 floors of a building. Because hoist-way doors are not located within the blind hoist-way except at the top and bottom, a stalled elevator within a blind hoist-way can present very difficult and challenging problems for fire departments.

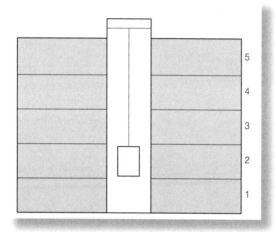

Fig. 6–5 A single hoist-way

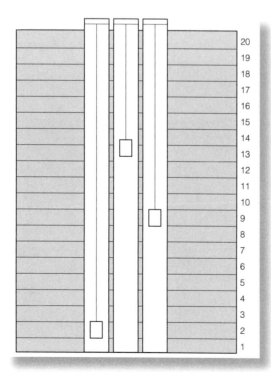

Fig. 6–6 A multiple hoist-way

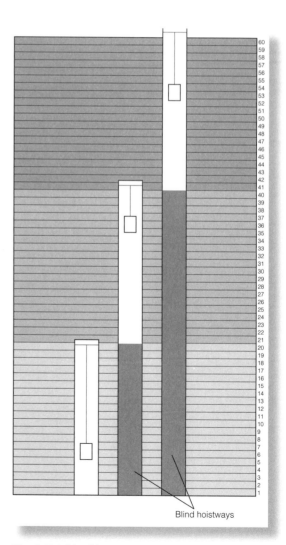

Fig. 6–7 A blind hoist-way

Elevator car

The actual elevator car has to be assembled within the hoist-way itself. There is an elevator frame, and the cab of the elevator sits inside this frame. The whole thing is then suspended by hoisting ropes. Roller guides, which are attached to the frame, are sets of wheels that roll against the rails to guide the elevator up and down within the hoist-way (fig. 6–8). Elevator car safeties are mounted on the frame and provide the critical safety for passengers by stopping the elevator from falling, should the hoisting ropes fail.

Several wire hoisting ropes connect to the top of the car and to the counterweight. Each of these ropes is extremely strong, typically providing enough strength to carry the entire load of the elevator. The multiple ropes provide an extra margin of safety, but their primary purpose is to increase the traction.

There is usually an inspection station provided at the top of the elevator car. At this location, controls are provided to operate the elevator during inspection, maintenance, and repair operations (fig. 6–9). During elevator emergencies, trained firefighters can also use the inspection station to control an elevator being used for rescue purposes.

Hydraulic Elevators

Hydraulic elevators are generally found in shorter buildings, typically those that are two to four stories tall. Some installations are slightly higher, but rarely over eight stories in height. A representative from Otis Elevator Company told me that the maximum height they will install a hydraulic elevator is seven stories, although it is possible that fire departments might encounter hydraulic elevators in much taller buildings. For example, the 20-story Flatiron Building in New York City has hydraulic elevators that are still in use today. These are *roped hydraulic*–type elevators.

A hydraulic elevator works on a very basic principle of using fluid, specifically hydraulic oil, to lift the elevator. A hydraulic pump, which is typically driven by an electric motor, pumps oil from a hydraulic fluid reservoir to a cylinder. As hydraulic fluid is forced into the lifting cylinder, the piston lifts and moves the car upward. The car descends by way of an electrically operated valve, which opens, and using the weight of the elevator car, allows the hydraulic oil to be forced back into the hydraulic fluid reservoir as the car descends down to the lower levels.

There are several different types of hydraulic elevators, including:

1. Holed hydraulic elevators
2. Holeless direct (1 piston) hydraulic elevators
3. Holeless direct (2 pistons) hydraulic elevators
4. Holeless roped (1 piston) hydraulic elevators

Fig. 6–8 These are elevator roller guides attached to the frame of the elevator car

Fig. 6–9 An elevator inspection station located on top of an elevator car

Hydraulic elevators are limited to the overall operating length of the hydraulic ram or piston. Most installations utilize a *telescoping* ram or piston, which retracts to a much smaller size requiring less space for installation.

The primary disadvantage to the use of hydraulic elevators is, obviously, their limited overall height capabilities. Furthermore, they travel at much slower speeds than the electric traction elevators, and therefore are not as desirable for most installations. They do, however, have a specific advantage in that hydraulic elevators don't require the heavier construction at the top of the hoist-way and the additional overhead space to accommodate an elevator machinery room. Most hydraulic elevator installations have the machinery room located at or below the lowest level of elevator service.

Hydraulic elevators are most commonly found in shorter, low-rise type buildings. However, a few hydraulic elevators can also be found in many high-rise buildings. In large high-rise buildings, with multiple banks of elevators, the elevators traveling shorter distances and servicing lower levels, such as parking garages, are often times the hydraulic type. Shuttle elevators of the hydraulic type, servicing only a few floors, and short distances, can also be found on the upper floors of some high-rise buildings.

Elevator Banks

In high-rise buildings, elevators are separated into individual banks. There may be one or several elevator cars located within a single bank. Larger high-rise buildings usually have multiple banks.

For example, a residential high-rise building of 20 stories will likely have one elevator bank, with two or three separate cars within that one bank (fig. 6–10). Due to the size of this building, and the number of occupants, there is no need for separate, multiple high- and low-rise banks.

On the contrary, a large commercial high-rise building, certainly those over 50 stories tall, will usually have multiple elevator banks, each serving different areas of the building. One particular commercial high-rise building in downtown Denver has 29 elevators. Most large high-rise buildings divide the elevator banks into at least three categories, such as: (see fig. 6–11 and fig. 6–12).

1. A low-rise bank
2. A mid-rise bank
3. A high-rise bank

Elevator Operations During Fires

When I was a very young firefighter, my mentors started me out on the right track with regard to elevators. They expressed their opinions and shared their experiences with elevators, both good and bad. It was easy to conclude that the elevator can be a dangerous tool that must be dealt with in a cautious and respectful manner.

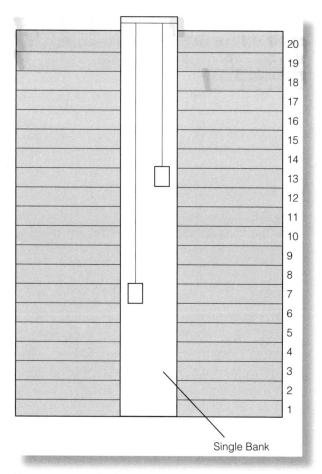

Fig. 6–10 A single elevator bank with two elevators in a 20-story residential high-rise

Throughout my fire service career I have had the opportunity to see for myself on numerous occasions just how dangerous and unpredictable elevators can be, especially once the elements of heat, smoke, and/or water are introduced. Heat from a fire can be very damaging to the myriad of sensitive, electronic components controlling an elevator. And, needless to say, the water used to stop the fire, either from handlines and or sprinkler systems can quickly cause the failure of an elevator.

It doesn't take a lot of water either. A very small amount of water that finds its way into an elevator hoistway or machine room can quickly render that elevator useless. If firefighters using an elevator as a logistical tool become trapped in a stalled elevator, they will no longer be able to complete their assignment, and the incident commander (IC) will not be able to use them to

Fig. 6–12 A lobby view of multiple elevator banks, note the floor numbers served by different banks.

accomplish the fireground mission. In fact, rather than being part of the solution, these firefighters will become part of the overall fireground problem. This problem is critical if an elevator stalls above the fire floor.

If you are going to use elevators during fireground operations in multistory and high-rise buildings, the most important strategic component is making sure your fire department and personnel are prepared to operate without the elevators at all. Elevator failure during serious fire conditions is a real possibility. If elevators are deemed too unsafe to use or if during the operation, building systems fail, such as the electrical power or the elevators themselves, you must have an alternate plan. This means ensuring that all members are physically and mentally prepared for a much more difficult and demanding logistical operation.

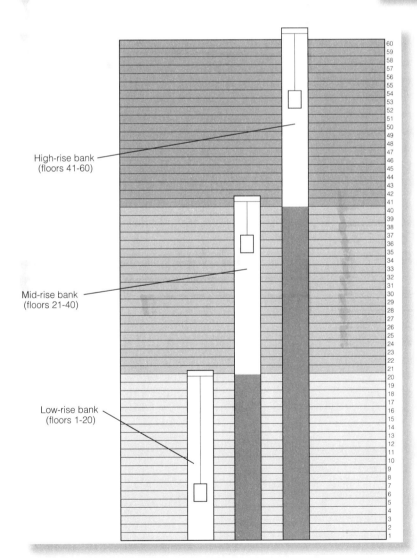

Fig. 6–11 Multiple banks of elevators inside a commercial high-rise building.

You must plan for the worst, and of course, hope for the best. If the elevators are out of service for whatever reason, or if the officer in charge has determined that they are unsafe to use or if they fail at some point during the operation, you must be prepared. Preparation means being physically and mentally prepared to climb up what may be several flights of stairs, loaded with heavy firefighting equipment, to get to the fire area, rescue victims, or make it up to staging, and so on. Strategically planning on the use of elevators for high-rise operations must begin with the area of firefighter physical fitness and strength. Someone, including your brother firefighters, may be counting on you to make it up there to help them.

I realize that the idea of climbing up several flights of stairs is probably not all that appealing to most of us. Anyone who has ever done it understands that it is quite simply an extremely demanding and arduous task, even for short distances. But we may not be left with a choice other than to climb stairs. You only need to review modern fire service history to see that elevator failure and stair climbing has been the case at several of the most serious high-rise fires and emergency incidents in American fire service history. I specifically address some of these incidents as well as stair operations and the lessons learned in chapter 7.

If elevators are working, specific procedures should be followed to increase firefighter safety. Once again, the elevator can be a valuable tool. A properly operating elevator can give a tremendous logistical advantage to fire department personnel operating at a high-rise building. The simple fact that we may be able to significantly reduce our overall reflex time, and thus get water on the fire several minutes sooner, can be important in the control of that fire.

Fire service recall and control

Most modern elevator installations are required by model fire and building codes, including the International Building Code, to be equipped with a *fire service recall and control* feature. This feature basically provides for an automatic recall of elevators to a designated location during fire emergencies. In addition, it gives firefighters the ability to control the elevator, everything from moving up and down to opening and closing the doors.

Phase I. Phase I fire service recall and control is the first step in achieving control of the elevators. Phase I is typically designed to automatically recall the elevator or elevators to a designated location when the fire alarm, or in many installations, specific zones of the fire alarm system is activated. The recall location is generally designated to be the lobby or main floor in most buildings.

In the event of a fire alarm from the lobby level, the recall destination defaults to a secondary, backup location. It is important to keep this in mind, as oftentimes fire companies choose to quickly recall the elevators by activating the fire alarm, sometimes by activating a lobby pull station. However, if a fast recall is desired, when the fire alarm system has not yet been activated, members must remember to activate a pull station other than one located in the lobby. This will ensure that the elevators recall to the desired location. Keep in mind that, in some installations, activation of a pull station may not recall the elevators.

In addition, most installations are designed to allow for a Phase I recall using a key to activate the system and recall the elevators to the lobby. The keyway and switch to complete this are usually located at the main floor elevator lobby near the elevators. However, in many buildings, the keyway and switch for elevator Phase I fire service recall and control is located inside the fire command center (FCC).

Phase II. Phase II of the fire service recall and control feature allows firefighters to control the movement and operation of a specific elevator from inside the elevator car. After Phase I, which usually happens automatically (if the building is in alarm), firefighters arriving at a building lobby will usually see all elevators recalled to the lobby level, with their respective doors open. Firefighters obtain elevator keys along with master keys, and use these keys to activate the Phase II service inside the elevator car. From here the elevator can be fully controlled by the firefighters.

Recommended operational procedures for elevator use

The following are recommended operational procedures for fire department use of elevators in high-rise buildings. The intent of these recommended procedures is to maximize firefighter safety and increase the probability for a safe and successful conclusion to incidents that require the use of elevators as a logistical tool.

1. **In low-rise buildings, use the stairs only:** Elevators should not be used in low-rise buildings, those less than 75 feet tall, or approximately 7 stories or less in height, during fireground operations. The time needed to retrieve elevator keys, recall an elevator, periodically stop the elevator, and test that elevator's performance on the way up, will take longer than simply walking up to a proposed staging location generally two floors below the reported fire floor. Furthermore, an added benefit of taking the stairs is that the initial Fire Attack Group doesn't lose one of their members, who would have to be assigned to the position of elevator operator. This is particularly important for those fire departments with limited manpower, which includes most fire departments, when the incident happens to be a high-rise fire.

2. **Heavy fire upon arrival:** If there is an obvious working fire upon arrival, with heavy fire venting from an upper floor, especially if it is possible that the fire has control over a large portion of the floor area and may have already compromised electrical power or other building systems, specifically the elevators, firefighters should use the stairs.

3. **Smoke or fire in elevator machine room:** If the fire alarm panel or other information indicates that the source of the smoke or fire may be inside an elevator machine room, do not utilize that bank of elevators. If that is the only bank of elevators in the building, firefighters should use the stairs.

4. **Multiple elevator banks:** If the building has multiple elevator banks, such as low-rise, mid-rise, and high-rise, make every effort to utilize an elevator within a bank that does not directly service the suspected fire floor or floor of alarm. Avoid using any elevator that provides direct service to a suspected fire floor or floor of alarm.

5. **Fire service recall and control (Phases I and II):** Check to see if the elevator or elevators in question are properly equipped with an approved fire service recall and control system. In other words, can firefighters control the elevator operations in a Phase II mode? If not, than you should consider this a major concern, enough to ban the use of these elevators, and opt for stair operations from the outset of the incident.

6. **Full PPE, SCBA, radio, can, and irons:** Any time fire department personnel are utilizing an elevator for fireground operations, all personnel should be equipped with full personal protective equipment (PPE), self-contained breathing apparatus (SCBA), portable radio(s), a portable extinguisher (2½-gallon water extinguisher *can*), and the appropriate forcible entry/exit tools (*set of irons*) when inside the elevator car. A set of irons should include at least one striking, one prying, and one pulling tool. The striking and prying tools (flat-head axe, TNT (Denver) tool, or sledge hammer coupled with a Halligan) can be used to force the elevator doors open or even breech some types of elevator hoist-way walls if an emergency exit is required. The pulling tool (6– to 8-foot pike pole) can be utilized to push open a roof hatch, or simply to extend a firefighter's reach within the hoist-way to, for example, trip a hoist-way door-locking mechanism. Anytime a team of firefighters must use an elevator, even an engine company team (whose primary responsibility is fire suppression) operating without truck company support, that team should never be without a basic set of forcible entry/exit tools. Under absolutely *no conditions* should a team of firefighters enter and operate within an elevator without a basic set of forcible entry/exit tools.

7. **Don't overload the elevator:** Elevators should not be overloaded by firefighters and equipment. A maximum of six personnel for a standard elevator should be maintained (fig 6–14). An average firefighter weighs close to 200 pounds. With close to 100 pounds of additional weight from firefighting tools and PPE, each member may add 300 pounds to the elevator. For six members, that's close to 2,000 pounds, which comes close to exceeding the capacity of many elevators, especially those in residential high-rise buildings. Besides the issue of weight, not overloading an elevator also provides enough room for forcible exit from the elevator car. An overloaded elevator doesn't allow members the

maneuverability needed for emergency egress. Members equipped with forcible entry/exit tools should board the elevator last, so as to be in position at the front of the elevator car near the elevator doors with enough room to swing a striking tool and operate a prying tool if it becomes necessary. Keep in mind that some elevators in larger commercial high-rise buildings have two elevator doors, one at the front, and one at the rear. The door at the rear is typically the egress door for the floors above, thus allowing the first occupants to board an elevator to be the first occupants off. Position your forcible entry/exit team accordingly.

8. **Designate and assign a member to be the elevator operator:** The officer in charge on the elevator should designate and assign a member to be the elevator operator. The elevator operator should remain in control of that elevator until relieved or reassigned. The elevator operator should be equipped with the appropriate full PPE, SCBA, portable radio, a portable extinguisher (2½-gallon water extinguisher can), and a set of irons (fig 6–14).

9. **Inspect the hoist-way for water, smoke, and fire:** Once an elevator is chosen for use, prior to leaving the starting location, one member should inspect the elevator hoist-way. Using a powerful hand light, direct the beam up the hoist-way through the gap located between the elevator car and the hoist-way wall (fig. 6–15). You are looking for anything unusual, but specifically any signs of water, smoke, or fire. This inspection practice should continue with each of the periodic stops made at designated locations on the way up as you test the performance of the elevator. This is critical, especially when traveling long distances, for example from the lobby level to the fortieth floor. It is recommended to stop the elevator at least once for short distances (less than 10 flights) and two or more times for distances greater than 10 flights. Obviously, any sign of anything unusual, but specifically water, smoke, or fire in the hoist-way, should

dictate that all members immediately evacuate the elevator and proceed to the nearest stairwell. The incident commander must also be notified of these facts and actions. It is extremely important to train all firefighters to stand all the way inside the elevator car when

Fig. 6–13 Firefighters should never overload an elevator. Six members with equipment should be the maximum for most elevators.

Fig. 6–14 One member should be assigned to the position of elevator operator and equipped with the appropriate gear.

completing this inspection. If a member has one foot in and one foot out, he could sustain serious or fatal injuries should the elevator car suddenly move in either direction, which has been known to happen. Also, members should not overlook the possible use of an elevator car hatchway door, which is sometimes located at the top of the elevator car. This door can sometimes be opened to provide a continuous inspection of the hoist-way while the elevator is moving upward. However, generally these top hatchway doors are locked from the outside. Once again, if your experience tells you that something just doesn't look right, immediately stop the elevator, get off, and get to the stairwell.

10. **Never take an elevator below grade:** Elevators should not be used to access areas below grade during fireground operations. This seems like a common sense recommendation, but failure to follow this procedure has had tragic results. It's simple: if you are above the fire when you enter the building, you're already in a very dangerous position. Don't exacerbate this danger by climbing into a transportation vehicle that can be very dangerous and unpredictable during fire conditions, and which travels inside a chimney that could already be filled with smoke and hot gases.

11. **Never take an elevator directly to a reported fire floor:** Elevators should *never* be utilized to gain direct access to a reported fire floor, floor of alarm, or to the location of any other potential fire-related emergency condition during the pre-control phase of fireground operations. This recommendation is also such a common sense principle that it might hardly seem worth mentioning. However, firefighters have been injured and killed because they failed to follow this very basic procedure.

12. **Stop two floors below the reported fire floor:** Anytime an elevator is utilized for fireground operations, fire companies should not take the elevator to a location any closer than two floors below the reported or suspected fire floor or floor of alarm. Fire companies should stop at least two floors below and walk up the remaining two flights. The company officer in charge should notify the incident commander as to which stairwell is being used to access the fire floor, and once a determination has been made, whether that stairway will be designated as the *attack stairway* or the *evacuation stairway*.

13. **Test the elevators performance:** If the elevator is equipped with fire service recall and control features, it should allow you to completely control all elevator operations on the way up. Unfortunately, elevator recall has been known to fail as well. Therefore you must test the elevator's performance on the way up in order to confirm that the operation is as safe as possible. When utilizing an elevator equipped with fire service recall and control features, fire companies should stop the elevator on the way up to the staging destination in order to evaluate that elevator's

Fig. 6–15 Inspect the elevator hoist-way for signs of water, smoke, or fire by shining a powerful hand light up between the hoist-way wall and elevator car.

performance, at least once for short distances (fewer than 10 flights) and two or more times for distances greater than 10 flights. The following items should be evaluated:

1. Does the elevator stop at the desired floor?
2. Do the elevator doors remained closed when the elevator stops?
3. Do the elevator doors open when the door open button is activated?
4. Do the elevator doors close when the door open button is released before the doors are fully opened?
5. Does the elevator car stop where the elevator car floor is level with the floor of the desired location?
6. Does the elevator perform normally and not in an erratic or potentially dangerous manner?

If the answer to any of these questions is no, then a complete evacuation of the elevator by all fire department personnel and switching to the stairs to gain access to the upper floors is justified.

Additional size-up considerations (for elevator use)

There are many considerations when deciding whether or not elevators are safe to use, for example, late night (after business hours) and early morning (midnight to 6 a.m.) reports of fire in a commercial building. This is time when there are few occupants inside a commercial high-rise building. Those few occupants are typically maintenance and custodial staff. Fewer occupants can result in a delayed report of smoke or fire within the building, especially if the smoke and fire detection system has not been properly maintained or is not being properly monitored. When dealing with residential high-rise buildings, there are generally numerous occupants in the building at night. However, as with any residential occupancy, most of the occupants will likely be asleep at late night and early morning hours. Nighttime and early morning hours are times when elevators should be used with extra caution.

On the other hand, during the daytime (normal business hours) at a commercial high-rise, there are countless occupants, several of who will likely call the fire department via 911 if smoke of fire is detected. Furthermore, as fire companies arrive at a building, they are frequently met by someone from the building engineer's staff, security personnel, or the fire safety director, who may have already investigated the fire alarm location. Of course we don't recommend that non-fire-department personnel investigate potential fire areas, but they will usually do it regardless. We can typically use elevators with less trepidation during daytime hours, in populated buildings, when there have been no second source calls or reports of fire or smoke.

The point is this, if you have not received any additional information due to few occupants, or if you're not met by someone with an explanation of the alarm, beware; you are now left to discover for yourself what is really going on above. Also, keep in mind, the information from occupants, security personnel, or members of the engineering staff may not always be accurate. A proper firefighting mind-set drives our good habits, and following all of the previously recommended operational procedures, each and every time will result in application of these procedures when it's really necessary.

Ultimately, I recommend using the procedures previously addressed for safe and effective elevator operations. At the very least, exercise good discipline by consistently and always stopping at least two floors below the reported fire floor or alarm location, get out, and walk up the remaining two or more flights every time! Keep in mind, once again, use caution when determining the safest way to get to a fire or floor of alarm on the upper floors of a high-rise building. If in doubt, use the stairs.

Elevator Rescue Operations

During the normal course of daily activity, elevators in high-rise and low-rise buildings occasionally do not operate properly and may stall. Stalled elevators are a very common occurrence in most high-rise buildings, especially buildings that contain multiple elevators. In fact, many large commercial high-rise buildings have their own elevator mechanic or a staff of several mechanics on duty during normal business hours.

Most of these elevator failures are minor and can be easily handled by the elevator mechanic. More often than not, the fire department is not even notified or requested to respond and assist. That procedure is appropriate only if there is a competent elevator mechanic on duty, the elevator failure is very minor in nature, and the elevator can be returned to normal operation in a short period of time. I have responded to several elevator emergencies where the occupants have been trapped inside a stalled elevator for several hours. Procedures outlined by the Denver Fire Department (DFD) require that buildings notify the fire department if the on-duty elevator mechanic cannot safely remove the occupants in a timely fashion.

Rescue or removal

Once again, I hate to use the word *routine*, but for lack of a better word, most elevator failures become very routine and minor events. However, we must always be mentally prepared for the worst-case scenario. After an initial size-up, information gathered can lead to a determination as to whether the situation is a true elevator emergency, which will require the *rescue* of occupants from inside the elevator car, or simply a *removal*. Most of our operations will be simple removals; however, fire departments should establish comprehensive procedures and maintain professional discipline whenever operating at any elevator incident, whether it is a true emergency and rescue, or just a simple removal. Failure to follow the necessary safety procedures and to operate in a disciplined manner can quickly turn a removal into a serious emergency, sometimes with tragic results.

Removal procedures

On a typical day, the DFD, like most city fire departments, responds to numerous elevator incidents. Most of these turn out to be an elevator that has stalled between floors. The occupant or occupants are normally not in any danger, and barring any medical conditions or serious claustrophobia and panic attacks, they typically consider the event to be a minor inconvenience. Once it has been established that the elevator car has simply stalled and the occupant or occupants are not in any danger, are not injured or suffering from any medical problems, and are not in a state of panic, the event becomes a removal.

Response assignment. On the DFD, a basic response assignment to an elevator incident in a low-rise building includes an engine company, a truck company, and the district chief. This represents a total of nine personnel and we believe that this assignment gives us the appropriate and necessary resources to safely and efficiently handle a typical elevator incident. A rescue company (heavy rescue) is added to the assignment if the building is a high-rise. The technical expertise of rescue company personnel is invaluable at elevator incidents that are true emergencies and rescues, especially those that involve rescues within blind hoist-ways or other complicated long-term operations.

Reconnaissance. As with all emergency incidents, at any elevator incident, one of the first steps is to complete a thorough size-up, and gather as much information as possible. The early reconnaissance by the first arriving company or companies will yield critical information and help the IC establish a plan of action.

A company, preferably a truck company, should be assigned to determine the location of the stalled elevator, identify how many occupants are inside the elevator, their condition, and if the elevator is simply stalled or if a more significant elevator failure caused damage to the elevator car, hoist-way, or any other operating components. All procedures from this point on are determined by these critical pieces of reconnaissance information.

Power control. A company, preferably an engine company, should be assigned to locate the elevator machine room and prepare to control the power to the affected elevator. Shutting down the main electrical power to the affected elevator prior to initiating any removal efforts is critical. Failure to maintain discipline and control the electrical power has resulted in injuries and deaths. You should never underestimate the potential of any elevator incident. Taking shortcuts and specifically failing to control the electrical power is a very dangerous gamble.

Experience on the DFD has proven many times that the cause of a stalled elevator happens to be a fire in the elevator machine room. These fires range from very minor to quite serious, sometimes requiring suppression efforts by fire department personnel. Further, smoke from these fires may affect numerous areas of the building. Because of these experiences, sending an entire engine company to control the electrical power is an excellent and effective procedure. The engine company should be prepared to take the necessary fire suppression equipment with them as part of their assignment to control the power.

Basic equipment should include full PPE and SCBA for each member of the engine company. High-rise/standpipe hose packs and the associated equipment, a CO_2 fire extinguisher, and a set of irons are the essential equipment. A set of building master keys, if available, will make access much easier, eliminating the need to force doors.

Assigning an engine company that is prepared for a fire to control the electrical power is good habit at all elevator incidents, but it becomes essential anytime there is additional size-up information that indicates a possible fire or smoke condition in the elevator machine room. For example, if the building has a built-in fire detection system and the fire alarm panel indicates an alarm with a smoke or heat detector activated from the elevator machine room, you need to assign an engine company and a truck company with fire suppression equipment to respond to the elevator machine room. Furthermore, when only a couple of companies are sent to a reported elevator incident, the IC must call for help anytime smoke or fire is encountered by those companies or reported to fire department personnel from occupants of the building. Reports of smoke or fire in a multistory building, especially a high-rise, calls for a large assignment of resources. If the early reports prove to be a serious fire, a lack of necessary resources will put the IC in a dangerous position of playing catch-up from the outset, not to mention the danger posed to firefighters operating shorthanded.

As part of our evolutionary process in the fire service, I believe we are responsible for continually progressing and developing new, more effective, and safer methods of operation. Over the years, it has been common procedure to send only one firefighter to control the power at elevator incidents. Accidents, injuries, and near misses involving firefighters operating in elevator machine rooms should be strong enough evidence to change a "that's the way we've always done it" attitude.

A brief survey of any elevator machine room, but especially those with machinery for multiple elevators, provides strong insight into the countless hazards in these machine rooms. A drive sheave and drum are designed to be powerful enough to move significant weight. A firefighter's protective clothing, tools, or other equipment can easily be caught up in the movement of the drive sheave and pull a firefighter into the machinery causing serious injury that could be potentially fatal. Electrical components and open circuitry on a control panel can electrocute an unsuspecting firefighter. There are numerous slip, trip, and fall hazards, including spilled oil, grease, and various other liquids found in a poorly maintained equipment room. The teamwork associated with an entire company, is paramount to firefighter safety and survival. If manpower is limited, at the very least, ensure that two firefighters are assigned to control the electrical power. There is truly safety in numbers, and having a partner to watch out for you and assist you if you encounter problems is a significant component to firefighter safety.

Lockout, tag out. Ideally, it is best to have a minimum of two radio-equipped firefighters operate and remain in the elevator machine room at the main electrical power control until the incident is brought to a safe conclusion. However, with limited manpower, many departments might not have the luxury of leaving members at this location. This can be addressed with the use of a lockout, tag out system. Many main electrical shutoffs for elevators can accommodate a pad lock, which will keep the shutoff in the off position. A tag, indicating that the shutoff has been locked in the off position and that it is not to be tampered with should also be used. Fire departments must be equipped with the appropriate equipment to complete an effective lockout, tag out operation. The lock should be a substantial padlock, and the fire department should hold the only keys to these locks (fig. 6–16).

If a lockout, tag out system was used during the operation, after any occupants have been removed or rescued from a stalled elevator car, and all fire department emergency operations associated with that elevator have been completed, the padlock can be removed. If firefighters were assigned to this position, they can leave this position. However, in both cases, the main electrical power should be left in the off position, and the property representative notified. Power should be left off until the problem with the elevator has been identified and corrected.

Methods of Removal and Rescue

After a complete size-up and assessment of the elevator failure and after the main electrical power has been shutoff and maintained in that position, members can proceed with the operation. Many times, some very simple solutions bring an elevator incident to a conclusion, quickly and safely. When completing a removal or rescue, members should attempt to use the safest and easiest methods possible, leaving the more complex, complicated, and hazardous operations as a last resort.

Once the hoist-way door closest to the stalled elevator has been opened and members can visualize the location of the elevator car, a determination can be made of how to best complete the removal or rescue of the occupants. When an elevator car is stalled between floors, generally, removing the occupants by bringing them down is better than taking them up. However, bringing them down will in many situations leave the hoist-way exposed at the removal location. It is critical to take the necessary steps to protect operating firefighters and civilians being removed from the stalled elevator from the open hoist-way.

1 Resetting the system

If the elevator is simply stalled, and no other significant event led to the elevator failure, one of the first solutions that can be attempted is to simply reset the system. This is most applicable to newer, modern elevator systems, which operate via computer. The team or company that is assigned to control the power can complete this operation, but only after given specific orders to do so from the IC. The operation simply involves shutting the electrical power off and than turning it back on. Sometimes, this resets the elevator and the elevator car will return to its normal operation, stopping and opening in a normal position. Once again, this will certainly not work every time, but it has been known to work on many occasions, and is often a good place to start.

2 Fire service recall

Another simple, but oftentimes, effective method to get a stalled elevator moving is to use the fire service recall. This method should only be completed after a thorough size-up and confirmation that the event is simply a stalled elevator. Furthermore, a team or company should be in position at the main electrical power control prior to attempting a recall. Once these procedures have been completed, a Phase-I recall of the affected elevator can be attempted. Sometimes this procedure works to bring a stalled elevator down to the lobby level, typically the destination for recalled elevators. The power shutoff method should be attempted first; many times the recall will only work in conjunction with first shutting down and then restoring power.

3 Hoist-way door

If the elevator will not reset after power control and recalling the elevator did not work, the next method usually involves attempting removal

Fig. 6–16 A lockout, tag out procedure can be used by fire departments to secure an elevator's electrical power in an off position.

or rescue via the hoist-way door. Most elevator installations have a panel above or near the elevator indicating that elevator's position. In buildings with multiple elevators and multiple banks, there is usually an elevator panel in the lobby, security office, or the fire command center. These panels are the fastest, easiest, and safest method to determine the location of a stalled elevator. If no panel exists, a hoist-way door will have to be opened initially to determine the approximate location of the stalled elevator. However, no hoist-way door should ever be opened until the main electrical power to the affected elevator has been controlled, and shut off by radio-equipped fire department personnel.

In order to open a hoist-way door from outside the hoist-way, the hoist-way door must have a keyway and members must be equipped with the appropriate key. A wide range of different types and styles of keys are used, depending on the type of elevator and the company that manufactured it (fig. 6–17). With a keyway access and the proper key, members can unlock the hoist-way door and open it.

Once the hoist-way door has been unlocked and opened, members can inspect the hoist-way in an attempt to locate the stalled elevator. One member can look into the hoist-way with a powerful hand light, while other members secure that member to protect him from falling into the hoist-way. From this position, the location of a stalled elevator can be approximated, narrowing down the search on the upper floors. The member looking into the hoist-way can also attempt to briefly communicate with occupants inside the stalled elevator, that is, if the elevator is reasonably close to his position. Once again, this operation should not be attempted until the main electrical power to the affected elevator has been controlled and shut down.

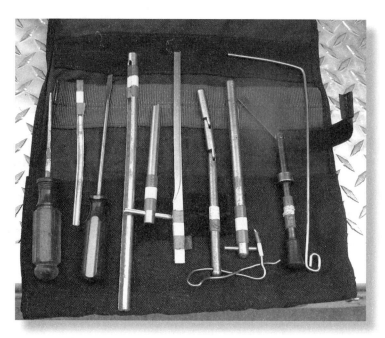

Fig. 6–17 There are a wide range of different elevator hoist-way door keys.

Fig. 6–18 Members using an elevator hoist-way door key to access a stalled elevator and effect rescue/removal of occupants.

Sometimes the location of the elevator is already known or can be determined from an elevator panel. In these situations, members can then go directly to the location where the elevator is stalled, typically between floors. From this location, a hoist-way door can be unlocked and opened up, providing immediate access to the stalled elevator car. Once again, there must be a keyway access in the hoist-way door, and members must be equipped with the proper key (fig. 6–18).

In many buildings, not all the elevator hoist-way doors are equipped with keyways, or only certain levels have keyways. In these situations, a different approach is needed to open the hoist-way door and gain access to the stalled elevator. If the location of the stalled elevator can be accurately determined, occupants may be able to assist firefighters in opening the hoist-way door. First, you must make contact with the occupants of the stalled elevator, and determine how many occupants are inside the elevator and the condition of all the occupants. If the occupant or occupants are physically able, firefighters can instruct them on opening the hoist-way door.

The occupants must be instructed to open the elevator car door first. The elevator car door is not typically equipped with any locking mechanism, and can be opened by simply applying lateral pressure in the direction that the door opens. Once again, firefighters will have to size-up the elevator, and can actually look at another elevator if the building has multiple elevators, in order to identify the type of door and how it opens. From this information, firefighters can communicate through the doors to the occupants of the stalled elevator, instructing them how to open the elevator car doors.

Once the elevator car doors are open, the occupants might have sufficient access to the hoist-way door and the locking mechanism. There will, however, be some instances when the occupants will not be able to reach, or in some cases, even see, the locking mechanism for the hoist-way door. If they do have access, firefighters can explain the locking mechanism to the occupants, and with minimal instruction, most persons will be able to trip the locking mechanism allowing the firefighters to open the hoist-way door (fig. 6–19). With both doors open, firefighters can begin the removal or rescue of the trapped occupants.

Before any of these procedures are undertaken, the power to the affected elevator must be shut off. Furthermore, once contact with the occupants of a stalled elevator has been established, the occupant or occupants should be instructed to immediately activate the emergency stop switch on the interior elevator car panel. This will provide an additional margin of safety when removing or rescuing occupants from a stalled elevator. If occupants are unable to perform this function, a firefighter should first be placed in the elevator to activate the emergency stop, before initiating a removal or rescue.

Fig. 6–19 With instruction from firefighters, occupants of a stalled elevator can trip the locking mechanism for the hoist-way door, allowing firefighters to open the hoist-way door and remove the occupants.

Poling down

Another method that can be used to gain access to a stalled elevator is called *poling*. The term *poling* comes from firefighters' use of a pike poles to perform this procedure. The poling method involves using a pike pole or other acceptable tool to extend a firefighter's reach, in order to trip a hoist-way door locking device. Over the years, the use of pike poles for poling has diminished in favor of smaller, lighter-weight, and narrower tools which are much easier to operate and fit into small openings.

Poling down operations begin with gaining access to the hoist-way of the stalled elevator car. This method is particularly useful in situations where there are no keyways on the hoist-way doors near the stalled elevator. Firefighters begin above the stalled elevator at the first hoist-way door location that has a keyway. After opening this hoist-way door, a firefighter lying on the floor will reach into the hoist-way with a poling tool and attempt to trip the locking mechanism for the hoist-way door on the floor below. One or more firefighters will assist that firefighter by securing him to keep him from falling into the hoist-way.

At the same time, a separate team of firefighters will be positioned on the floor below, applying a slight lateral pressure on the outside of the hoist-way door, so they can push it open when the lock has been released (fig. 6–20). It is important to point out that the members attempting to open the door from the outside must use some finesse. They should only apply a slight amount of pressure, as too much lateral pressure will cause the locking mechanism to bind up and the firefighter attempting to trip the mechanism will have a difficult time.

Several teams of firefighters are needed to complete this operation quickly and efficiently, especially if poling is required for a long distance more than 10 flights. Two teams of at least two firefighters per team is the minimum; however, additional teams are ideal. Separate teams are set up on the floors below, and as one hoist-way door is opened, the firefighter from above, who just tripped the locking mechanism, can then pass the poling tool down within the hoist-way to the next team. They will then trip the locking mechanism on the next floor below as firefighters arriving at that location open the hoist-way door. After tripping the lock and passing the poling tool down to the next team, firefighters should close the hoist-way door, and can then use the stairs or a separate elevator on independent or fire service control to access the level below. The separate teams of firefighters continue this process, moving from floor to floor, until the stalled elevator car has been reached.

Poling across

Another method used to open a hoist-way door and access a stalled elevator involves *poling across*. This method is completed using the same basic principles as poling down. Because the

Fig. 6–20 Firefighters on the floor below, applying slight lateral pressure to the hoist-way doors can open this door after the locking mechanism has been tripped by members poling down from above.

firefighters are not exposed to an open hoist-way, poling across is generally considered to be a safer operation; furthermore, it's typically a much faster and more efficient method than poling down. However, this method can only be used when the stalled elevator is close to floor level, as opposed to stalled between floors. If possible, this method should be attempted before poling down. Poling across can only be attempted and completed if there is another elevator that operates next to the stalled elevator within the same hoist-way. This is not uncommon in many buildings that have multiple elevators.

To begin a poling across operation, firefighters must first locate and establish the location of the stalled elevator. Firefighters should gain control of an elevator that operates next to the stalled elevator and within the same hoist-way via independent service or fire service Phase II. This elevator is the *rescue elevator* and should be positioned at a landing that is closest to the location of the stalled elevator. The doors to the rescue elevator are then opened and the elevator is kept in this position to complete the poling operation. For safety purposes, the emergency stop switch should be activated; however, activation of the emergency stop is frequently accompanied by an audible alarm, usually a bell. This can interfere with the operation, limiting communication between firefighters and communication with the occupants of the stalled elevator. Therefore, a better operation involves using the team of firefighters assigned to control the electrical power, as they can also shut off the main power to the rescue elevator once it is in position. If the rescue elevator has to be moved, power can be quickly restored, but members in the machine room should be certain which power switch controls which elevator. Inadvertently restoring power to the stalled elevator during the rescue operation could be very dangerous.

Once the rescue elevator is in position next to the stalled elevator, firefighters using a powerful hand light can attempt to locate the locking

Fig. 6–21 Firefighters performing a poling across operation

mechanism for the hoist-way door of the stalled elevator. With a poling tool, preferably a very narrow poling tool that can easily fit into this space, firefighters can attempt to reach across and trip the locking mechanism. A separate team of firefighters is once again positioned outside the hoist-way door applying slight lateral pressure to open the door once the lock has been tripped. This operation places both operating teams in the same general location, easing communications and making the overall operation much more efficient (fig. 6–21).

Poling tools

The original poling tool was a pike pole, hence the term *poling*. As firefighters gained experience performing poling operations it quickly, however, became apparent that the conventional pike pole was a rather cumbersome, large, and sometimes impossible tool to use for many poling operations, especially poling across. Most pike poles come in standard sizes and lengths and cannot be adjusted. The larger diameter of the pike pole itself can be a limiting factor, especially for poling across.

If we weren't called firefighters, the next best term might be *emergency problem solvers*. Most of the best equipment used by firefighters today originated as homegrown ideas conceived, designed, and built by firefighters themselves. Many firefighters probably don't realize as they carry a set of irons into a building that the Halligan bar was invented by an FDNY captain named Hugh Halligan.

As a young firefighter on DFD Truck Co. 8, the engineer (apparatus operator) on my crew was Mark Trujillo and the senior firefighter was Bob Terry. I learned a lot from these two guys, especially with regard to hand tools. They went on to invent the now widely used TNT (Denver) tool. There is a long list of firefighters who have developed valuable tools and techniques over the years.

In recent years, various tools have been developed, or adapted from other tools, to be used for elevator rescue, particularly poling operations. A close friend, District Chief Dave Frank, DFD, adapted an excellent poling tool from a painter's pole several years ago when he was a company officer. The lightweight fiberglass pole was purchased at a local hardware store, and was designed and marketed as a painting tool. It is a telescoping pole, which, when retracted, is very short and compact, but can also be easily extended several feet for poling operations. The length can be adjusted as needed within a wide range. Dave joking says "what was once a $15.00 painter's tool is now an elevator rescue tool, available at the new low price of $49.95, while quantities last." Since his discovery several years ago, these or similar type tools are now used by all DFD Truck and Rescue Companies for elevator rescue operations.

There are a wide variety of tools available for poling operations. Some have patents and are marketed in various fire service publications; others are cheap, but very effective homegrown tools. The latter is by far the most popular, as firefighters generally have to buy their own tools, that is, unless they want to fight the bureaucracy and write letters requesting the equipment, only to oftentimes end up with a low-bid piece of garbage, designed, specified, and purchased by someone who doesn't understand its use.

Roof hatchway door

Many situations require that occupants of a stalled elevator must be evacuated by a means other than the elevator and hoist-way doors. For example, if the elevator and or hoist-way doors have been damaged and cannot be opened, occupants will have to be removed or rescued another way. Also, anytime an elevator becomes stalled in a blind hoist-way, removal or rescue will have to be completed via a means other than the elevator and hoist-way doors. One of the possible alternative avenues of egress from a stalled elevator is the roof hatchway door (also referred to as the top emergency exit) (fig. 6–22).

Most elevators are equipped with a roof hatchway door that can be used for emergency egress. The roof hatchway door is typically locked from the outside, thus requiring firefighters to first access the top of the elevator car in order to unlock the roof hatchway door. In many elevators, there is also a framework of panels on the ceiling of the elevator car that have to be removed before occupants can access the roof hatchway door. These panels are usually lightweight, plastic panels, that cover the elevator car lighting and ventilation components.

Anytime firefighters must enter and operate within the hoist-way, the operation truly becomes a *rescue* operation rather than a simple *removal*. An elevator hoist-way is a very dangerous place, for both firefighters and the occupants being rescued from a stalled elevator. Firefighters must remain alert and operate in a most disciplined manner.

Many times, elevators become stalled between floors. Oftentimes the elevator is close enough to a floor that access to the occupants is relatively easy once the hoist-way and elevator doors have been opened. However, in some situations, there may be a greater distance between floors, especially in commercial buildings with mezzanines and mechanical floors. After the hoist-way door is opened, only the top or bottom of the elevator can be seen by firefighters. It quickly becomes obvious that access to the elevator car doors will be impossible with the elevator car stalled in such a position.

Depending on the distance, often the roof of the elevator car can be accessed from above. Firefighters begin by accessing the hoist-way from above, using keys if there is a keyway in the hoist-way door or a poling operation. Once the hoist-way door is open, firefighters can get a precise idea as to the location of the stalled elevator car. Firefighters should size-up the roof and determine if there is a roof hatchway door. A short ground ladder, and in many situations, a folding or Little Giant®-type ladder can be used to access the roof of the elevator car.

Firefighters entering the hoist-way must have the necessary personal protective equipment, specifically a helmet, gloves, and work shoes with a steel toe and shank. A powerful hand light, and preferably a separate, hands-free light, such as a helmet light, will be essential. Firefighters must be very careful and alert to any trip or fall hazards on top the car and within the hoist-way. If the hoist-way has multiple elevators, they should all be stopped and taken out of service prior to placing firefighters into the open hoist-way for rescue operations. This can be done by placing all of the cars into fire service recall Phase I, or shutting the power down at the elevator machine room. It would be far too dangerous for firefighters operating within the hoist-way, and occupants being rescued from the stalled elevator car, to allow for the continued operation of other elevators within the same hoist-way. Fast-moving elevators and counter weights are difficult to see in a dark hoist-way. If a firefighter or occupant being removed from the stalled elevator were to be struck by any of these moving components during rescue operations, it would likely be fatal.

Side emergency exit door

When a building has multiple elevators within the same hoist-way, the elevator cars are often equipped with side emergency exit doors. The side emergency exits are typically larger than the roof hatchway door. Like the roof hatchway door, the side emergency doors are sometimes locked from the outside, and must be accessed and unlocked by firefighters. In addition, many require special keys to unlock them.

Rescuing occupants from a stalled elevator using the side emergency doors requires using a separate *rescue elevator*. Many elevators that have side emergency exit doors only have one. Firefighters must complete a thorough size-up to determine the location of the side emergency exit door on the stalled elevator and on the rescue elevator. Once this has been determined and a rescue elevator has been selected, that rescue elevator must be placed on independent service, preferably using fire service control Phase II.

Ideally, it works well to utilize the independent service control features located on the top of the rescue elevator at the inspection location. From this position, firefighters on top of the rescue elevator car can stop the rescue elevator in a position precisely in line with the stalled elevator. Firefighters inside the rescue elevator car can then open the side door on the rescue elevator, access the side door on the stalled elevator, open that door, and rescue the occupants.

Fig. 6–22 Some elevator rescue operations may have to be completed using an emergency roof hatchway door.

There will be a need for some sort of a stable platform to be placed between the two elevators for firefighters and occupants of the stalled elevator to walk from one elevator to the other. After such a platform has been placed in position, a firefighter from the rescue elevator should first go into the stalled elevator. That firefighter must access the elevator panel, and activate the emergency stop, which might be a button, but usually requires a key to activate on newer elevators, as perpetrators have used the emergency stop switch to trap their victims. The firefighter can then assess the occupants for any injuries or problems, and prioritize the removal of multiple occupants. That firefighter will remain inside the stalled elevator car assisting the occupants out of that car, across the hoist-way, and onto the rescue elevator. At the same time, firefighters inside the rescue elevator are assisting and receiving the occupants coming from the stalled elevator car. Depending on the number of people inside the stalled car, firefighters may have to make multiple trips to complete the rescue and removal of all occupants, in order not to overload the rescue elevator.

Summary

Once again, although they can be valuable logistical tools, elevators present a wide range of potential problems and dangers. Elevators must be used and treated with respect. Their use during fireground operations should only be done within a comprehensive list of appropriate parameters and operating procedures. Rescuing occupants from elevators presents a long list of potential dangers and numerous associated problems. Elevator operations cannot and should not ever be thought of as *routine*. Elevators are valuable and powerful pieces of machinery that require a cautious and respectful approach, every time!

References

1. Otis Elevator Company, 1974. Tell Me About Elevators. Farmington, CT: Otis Elevator Company, 8, 10.
2. Ibid., 16.
3. Ibid., 17.
4. Ibid., 19.

CHAPTER 7

Stairway Operations

When elevators are out of service, or it has been determined that they would be unsafe to use for fireground operations, that leaves us with only a few options. As I stated in chapter 5, there are four basic logistical options in a multistory or high-rise building. With the elevator option gone, that leaves us with fire department ladders from the exterior, or the use of aircraft, specifically helicopters to access the roof. Obviously, using fire department ladders on the exterior can be an impractical and limited option, especially if we need to gain access to the upper floors of a tall high-rise building. The use of a helicopter from the outside will, for most fire departments, not be a practical option early in the incident, but it could be a beneficial logistical tool for long-term incidents. Our only remaining internal option is the stairs.

As I begin to address the critical lessons and procedures to utilize when dealing with stairs, I will again reemphasize the importance of firefighter physical fitness and strength. Planning for the worst, and of course hoping for the best, places us in an appropriate mind-set that will compensate for a situation where the elevators are out of service, if the officer in charge has determined that they are unsafe to use, or if they fail at some point during the operation. Preparation means being physically and mentally prepared to climb up what may be several flights of stairs. Doing this is much easier said than done, but it can be accomplished through physical and mental preparation, along with training and development on specific techniques.

Resorting to a logistical operation that involved stair climbing has been the case at several of the most serious high-rise fires and emergency incidents during the 1980s and 1990s. For example, the 1988 high-rise fire in the First Interstate Bank Building located in downtown Los Angeles was a fire where the majority of elevators inside that building failed. In fact, the one fatality at that fire was a building security officer who took an elevator directly to the fire floor to investigate what he assumed would be a false alarm.

The 1991 fire in the One Meridian Plaza Building located in downtown Philadelphia was a fire where most of the building systems, including the elevators, failed to operate. This was based primarily on the failure of the primary electrical power and shortly after that, the failure of the emergency (secondary) electrical power.

The 1991 fire in the Polo Club Condominiums Building located in south Denver was a fire where the elevators failed early in the operation, and the entire logistical operation had to be completed using the stairs.

The 1993 bombing of the World Trade Center and the events of 9/11 both serve as incredible examples of overwhelming logistical operations. Both events required that firefighters use stairs as the primary method to gain access to the upper floors of the buildings. Both towers stood well over one thousand feet in the air with an acre of territory on each floor that had to be searched by firefighters. At the 1993 bombing, searching the 100-plus elevators in one tower was an awesome undertaking in itself. Firefighters given that assignment gained access to these countless elevators only by first climbing 25, 50, 75, and, in some cases, more than 100 flights of stairs.

The use of stairs to gain access to the upper floors of a high-rise building is not a very appealing option. Plain and simple, it means hard work. An average firefighter weights approximately 200 pounds. Add to that his personal protective equipment (PPE), self contained breathing apparatus (SCBA), hand tools, fire hose, and spare air cylinders, and the firefighter quickly has an extra 100 pounds or more of weight to carry. When we start talking about climbing stairs for distances of 20, 30, or 40 flights, the logistical portion of the high-rise operation rapidly becomes a very arduous and exhausting task. And don't forget, these firefighters have to fight a fire, search for occupants, and check for extension once they get up to their destination. They might even have to climb several more flights of stairs, checking the upper floors, stairwells, and elevators for occupants and extension of fire. Several more flights might be another 10, 20, or more flights.

Like it or not, the use of stairs as a logistical component of high-rise operations is a fact of high-rise firefighting. That's why firefighters in high-rise districts have to be in superior physical condition. The high-rise operation is not going to involve a 200-foot preconnect, a couple of turns, and voilà, you've arrived. Furthermore, the comfort of knowing you are just a few feet from the safety of being outside, or being able to count on that window that's just a few feet back for emergency egress, are not viable options on the upper floors of a high-rise.

The tactical problem of logistics, once again, coupled with the issue of exposures, are the two most significant overall problems on the fireground with regard to operations at high-rise fires. If logistics weren't difficult enough already, the issue is frequently compounded by the loss of exterior access due to height, the loss of elevators due to electrical or system failure, the impractical use of air support early in the event; which leaves us with the extreme physical and mental demands of gaining access to the upper floors via stairs.

There is, however, some good news associated with the use of stairs for a logistical tool. For fire department operations, although very physically demanding, the use of stairs provides a significant margin of safety for operating forces. The simple concept of stopping an elevator at least two floors below a suspected fire floor or floor of alarm and walking up the remaining two flights via stairs if based entirely on firefighter safety and survival. Barring a massive structural collapse or some sort of explosion, we have great control over our own destiny when operating inside a stairwell. The stairwells of most high-rise buildings, at least below the fire floor(s) in most situations, can be a safe haven for firefighters to launch their attack and other fireground operations. So, although there is a tremendous physical challenge associated with the use of stairs, oftentimes, the benefit of the safety afforded to members far outweighs the disadvantage of the significant physical demand of climbing the stairs.

Types of Stairs

Before we can begin to use stairs as an effective logistical tool, it is important to identify the different types of stairs that can be found in most high-rise buildings, and the various components associated with stairs. A comprehensive understanding of stairs and everything associated with them will pay significant dividends on the fireground. It's not just a simple matter of locating the stairs and starting to climb. There is a definite art to this aspect of high-rise firefighting, as is true with most everything associated with high-rise operations.

Most high-rise buildings have at least two stairways that travel the entire vertical distance of the building. Typically, only one of the two provides access to the roof, and in many circumstances this access is not directly to the roof, but rather to another area, such as a mechanical room that in turn leads to the roof. The stairs are generally designed in one of two standard ways: 1) return stairs, and 2) scissor stairs.

Return stairs

This design, also sometimes referred to a *U return stairs*, is commonly found in many high-rise buildings, especially the older generation of high-rises. Just as the name implies, the return stair actually returns to the same geographic location of the building at each floor landing. In other words, if you enter a stairway with return-type stairs on the 32nd floor, the entrance/exit for this stair on the 33rd floor will typically be in the same vertical location within the building. The only exception to this is when there are horizontal portions to the stairway adjusting for building construction features such as setbacks.

Return stairs are laid out in such a manner that upon entering a stairwell with return-type stairs, one can walk up a half flight (approximately 10 steps in a typical commercial building) and arrive at a half landing. This half landing is a typically a smaller landing and is located half way between floors. It is at the half landing that the stairs turn, or make a U-turn back the other direction. There is once again a half flight of stairs, which thus returns back to the same vertical geographic location, and the landing for the next floor (fig. 7–1).

✷ The return-type stair design provides a distinct advantage to firefighters, especially with regard to maintaining a good orientation. This is extremely important for our operations in the low visibility of smoke conditions. Furthermore, the officer in charge of an attack team can take a look at the floor layout on the floor below the fire floor prior to commencing with the fire attack. By doing this, he can gather critical information regarding the probable floor layout of the fire floor, which is more reliable in residential high-rise buildings than it is in commercial high-rise buildings (fig. 7–2).

The location of many building features and components are stacked during the construction of a high-rise building. This is done simply as a matter of economy, as it significantly reduces construction costs. Items such as restrooms, electrical and custodial closets, standpipe cabinets/outlets, and so on, will generally be in the same location on floor 33 as they are on floor 32. In many cases, but certainly not all, hallways and actual rooms will be in the same vertical location from one floor to another.

Information regarding floor layout is particularly reliable when operating in residential high-rise buildings, (high-rise multiple dwellings). For a fire reported in apartment 3312 on the 33rd floor, a quick orientation of the 32nd floor and identifying the location of apartment 3212, will likely give the officer in charge reliable information on the best access and route to the fire apartment.

Fig. 7–1 An example of return stairs

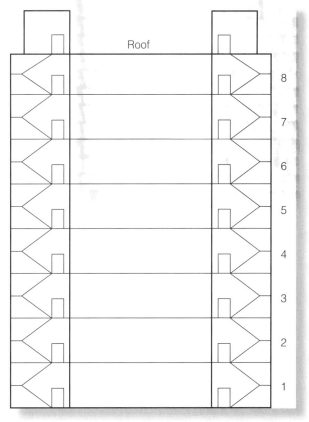

Fig. 7–2 The layout of typical return stairs

As for commercial high-rise buildings, certainly the floor layout of some floors is similar, but you can also have one floor on top of another that is laid out completely differently. This is common in buildings with multiple tenants. However, if floors 32 through 36 of a commercial high-rise building are occupied by the same company, it's very possible that the floors might have a similar layout and design. This information is best gained through comprehensive pre-fire planning.

The time needed to complete a good pre-fire plan of a large high-rise building is significant. Busy fire companies, already stretched thin with little or no extra time will not likely have the time needed to conduct comprehensive pre-fire plans on all of the buildings in their first due response area, nor will most fire departments have the necessary resources. Those fire companies with a significant number of high-rise buildings in their first due area will certainly not be able to pre-plan them all. Companies should focus on the most significant target hazards first, placing a high priority on the more complicated buildings that might prove particularly challenging and dangerous to operating firefighters.

It is important to also take every opportunity to learn about the buildings in your district during your daily activities. The countless times you're inside these buildings on automatic alarms, EMS calls, elevator emergencies, and other daily calls for emergency service are excellent opportunities to glean little bits and pieces of good information. Eventually, all these little pieces of the puzzle can add up to quality pre-fire information.

Scissor stairs

The *scissor stair* design is much more common in newer high-rise construction. The reason, as with most high-rise design features has to do, once again with cost. The design of scissor stairs allows for two separate stairwells to be contained in one vertical shaft within the building. The stairwells are typically separated by a rated sheetrock wall, but they are in the same geographic location of the building.

The biggest drawback or concern for firefighters with regard to scissor-type stairs is the issue of orientation. Unlike the return-type stair, scissor stairs do not return to the same vertical geographic location at each floor landing. In most scissor stair designs, the stairs exit on the opposite side of the building at each landing. For example, you enter a stairwell with scissor-type stairs on the 32nd floor and the entrance to that stairwell happens to be located on the west side of the center core. From the 32nd floor landing there is one continuous run of stairs going directly up to the 33rd floor landing with no half landing or turns. At the 33rd floor landing, you exit out onto the 33rd floor, but you are now located on the east side of the center core (fig. 7–3).

Fig. 7–3, A scissor stair

✱ In a building with scissor-type stairs, if the officer in command of an attack team attempted to orient himself with the fire floor by going to the floor below, he would actually be looking at a different geographic location of the floor below than the location where his attack will commence on the fire floor. In many cases, this can be corrected by having the officer check two floors below and orient himself at that location. Typically, in most buildings with scissor stairs, the stairs come out on the same side on every other floor (fig. 7–4).

I recently pre-planned a high-rise building on the south end of my district. This is a building where two of my companies run stairs on a daily basis for physical fitness. My good friend, Firefighter Sean Roeper, DFD Tower Co. 4, told me about some interesting features in the building, including automatic pressure reducing valves (PRVs) set to 70 psi, and scissor stairs. I decided to take a closer look for myself.

One of the things I found was that the labeling of the scissor stairs was actually incorrect, and could be very confusing to occupants and firefighters alike. In this building there are two stairwells, both of scissor-type configuration, and both located within the same vertical shaft. They are separated internally by a fire rated wall of sheetrock. The stairwells are labeled "South Stair" and "North Stair," respectively. However, on the actual floors of the building, all of the doors on the north side of the building, which lead to the stairs are labeled "North Stair Access," and all of the doors on the south side of the building that lead to the stairs are labeled "South Stair Access."

The problem lies in the fact that on both sides of the building when you enter a stairwell, you are actually entering a different stairwell on the even floors than the stairway you are entering on the odd floors. Specifically, entering a stairwell on the south side of the building at the 14th floor requires you to pass through a door labeled "South Stair Access." The same thing is true when entering a stairwell on the south side of the building at the 15th floor; you pass through a door labeled "South Stair Access." However, entering from the 14th floor actually places you inside a different stairwell than the stairwell accessed by at the 15th floor.

For this building, a much better and safer labeling system would be to have these stairs labeled as Stair A, and Stair B respectively, or Stair 1 and Stair 2. Under the current labeling system, an occupant trapped in a stairwell might have entered on the south side of the building and would be referring to that stairwell as the "South Stair." Firefighters would likely enter the stairwell on the south side at the lobby level in order to conduct a search, but the occupant might not actually be inside that stairwell. Firefighters reading this can probably identify many buildings in their own districts with similar problems. The owners and management of such buildings should be advised that changes need to be made. In some cases, the building and fire codes used by the authority having jurisdiction can mandate such changes.

Scissor stairs are being found more and more often as the cost associated with this type of stair design is frequently much less than the cost for return-type stairs. This is once again based on the available space inside the building. A little extra rentable space on each floor adds up, and ultimately means more money in the owner's pocket. As long as such stairs are allowed by building codes, owners will likely use them in the design of their buildings. Ultimately, the code is essentially going

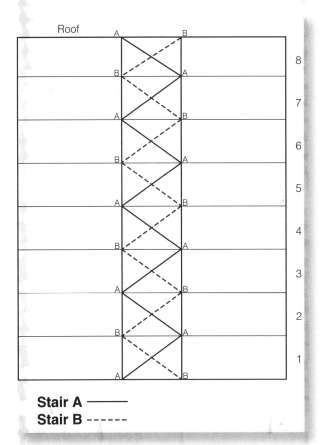

Fig. 7–4 The layout of typical scissor stairs

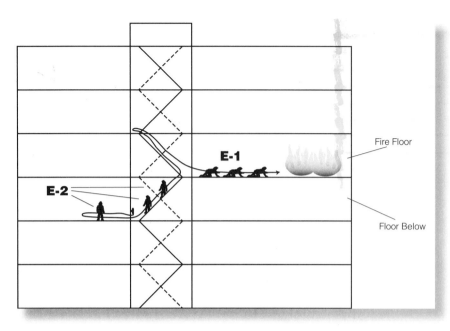

Fig. 7–5 Stairwell hose stretch, scissor stair

to call for two exits from any given floor, with minimal distance or separation between exits. In fact, the distance between exits, which in older generation high-rise buildings was much greater, is now significantly less. Most fire codes relaxed the distance required, which allowed scissor stairs and the exit doors on each floor became much less remote from one another.

✴ Although not ideal from a fire protection standpoint, and sometimes confusing with regard to orientation, the scissor stair design does have some definite advantages for fireground operations. Most notably, the long run of stairs between floors without interruption and turns can be a benefit to the engine company for stretching hoselines. For example, sometimes our attack has to commence from the stairwell (*stairwell stretch*) because of a well-involved fire floor, or the need to at least have a charged handline prior to entering the fire floor due to high heat and smoke conditions. In these situations, it is standard procedure to stretch hose up past the fire floor landing to the next landing above before charging the handline. This basic procedure then allows our advancement out onto the fire floor to be assisted by gravity. In other words, we are pulling a charged hoseline down rather than up, at least for a short distance, which is typically 25 feet (1/2 length) in buildings with return-type stairs, and 50 feet (one length) in buildings with scissor-type stairs.

The long runs associated with scissor stairs allow for, in many cases, at least one full section (*length*) of hose, or 50 feet, to be stretched up past the fire floor and back down (fig. 7–5). This advantage doesn't exist with return-type stairs. When stretching up past the fire floor, we can only stretch up to the next landing, and we can't go past that location because it would require a turn (fig. 7–6). That turn becomes a significant friction point, and the hoseline will become hung up and caught on the corner of that upper landing.

✴ In this scenario, we are preparing a charged hoseline before we enter a suspected well-involved fire floor (stairwell stretch) or one that at least contains high heat and smoke conditions. While the door to the fire floor is closed, members can operate above the fire floor in the stairwell stretching the dry line. In fact, one member must be assigned to hold the hoseline in place at the landing above the fire floor as the hose is being charged. Failure to do this will result in the hoseline being pulled down the steps as the weight of the water fills the hose. Once the line is fully charged, the weight of the static water in the hoseline will hold it in place.

✴ Although a firefighter or firefighters could operate above the fire floor landing in the stairwell initially, while stretching the dry line and prior to the fire floor door being opened up, they cannot operate above after the door is opened when conditions include high heat and smoke (fig. 7–7). Therefore, the additional hose stretched up past the fire floor landing can only go as far as the next landing. For scissor stairs, this is a longer distance; typically twice that of the return stair design. When advancing a charged 2½-inch attack line, pulling an additional 50 feet of hose out onto the fire floor from above, gravity assisted, is a significant benefit.

✴ I have climbed a lot of stairs over the past 25 years for physical fitness and preparation for high-rise operations, as well as for actual fire and emergency operations in high-rise buildings. I have to admit, I actually prefer the scissor-type stair design to that of the return-type. To me it seems a little less fatiguing to climb scissor stairs than return-type stairs, based primarily on fewer turns. Also, I find that having the ability to stretch additional hose up past the fire floor landing is extremely beneficial during operations. The building that I do most of my physical training in has scissor-type stairs, so perhaps I am a creature of habit and have gotten use to this type of stair. Every time I operate in a return-type stairwell, I find it to be much more difficult. Other firefighters have commented to me that they prefer the return-type stairs because the half landings provide an additional resting point between climbing the stairs, even if it happens to only be for a few seconds.

There are many pros and cons to both of these types of stair designs. The most important point for firefighters to remember is they must understand and be able to identify both. Furthermore, a good data base, perhaps located within a computer-aided dispatching system, identifying these features in the various buildings of a fire company's district, will be very beneficial when smoke and fire are showing.

Access stairs

Another type of stair located in many high-rise buildings is called the *access stair*, also sometimes referred to as tenant stairs, or convenience stairs. Access stairs, as the name implies, provide access between floors of individual occupancies. For example, larger businesses that occupy multiple floors in a high-rise building frequently request that the building owner install access stairs for their employees. One particular company might have leased office space on the 42nd and 43rd floors of a particular building. Access stairs cut into the building provide immediate access for employees to move within the tenant space between the 42nd

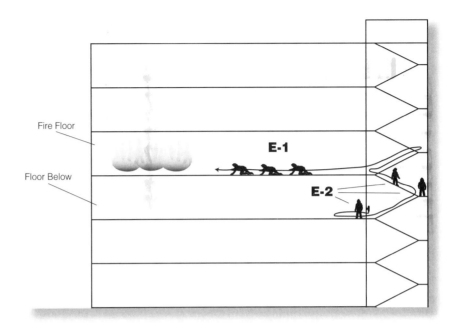

Fig. 7–6 Stairwell hose stretch, return stair

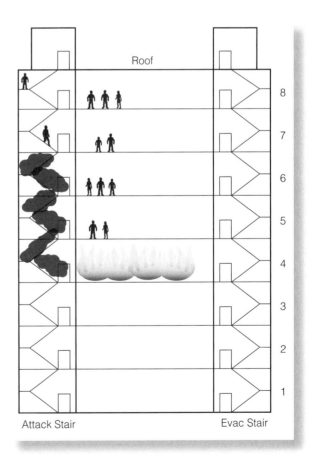

Fig. 7–7 This diagram illustrates the danger associated with smoke extension into the stairwell above the fire floor.

and 43rd floors without having to go out into the public elevator lobby and use the elevators, or the public stairwell (fig. 7–8).

These access stairs are of particular interest to fire departments, as there existence can create significant problems during fireground operations. Many of these type stairs are wide open without any specific fire protective barriers, such as fire doors, or alarm activated shutters. Therefore, it is clear to see, a fire on the 42nd floor could quickly lead to a smoke condition, or the extension of fire up onto the 43rd floor by way of this open, interior access stair.

I have found buildings in Denver where some access stairs are not recognized by the building's built-in, automatic fire protection systems. In many of these examples, the buildings in question are modern high-rises with sophisticated fire protection systems, including smoke management systems. Many of these type smoke management systems are designed to isolate the fire and smoke by way of exhaust and pressurization. In one particular building, an access stair was in place between the 36th and 37th floors. There were no fire doors or shutters to isolate one floor from the other in the event of a fire upon activation of the fire alarm.

After reviewing this with building engineers, it was discovered that a fire alarm on the 37th floor would cause a pressurization of the 36th and 38th floors, with the intention of sandwiching the smoke and fire to keep it from spreading. As one can clearly see, this could actually lead to accelerating the fire, as the pressurization of the 36th floor would send a flood of positive pressure air currents up the access stairs and onto the 37th floor, potentially feeding the fire (fig. 7–9). Once this was pointed out, building engineers quickly reprogrammed their smoke management system to recognize the 36th and 37th floor as one floor in the event of any alarms in that portion of the building (fig. 7–10).

It is interesting to note that at the One Meridian Plaza fire in Philadelphia, in February 1991, two of the tenant spaces included open interior access stairs between the floors. Specifically, an access stair between the 21st and 22nd floor added to the numerous problems the Philadelphia Fire Department was already attempting to deal with at this fire. Fire that originated on the 22nd floor, communicated to the 21st floor via this open access stair between the two floors.

Fig. 7–8 Typical access stairs in a high-rise building

"Access stairways between floors were a factor in this fire and reiterated that they are as convenient for the fire as they are for the tenant. Their use should be limited to fully sprinklered buildings. In most circumstances, the possibility of rapid fire spread up (or down) these unprotected vertical channels far outweighs any benefits derived from their tactical use during fire attack."[1]

Tactical use of Stairs

I must give credit to my associates from the FDNY, specifically, Battalion Chief Jerry Tracy, for first introducing me several years ago to specific tactical uses for stairwells during high-rise operations. There is, in fact, a great deal of finesse and coordination necessary with regard to the use of stairs at high-rise fires. Stair designations used by the FDNY have been adopted by many fire departments across the American fire service. The importance of designating a stairwell for attack purposes, and another, separate stairwell for evacuation purposes, at the outset of an operation cannot be overemphasized. The success of the entire operation, along with firefighter safety, and the survivability of civilian occupants, especially those located above the fire floor(s), rests heavily on the disciplined use of stairs.

Attack stair

The first arriving fire companies to a high-rise fire, specifically the officer in charge of the company, fire attack group, or team that arrives at the fire floor first, must make a determination as to what stair will be best suited for attack operations. After a stair has been chosen, the officer in charge of the attack must communicate this information to the incident

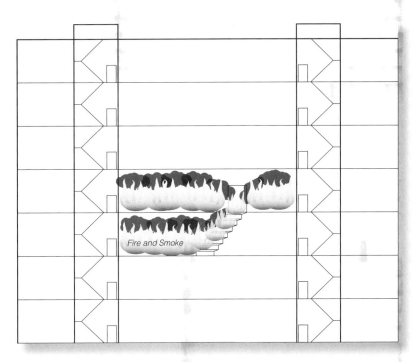

Fig. 7–9 This diagram illustrates how smoke could easily travel from one floor to another via an open access stair.

Fig. 7–10 This smoke control panel shows the location of an access stair.

commander. Furthermore, the IC must make it clear to all incoming fire companies which stair is being used as the attack stair.

Selecting the appropriate stair for attack purposes involves several critical pieces of information. It's not just a matter of using the first stair you find and automatically designating that one for attack. This is a conscious decision based on size-up information and specific characteristics of the building, the stairwells, and the fire conditions. The attack stair is selected based on the following information:

1. Location of the fire in relation to the stairs
2. Location of the fire in relation to the standpipe
3. Number of stairways serving the fire floor
4. Stairway(s) being used by occupants for evacuation
5. Types of stairways, specifically fire/smoke towers
6. The best, fastest, and safest access to the fire

Ultimately, when selecting an appropriate stair to be used for attack, consider a stair that will provide the best, fastest, and safest access possible to the fire area. In addition, consider the location of standpipe outlets and cabinets, and select an attack stair with as short a distance as possible from the standpipe outlet to the fire area. Fire companies should not use a stairway that is being actively used by building occupants as an evacuation stair. Also, fire/smoke towers should be avoided for use as attack stairs. Many fire departments have encountered severe problems with the fire/smoke tower drawing large currents of forceful air, with heat and smoke from the fire drafting toward the fire/smoke tower, directly into the path of attacking firefighters. In other words, fire/smoke towers tend to exacerbate the natural stack effect, creating conditions similar to a wind-driven fire, or making a wind-driven fire much worse, and making advancement extremely difficult (fig. 7–11). This problem is especially severe during wind-driven fires, where windows have failed on the windward side of a fire building. Fire/smoke towers should be used primarily for evacuation, and thus, should be designated, whenever possible, as the evacuation stair.

After an attack stair has been chosen, and prior to opening the door to the fire floor and commencing with the attack, every effort must be made to ensure that the attack stair is clear of all occupants above the fire floor. While the engine company members are stretching out the line, the officer in charge of the fire attack group should have truck company members check above, preferably a minimum of five floors above the fire floor, to ensure the attack stair is clear of any occupants. At a point at least five floors above the fire floor, a firefighter should call out in a loud voice asking if any one is above. If someone responds, a determination must be made as to whether or not they need help. Ultimately, they must be physically contacted and removed to an area below the fire floor, or protected in place in a safe refuge location on an upper floor, prior to commencement of the

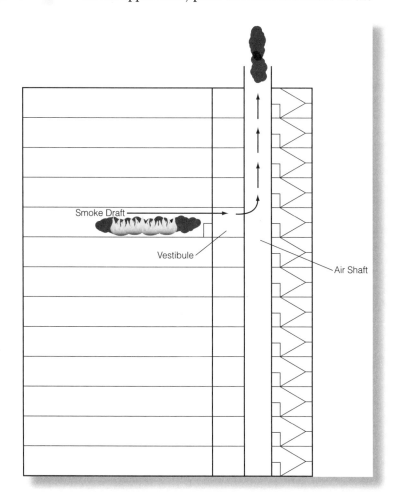

Fig. 7–11 This diagram illustrates how a fire/smoke tower can greatly increase the stack effect draft.

attack. If after calling out, there is no response, the firefighter should yell out in a loud voice, advising anyone above to get out of the stairwell, if possible, and if the doors leading from the stairwell to the floors are not locked. Any known occupants in the attack stairway above the fire floor(s) must be removed from the attack stairway before the door to the fire floor is opened (fig. 7–12). The buildings public address (PA) system can sometimes be particularly beneficial to communicate with occupants, and direct them out of and away from the attack stairwell, or preferably, to keep them out of it to begin with.

Evacuation stair

Making every effort to maintain at least one stair free of smoke and heat for the purposes of occupant evacuation can save countless lives. In most high-rise buildings, there are only two stairwells that travel the entire vertical distance of the building. One stairwell will have to be used for attack, therefore, that leaves only one reasonably safe escape route for occupants located, or trapped above the fire floor(s).

Just as careful consideration is given to the selection of an attack stair, simultaneously an evacuation stair must be selected and designated as well. This must also be a conscious decision based on careful deliberation. As previously mentioned, fire/smoke towers are not desirable for fire attack, but are typically ideal for evacuation stairs.

Once again, the value of conducting comprehensive pre-fire plans is critical with regard to the tactical use of stairs. In some buildings, pre-fire information will already establish the answers with regard to stairway designation. In one particular building in lower downtown Denver, the decision as to which stair will be used for attack, and which will be used for evacuation has, for all practical purposes, already been determined.

This particular building is a high-rise multiple dwelling. It is occupied primarily by mentally and physically disabled persons, many of whom are nonambulatory and need wheelchairs for mobility. On the northeast end of the building, there is an enclosed stairwell with access to the roof. On the southwest end of the building, there is a large enclosed area with long ramps running back and forth from the front of the building to the back. This enclosure is not a true fire tower, but it does have automatic dampers to vent the air if it becomes contaminated with smoke. The ramps in this enclosure are obviously designed for the evacuation of numerous wheelchair bound occupants. It is clearly obvious that this will be designated as the evacuation stair, and the stairwell on the northeast end of the building will be the attack stair. Experience with several fires in this building in the past has proven that this operation works very effectively.

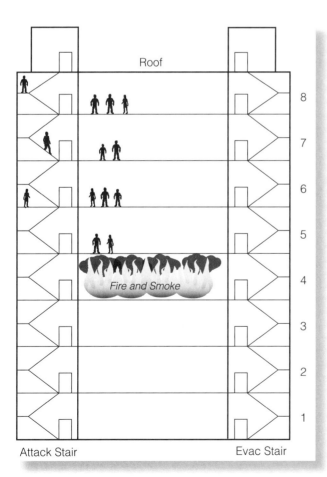

Fig. 7–12 Any occupants in the attack stairwell above the fire floor must be removed before the attack begins.

Stairwell Pressurization

Most modern high-rise buildings are equipped with a fire/life safety feature that pressurizes the stairwells in the event of a fire, upon activation of the fire alarm system. The stair pressurization is designed to keep the stairwells clear of smoke and heat when doors at the fire floor are opened by occupants escaping the fire area, or by firefighters commencing with an attack. Although this is a very good idea in theory, it can be subject to problems on the fireground. If numerous occupants are attempting to escape, and multiple doors are opened, or left open, the positive pressure within the stairwell may be lost. Furthermore, firefighters will hold doors open with wooden door chocks and other devices during an attack. If this involves multiple doors and floors, it could potentially cause a loss in positive pressure within the stairwell (fig. 7–13 and fig. 7–14).

Climbing Stairs

As a district chief, one of my responsibilities, along with the other district chiefs on my shift, is to coordinate the monthly training for our respective districts. At the time of this writing, I was preparing my district training for the month of September—the topic: high-rise and standpipe operations. This training was to include a hands-on training drill, with a simulated fire on the 29th floor of a commercial high-rise building. The elevators were not a logistical option. Most of the members were excited and enthusiastic about the upcoming training, but of course, several were not very happy about it at all.

I don't like climbing stairs anymore than the next guy. In fact, I hate it! But I did raise my right hand and swear to protect the citizens of the city of Denver. My city, like many others, has a fair number of high-rise buildings. Over half of them are not fully sprinklered, and, therefore, in the event of a fire, require manual firefighting. Elevators are not always 100% dependable or safe to use, so we are left with a professional responsibility and obligation to be physically and mentally prepared to climb stairs when necessary. Because of those facts, we also must conduct regular training drills on this topic.

I have been torturing myself by climbing stairs for the past 25 years. In that time, I have discovered a fair number of techniques that when applied properly, make for a much less fatiguing operation. The old adage, "work smarter, not harder" truly applies to climbing stairs. I address some specific techniques in this chapter.

A wide range of crazy ideas have been presented over the years in regard to high-rise operations. One in particular talked about having a group of firefighters dressed in shorts and tee-shirts run up the stairs to a staging location, and then put on their PPE, SCBA, and gather tools, and go fight the fire. They would be much less fatigued not having had to carry all that equipment up the stairs. That's certainly true, however, the one thing no one was ever able to answer, was, who took the equipment up to staging. Did it just mysteriously arrive there on its own? Not likely.

The fact is that when firefighters respond to a reported fire in a high-rise building, they put on full PPE, SCBA, and respond on fire apparatus. They get off, carry hose, tools, and other necessary equipment into the building, prepared to fight a fire. We simply don't have a fun bus, with 10 or 20 people dressed for basketball camp arriving at our fires. So, although interesting, it was not a very practical idea,

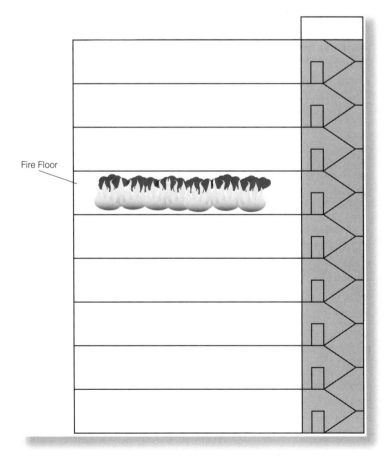

Fig. 7–13 This diagram illustrates a typical stairwell pressurization.

simply based on the fact that someone has to take the firefighting equipment up stairs. We must be practical, and focus on real life techniques that can truly be applied on the fireground.

Heat buildup

Heat buildup is the one item that I consider to be the most debilitating aspect of climbing stairs during high-rise operations. I am not talking about heat from the fire, but rather heat buildup from the physical exertion of climbing stairs in full turnouts, with a significant amount of additional heavy equipment being carried. A couple of techniques can be applied to help alleviate some of this heat buildup. The easiest method involves simply opening up the turnout coat, removing your gloves and hood and placing them in pockets, and carrying your helmet by clipping it onto your SCBA harness. This simple procedure takes little time to complete, but really makes a big difference in the long haul; pun intended. A good rule of thumb is that if you have to climb more than 10 flights of stairs, take a few seconds to appropriately adjust your PPE in order to reduce the heat buildup.

Carrying equipment

In addition to the heat buildup, many firefighters quickly become fatigued during a stair climb because they failed to preplan their operation. For example, try carrying a spare air cylinder up 30 flights by simply holding it by the valve and letting it ride low, at your side, next to your leg. This is extremely fatiguing, and if you're lucky enough to make it the entire distance, you will likely need another firefighter to pry your hand open in order to release the air cylinder from your grip. I find that firefighters attempting to simply muscle equipment such as an air cylinder, will rapidly increase their level of fatigue by the countless additional movements they will make adjusting the equipment every other floor on the climb up to staging. Passing the air cylinder from one hand to the other, bringing it up and resting it on the right shoulder, then the

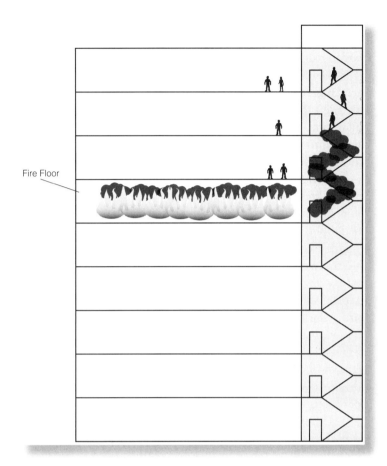

Fig. 7–14 If multiple doors are opened, stairwell pressurization can be quickly lost.

left shoulder, and then attempting to carry it like a baby, cradling it against the chest. These are all fatiguing movements.

There are good techniques to carrying all of the different types of equipment necessary for high-rise operations. As for the air cylinder, like all of our equipment, find some way to carry the load without having to adjust it multiple times on the way up several flights of stairs.

Something that I stumbled across many years ago works very well. About 10 years ago, all the fire stations on the DFD were issued a small supply of canvas bags to use for grocery shopping. The thought was that the fire department, having to shop for and buy groceries for the firehouse meals everyday, could help protect the environment by using the same canvas bags, over and over again, to transport the groceries, rather than plastic bags, which would be thrown out every time and just end up in a land fill. It's interesting what some people establish as the priorities. We asked for a couple of new nozzles and a saw for the truck, and we got canvas grocery bags instead.

Well, the canvas bags had a short life, that is, in regard to their intended function. One day, I decided to dig a couple of these bags out of the cellar, and try carrying an air cylinder in one of them during my daily workout. It worked very well. I simply placed the spare air cylinder inside the canvas bag, tied the carrying handles in a knot, and hooked a lightweight, aluminum carabiner on them. I then attached the carabiner to the D-ring on the shoulder strap on the SCBA harness, and I had a hands-free method to carry the spare air cylinder. Yes, the weight was still there, and I had to carry that weight up several flights of stairs, but I didn't have to adjust and re-adjust time and time again during the climb. From the time I discovered this method, I started carrying a couple of these canvas bags with me on the rig. I have used this method at actual working fires with complete success. Even during the fast-paced nature of a real fireground operation, setup time is minimal.

Other tools can be carried by applying the same basic principles. A large carabiner, also lightweight aluminum, provides an excellent means of carrying a hand tool. Attaching the carabiner to the other D-ring on the shoulder strap on the SCBA harness (opposite from the one being used to carry the spare air cylinder) and sliding a hand tool in, or clipping the hand tool into the carabiner, and you have another hands-free method to carry a Halligan, axe, or other hand tool. I have also found that by using the waist strap of the SCBA to hold the handle of the hand tool close to the body, the tool doesn't bounce and strike against your body or leg with every step, once again, reducing fatigue.

That brings up an important point. Whatever method you choose to utilize for carrying tools and equipment, you must remember to get the equipment up high—above your waist. I have seen firefighters using long pieces of webbing, attaching one end to an air cylinder and then placing a big loop over their head. A few flights up the stairs and the air cylinder is striking against their leg with every step. Furthermore, the webbing around your neck will tighten up, eventually cutting off your circulation at the carotid artery. This becomes a very miserable and highly fatiguing experience. The same is true for hand tools. If you can get them up high and secure the handle with something like the SCBA waist strap, the tool won't be striking against your body with every step, which means less fatigue.

As for hose packs, once again, you have to be practical. I recommend the use of 2½-inch hoseline for standpipe operations. I will discuss hoseline selection in depth in chapter 8. We developed an excellent method for assembling, storing, and carrying high-rise/standpipe hose packs on the DFD. Many departments have adopted the *Denver hose pack*, or are using a similar method. It is basically assembled in a horseshoe type configuration. This allows it to be carried either over a shoulder or over the air cylinder of the SCBA harness. I recommend the latter method, as once again, this reduces fatigue. When carrying the hose pack on the shoulder, it has a tendency to move, and will usually slide off the shoulder from the movement of climbing stairs.

Teamwork

Use teamwork when placing the hose packs on firefighters' SCBA. Have another firefighter place the hose on the air cylinder of your SCBA harness. It should be placed on carefully, in a well-balanced fashion, and pushed tightly against the carrying firefighter's back. If an extra few seconds is taken to properly load the hose pack, it will remain in place during a long stair climb, without moving or falling off. Once again, if you don't have to re-adjust it, or pick it up off the ground because it fell off, you reduce fatigue.

There you have it; an effective and efficient method to carry at least three different pieces of equipment necessary for high-rise firefighting—a hand tool, a spare air cylinder, and a length of hose. All totaled, with your full PPE and SCBA, this is a lot of extra weight. Nothing can change that fact. But taking the extra time to open up your coat, remove and stow gloves and hood, remove the helmet and clip it to the SCBA harness, along with applying well thought-out and previously evaluated methods of carrying hose, hand tools, and spare air cylinders, will make for a much more effective operation.

Additional Logistical Techniques

Let me point out just a few more recently developed equipment carrying techniques. First, for long climbs, certainly anything over 10 flights, it is a good idea to simply remove your turnout coat entirely. This involves a few minutes of preparation time, so it would not pay off for short distances, but the extra time used to adjust equipment, will save time in the long run, especially for extremely long climbs of 20, 30, 40, or more flights.

At lobby level, members remove their SCBA and turnout coat. Don the SCBA pack without your turnout coat on, and then place your turnout coat over the SCBA air cylinder (fig. 7–15). Placing the hose pack on top of this will hold the coat in place. Load all of the other equipment you are carrying in the best positions, and you're ready to start climbing. This procedure takes a physically and mentally prepared team less than 2 minutes to complete.

All of these methods to carry equipment have been developed, modified, and improved during the past 25 years. If you find any of these methods unacceptable, or undesirable, that's okay. Ultimately, individual firefighters must develop methods that best fit their needs and work most efficiently for them. Whatever that happens to be, figure it out well ahead of time, practice it, and be prepared the day you must use it for real.

Stairway Support Unit

Unfortunately, not everybody on the fireground gets to be the nozzleman. As firefighters, I think we all want to be where the action is: at the nozzle, performing a search, operating on the fire floor or at least the floor above. Those positions are all right in the heat of battle. The fact remains there are countless so called "unglamorous" support functions on the fireground, that, if left incomplete, the entire operation will likely fail. In fact, most fireground operations, especially during high-rise operations, are meant to support the fire suppression effort. One of those unglamorous assignments at a high-rise fire is the stairway support unit.

At a serious, long-term high-rise fire, a stairway support unit (SSU) will need to be established early and maintained throughout the duration of the incident. It is critical to be proactive with regard to establishing an SSU. Even if elevators are functioning properly at the beginning of an incident, relying on their use in the long run can be a very serious gamble. The tremendous heat associated with an ongoing and uncontrolled fire, along with tens of thousands of gallons of water used to stop that fire, can both find their way to mechanical areas and elevator machine rooms, causing system failure. Therefore, part of a proactive operation, involves establishing a logistical redundancy using the stairs, even if it is not immediately needed.

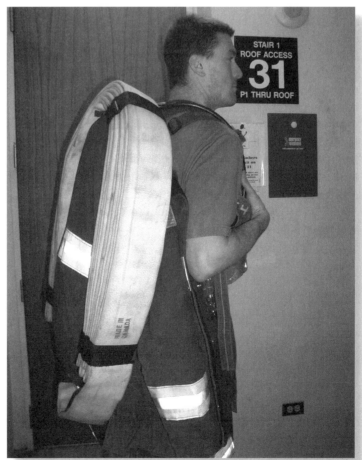

Fig. 7–15 For long climbs, it is best to take the turnout coat completely off, and carry it over the air cylinder of the SCBA harness.

The resources available to most fire departments will be significantly compromised by a serious fire in a high-rise building. Long before an SSU is established, numerous tactical priorities will have to be addressed. However, the benefit of a proactively established SSU is so crucial to the overall operation, that additional resources, although scarce, must be committed to this component of the operation as early as possible. Perhaps only one or two companies can be used initially for the SSU, but that's better than nothing.

For a fire that occurs on the 28th floor of a 50-story commercial high-rise building, you might establish an SSU in the following manner. The staging location will be on the 26th floor. In this hypothetical building, there is no floor 13, so you need to transport tools and equipment a total of 25 floors. If you could assign a SSU firefighter to each floor that would be ideal. However, that would require another full alarm assignment (in most cities) consisting of five or more companies. If a particular city or jurisdiction can muster up that kind of additional resource, on top of what is likely at least a second, perhaps a third alarm already, than fantastic, do it. Once again, if you can place one firefighter per floor for an SSU, that's ideal.

For most departments, the available manpower to be utilized for an SSU will likely be limited. Therefore, start by attempting to place one firefighter every two floors. A department with four-man companies could establish an SSU at this fire with only three four-man fire companies. One firefighter would be placed at the lobby level, with an additional firefighter placed at floors 3, 5, 7, 9, 11, 14, 16, 18, 20, 22, and 24. With some manpower assigned to staging to receive the equipment and some manpower assigned to the lobby to gather it up and place it near the stairway entrance, this SSU of three companies could be very effective (fig. 7–16).

Equipment arriving in the lobby would be carried by one firefighter up to the third floor landing. The firefighter on three would then carry the equipment to five, and so on, until it reached staging on the 26th floor. Firefighters assigned to the SSU would be carrying one or two pieces of equipment in one trip, such as two air cylinders, one in each hand. The firefighters would have a chance to rest as they descended the stairs back to their respective assigned landings to pickup more equipment. After an initial rush of equipment is successfully placed into staging, work for the SSU will likely slow down and periodically stop while waiting for additional equipment to arrive inside the building at the lobby level. During these periods, SSU members should stop working, rest, and drink fluids to maintain hydration.

Members of the SSU need to dress down, and shed any unnecessary personal protective equipment (PPE), specifically their turnout coat, and pants. The additional weight, along with the heat retention associated with PPE, will quickly lead to fatigue and potentially to heat exhaustion. The firefighters on the SSU are supporting the firefight, but from remote areas, not likely to be affected by

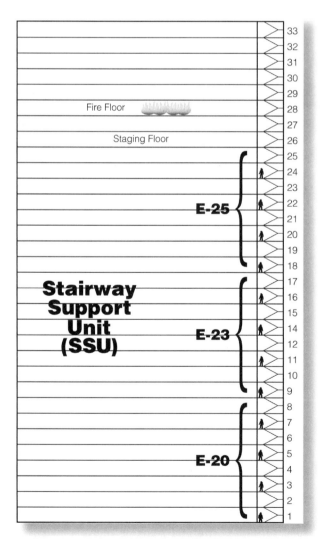

Fig. 7–16 This diagram illustrates the positions of a 12-member stairway support unit (SSU) delivering equipment to staging on floor 26.

smoke and heat conditions. Therefore, members of the SSU can remove whatever equipment is not necessary for their immediate personal protection.

All firefighters should plan ahead and be prepared for such support operations at high-rise buildings. A gym bag carried on the rig with a pair of gym shorts, a tee shirt, and light weight work shoes or tennis shoes can be very valuable when you're part of this type of support operation. Steel toe work shoes and leather gloves are the only PPE that may be needed by a firefighter assigned to a SSU. All of his other PPE, including his helmet, SCBA, and turnout coat and pants, can be stowed in the corner of the floor landing to which he is assigned. It's immediately available should conditions in his area change, or if that member's company is reassigned to actual firefighting duties.

Here are some additional suggestions and thoughts regarding the use of a SSU. Company officers should be placed at the bottom, top, and in the middle of the SSU to monitor the equipment movement, and provide communications via portable radio from all locations of the chain. Furthermore, company officers need to monitor the condition of their firefighters, keeping in mind that this operation, although not directly part of the firefighting, will still be a very arduous and physically demanding assignment. All members assigned to the SSU must be monitored for signs of fatigue and heat exhaustion. Bottles of water must be sent up the chain periodically to rehydrate SSU members.

With scarce resources, relief of SSU members will likely be difficult to accomplish as often as is necessary. However, fire departments, especially smaller organizations, must be creative with regard to obtaining additional resources. Automatic and mutual aid agreements with other departments can increase the pool of resources significantly. Five small- to medium-sized suburban fire departments within a metropolitan area, operating alone, will struggle to effectively fight a significant high-rise fire. However, collectively, these five departments may become a much more powerful force with reasonable resources to deal with larger incidents, such as high-rise fires. Small, medium, and many large city fire departments must also look outside their organizations for help with major incidents. The Chicago Fire Department is one of the largest in the nation. At the La Salle Bank Building fire which occurred in 2003, numerous suburban area departments provided mutual aid by covering other areas of the city. If necessary, fire companies from automatic and mutual aid departments are also excellent resources to perform many of the support functions of high-rise operations, such as that of an SSU. The terms *strike team* and *task force*, typically associated with wildland firefighting, should be adopted and utilized on a regional level for high-rise operations.

Summary

The stairs in a multistory or high-rise building are an important logistical component. A comprehensive understanding of stair types and configurations is essential. Quality pre-fire planning of buildings in your first due response area with information on stairs and their layout is critical. Most importantly, a firefighter mind-set, with mental and physical preparation, is the number one key to success during operations that involve the use of stairs. Exceptional firefighters stay in shape and prepared by climbing stairs as part of their routine physical fitness program.

References

1. Eisner, H and Manning, B. 1991, *Fire Engineering Magazine*, August: 69–70.
2. Manning, Bill and Harvey Eisner, "One Meridian Plaza Fire," *Fire Engineering*, August 1991 (Vol. 144, No. 8), pp. 50-70.

CHAPTER 8

Hoseline Selection

Hose is one of the most basic and fundamental components in the engine company's arsenal of various tools. The so-called "original Men in Black" have the basic fireground responsibility of delivering water, in the proper amount and form, to the seat of the fire in order to extinguish it. At the heart of this most critical fireground operation are fire hoses, along with nozzles and the numerous other essential water-delivery appliances. The size of hose chosen for firefighting operations in high-rise and standpipe-equipped buildings is of utmost importance.

Hoseline Size

Firefighters have several different sizes of hose to choose from for firefighting purposes in high-rise and standpipe-equipped buildings. For all practical purposes these are also the same basic choices for fighting fires at any other structural fire. Although fire hose can be specified and purchased in a wide range of different sizes, the most commonly used for attack handlines can be narrowed down to the following four sizes:

- 1½-inch handline (125 gpm)
- 1¾-inch handline (175 gpm)
- 2-inch handline (200 gpm)
- 2½-inch handline (300 gpm)

At this point in time, the vast majority of the American fire service seems to have settled on the 1¾-inch handline as the minimum size to be utilized for structural firefighting. Unfortunately, some fire departments also consider this to be the maximum size handline as well.

A car fire, a house fire, a fire in a hardware store, and a fire in a high-rise building are four separate and distinct fire problems. Yet, many fire departments will select and attempt to utilize the exact same weapon to combat these four very different fires, and that weapon is typically the 1¾-inch attack line.

There are unfortunately still a few fire departments out there that may utilize a 1½-inch handline for attacking some structural fires, but that weapon is not recommended for use on anything larger than a very small vehicle fire or wildland fire operations. It is definitely not recommended for high-rise and standpipe operations; to use it for that application would be down right dangerous. A 100-foot length of single jacketed 1½-inch hose with a small, plastic forestry nozzle is very lightweight and highly maneuverable. However, it is *not*, and I can't emphasize this enough, *is not* an appropriate tool for high-rise firefighting and standpipe operations.

In order to narrow our discussion on attack hoseline sizes down, as it relates to high-rise and standpipe operations, we will eliminate the 1½-inch handline from the field due to the extremely low volume and high friction loss associated with this weapon. I would strongly encourage the leaders of those fire departments who are still using 1½-inch handline for any type structural firefighting, especially if it is part of a high-rise/standpipe hose pack, to change the way you're doing business, quickly. It's a critical issue directly related to firefighter safety and survival.

The War Years

The decades of the 1960s and 1970s, the so called "war years" will be remembered as some of the busiest in the history of the American fire service. Fire departments across the country, especially those in the larger urban centers frequently found themselves overwhelmed with the amount of fire activity. It was not unusual for some urban fire companies to respond to multiple fires, back to back, and with little rest in between. Once again, a tactical dilemma presented itself to busier fire departments. The problem: so many fires, so little time. Some of the fires required the use of the 2½-inch handline, some could be handled with a 1½-inch handline, but much of the work seemed to fall somewhere in between.

Due to the heavy workload, time became a much more significant factor than ever before. The question was how could a fire department achieve higher flows appropriate for some fires, but still provide ease of use and maneuverability? The answer for many was the birth of the 1¾-inch handline in the late 1970s.

Because of the high fire workload, the '60s and '70s provided many lessons for the American fire service, including the fact that a 1¾-inch handline could be a very valuable tool. In fact, the value of this tool is evident everyday on fire departments across the country, as it is the appropriate weapon for most of our operations. Unfortunately, as can happen with any tool, the 1¾-inch handline can and has also been misapplied at times. Because it is easier to stretch, advance, and operate than the 2½-inch handline, the 1¾-inch handline is all too often selected as the weapon of choice for most of our fires, including many of those that are clearly beyond its capabilities.

Proper Weapon Selection

Many firefighters, company officers, and chief officers have likely witnessed, or perhaps even played a part in, the fireground blunder that I refer to as "poor weapon selection." It centers around the well-intended selection of a fireground weapon that is based solely on the "ease of use" and maneuverability criteria. It might be a fast moving fire in a lumberyard, a heavily involved garage fire that is attached to an occupied dwelling, or even a fire on the upper floors of a high-rise building. As the water reaches the nozzle, and the battle gets under way, it quickly becomes obvious to those involved that this "house fire" weapon is not having a positive impact on this unusual and underestimated fire.

GPM versus Btu's

We are talking about good weapon selection, specifically for those rare, high-risk, low frequency events. It boils down to a very simple equation; gallons per minute (gpm) versus British thermal units (Btu's). In the fireground battle we must deliver enough gpm to overcome the Btu's being produced by the fire. If we don't accomplish this basic mission, we lose the battle, plain and simple.

Many factors need to be taken into consideration when selecting the proper size handline for fireground operations, yet, by far the most important of all those factors is that of *flow*, or gpm. We must ultimately deliver a sufficient quantity of water and apply that water to the burning solid fuels in order to stop the fire. Numerous other factors will affect our decision of what size handline to utilize. Factors such as ease of use and maneuverability are certainly very important. However, quickly placing a smaller, more maneuverable handline into operation becomes a moot point if the water being delivered is not sufficient enough to stop the fire.

When poor weapon selection occurs at our most frequent fires, and the fire cannot be stopped, we are usually able retreat with our more maneuverable weapon to an area outside of the fire building, where a defensive operation becomes necessary to confine the fire to the building of origin. Unfortunately, this tactic is not a realistic and viable option at most high-rise fires, specifically those fires that occur beyond the reach of aerial-mounted, master stream appliances. Hence, this is one of the reasons why proper and proactive weapon selection is so critical at high-rise fire events, and all fireground operations in any standpipe-equipped buildings. We have a very short window of opportunity to stop these fires, and once that opportunity is gone, it's usually very difficult, if not completely impossible, to recover.

1¾-Inch Handline

Before addressing the pros and cons associated with the selection of 1¾-inch handlines for high-rise and standpipe operations, one must first have a basic understanding of the history associated with this size handline.

Throughout the first half of the 20th century the 2½-inch handline was the predominant size of hose used by firefighters for manual firefighting. As time went on, though, the need grew for a smaller, more maneuverable handline and so the 1½-inch handline, was introduced. For several years these two handline sizes, 1½-inch and 2½-inch, provided the two basic options for manual firefighting. Simply put, the old motto of *"big fire, big line"* and *"little fire, little line"* became the basic rule of thumb.

Realistic flow

In terms of the 1¾-inch handline we must identify and understand what the *realistic* flow truly is from this weapon. When it was first introduced, and for years since then, many individuals in the fire service proclaimed that the 1¾-inch handline could flow as much water as a 2½-inch handline. Without changing the laws of physics and hydraulics, this is impossible. The 2½-inch handline is a tube with an inside diameter that is a full ¾ inch larger than the 1¾-inch handline. The statement "I can get as much water out of my 1¾-inch line as I can the 2½" is 100% false, and simply impossible.

Unfortunately, based on that false premise, many fire departments did away with their 2½-inch handline, relying solely on the use of 1¾-inch for all manual firefighting operations.

Some fire departments that lack this extremely valuable tactical weapon (2½-inch attack line) can occasionally be seen attempting to place a 3-inch supply line in-service and operate it as an attack line when conditions are entirely beyond the capabilities of 1¾-inch line. The 3-inch supply line, is just that, a supply line. It is *not* an attack line, and should not be used as such.

Ultimately, it is of critical importance that firefighters and fire departments take a very careful look at the realistic flows achieved by all of their weapons. By realistic, I mean how much water a given weapon delivers as an attack team proceeds down a well-involved hallway, after making one flight of stairs and turning a corner. In other words, realistic water flow is based on an offensive operation with positive forward progress. Other than that of a defensive operation, specifically when utilizing master stream appliances, proactive water delivery is not, and should not be based on a stationary proposition.

From a purely theoretical approach, a 1¾-inch handline could potentially flow well in excess of 200 gpm. In fact, by utilizing the formula for flow from a given sized orifice:

$$(29.7d^2 \sqrt{p})$$

you can see that some extremely high flows could possibly be achieved with 1¾-inch hose. For example, using that formula, the 1¾-inch smoothbore nozzle used on a master stream appliance is rated to deliver 814 gpm at 80-psi nozzle pressure. That's a lot of water, and a very realistic flow from a properly supplied, master stream appliance. Unfortunately, it would not be realistic to deliver that same flow through a 1¾-inch attack hoseline, due to the excessive amount of friction loss in the hose. Remember, the fireground is not a theoretical place, it's real and the fire is real. So too is the friction loss in a hoseline, the nozzle pressure, and the nozzle reaction. All of these components play an important part with regard to realistic flow.

So, based on many of the previously addressed factors, what is the realistic flow from a 1¾-inch handline? First, a good rule of thumb to utilize when trying to quickly identify what a realistic potential flow is from various sized hoselines is to convert the hose size from a fraction to a whole number. With that done, simply remove the decimal and you have a whole number, which represents a realistic potential gpm. Hose size can easily be converted to realistic flow as follows:

1¾" = 1.75 = 175 gpm
2" = 2.00 = 200 gpm
2½" = 2.50 = 250 gpm

Those pros and cons of 1¾-inch handline can basically be narrowed down to the advantage of a smaller, more maneuverable handline, and the disadvantage of more friction loss and a greater overall pressure required in order to achieve higher flows. Once flows exceed 200 gpm, the maneuverability advantage of this weapon is lost. For high-rise firefighting and standpipe operations, the low pressures typically associated with standpipes, coupled with the potential for encountering various types of pressure-regulating devices (PRDs), makes it difficult when using 1¾-inch attack line to achieve flows that would be effective and safe for interior structural firefighting.

First and foremost, a good fire officer must thoroughly understand all of his weapons before he goes into battle. I consider 175 gpm to be a realistic top end flow from a 1¾-inch handline. You can achieve slightly higher flows and or less nozzle reaction force with low-pressure nozzles, specifically the smoothbore/solid stream nozzle. But if you want to maintain that advantage of maneuverability, forget about flows in excess of 200 gpm with this handline. Remember, we're talking about realistic water delivery inside a fire building, not a theoretical show and tell, typical of a parking lot demonstration.

Once again, we must fully understand that the low pressures associated with most standpipe systems will limit our ability to overcome the friction loss associated with the smaller- to medium-sized handlines, such as 1¾-inch. For the fire in a one-story dwelling within reach of a 200-foot preconnect, it will be easy for the pump operator to throttle up and give the attack team a boost in the pressure as necessary. However, this is simply not that easy when the fire is on the twentieth floor of a high-rise building. In fact, even if fire department pumpers can overcome the head pressure and friction loss, there may still be a pressure-reducing valve (PRV) installed at the outlet, which will further restrict the attack teams' ability to achieve an appropriate flow.

Guidelines for using 1¾-inch handline

Ultimately, it boils down to the company officer's ability to exercise good judgment and have the appropriate discipline to make a good weapon selection. My overall weapon selection recommendation is to utilize 2½-inch handline for all high-rise firefighting and standpipe operations. However, if 1¾-inch handline must be used, I would strongly recommend a very close adherence to the following:

1. If 1¾-inch handline is used, it should only be used for small fires in compartmentalized buildings such as residential high-rises. Keep in mind, a fire on the twentieth floor may be perceived as small from the street, but after 10 to 15 minutes of reflex time it will likely be anything but small. Proper weapon selection is based on what the fire will potentially look like five minutes from now, not what it looks like right now.

2. If 1¾-inch hand line is used, it should only be used in maximum lengths of 50-feet, used as a lead length (nozzle section) attached to one or more lengths of 2½-inch handline.

3. If 1¾-inch handline is used, the second handline (the backup line) must be a larger attack line, preferably a 2½-inch handline.

4. A 1¾-inch handline should not be utilized for fires in commercial high-rise buildings, due to the greater fire loading, uncompartmentalized areas, and a need to achieve a minimum flow of 250 gpm in these occupancies.

5. A 1¾-inch handline should not be utilized for fires in high-rise and standpipe-equipped buildings where the standpipe system is equipped with pressure-regulating devices, especially those that are non-removable, and non-fireground adjustable.

Hydraulic calculations for 1¾-inch hoseline

Let's take a look at the potential flow from various different a combinations of weapons based on hydraulic calculations, starting with 150 feet of 1¾-inch attack line, supplied from a standpipe for the purposes of combating a fire in a high-rise building. We want to achieve a flow of 150 gpm, which is considered the minimum for structural firefighting in residential occupancies.

The following flows were calculated using the friction loss formula for 1¾-inch hose, from *Fire Department Hydraulics*, by Eugene F. Mahoney.[1]

$2Q^2 + Q(6)$

$Q = gpm/100$

150 feet, 1¾-inch, 150 gpm (100 psi automatic combination fog nozzle)

1. Approximately 54 psi friction loss in 150 feet of 1¾-inch hose flowing 150 gpm
2. 100 psi nozzle pressure in order to deliver 150 gpm
3. 76 pounds force nozzle reaction
4. Required standpipe outlet pressure = 154 psi

150 feet, 1¾-inch, 150 gpm (low pressure 75 psi, combination fog nozzle)

1. Approximately 54 psi friction loss in 150 feet of 1¾-inch hose flowing 150 gpm
2. 75 psi nozzle pressure in order to deliver 150 gpm
3. 66 pounds force nozzle reaction
4. Required standpipe outlet pressure = 129 psi

150 feet, 1¾-inch, 150 gpm (low pressure, 50 psi, 15/16-inch smoothbore nozzle)

1. Approximately 54 psi friction loss in 150 feet of 1¾-inch hose flowing 150 gpm
2. 33 psi nozzle pressure in order to deliver 150 gpm
3. 46 pounds force nozzle reaction
4. Required standpipe outlet pressure = 87 psi

Next, we want to achieve a flow of 200 gpm, which is considered by many (but not me), to be a realistic and achievable flow from 1¾-inch handline.

150 feet, 1¾-inch, 200 gpm (100 psi automatic combination fog nozzle)

1. Approximately 90 psi friction loss in 150 feet of 1¾-inch hose flowing 200 gpm
2. 100 psi nozzle pressure in order to deliver 200 gpm
3. 101 pounds force nozzle reaction
4. Required standpipe outlet pressure = 190 psi

150 feet, 1¾-inch, 200 gpm (low pressure, 75 psi, combination fog nozzle)

1. Approximately 90 psi friction loss in 150 feet of 1¾-inch hose flowing 200 gpm
2. 75 psi nozzle pressure in order to deliver 200 gpm
3. 87 pounds force nozzle reaction
4. Required standpipe outlet pressure = 165 psi

150 feet, 1¾-inch, 200 gpm (low pressure, 50 psi, 1-inch smoothbore nozzle)

1. Approximately 90 psi friction loss in 150 feet of 1¾-inch hose flowing 200 gpm
2. 45 psi nozzle pressure in order to deliver 200 gpm
3. 71 pounds force nozzle reaction
4. Required standpipe outlet pressure = 135 psi

Next, we want to achieve a flow of 250 gpm, which is considered the minimum for structural firefighting in commercial occupancies.

150 feet, 1¾-inch, 250 gpm (100 psi automatic combination fog nozzle)

1. Approximately 135 psi friction loss in 150 feet of 1¾-inch hose flowing 250 gpm
2. 100 psi nozzle pressure in order to deliver 250 gpm
3. 126 pounds force nozzle reaction
4. Required standpipe outlet pressure = 235 psi

150 feet, 1¾-inch, 250 gpm (low pressure, 75 psi, combination fog nozzle)

1. Approximately 135 psi friction loss in 150 feet of 1¾-inch hose flowing 250 gpm
2. 75 psi nozzle pressure in order to deliver 250 gpm
3. 109 pounds force nozzle reaction
4. Required standpipe outlet pressure = 210 psi

150 feet, 1¾-inch, 250 gpm (low pressure, 50 psi, 1⅛-inch smoothbore nozzle)

1. Approximately 135 psi friction loss in 150-feet of 1¾-inch hose flowing 250 gpm
2. 44 psi nozzle pressure in order to deliver 250 gpm
3. 87 pounds force nozzle reaction
4. Required standpipe outlet pressure = 179 psi

Of the previous different weapons combinations, most would not be capable of achieving the target flow when operating off a low-pressure standpipe. In fact, if a factory preset pressure reducing valve is in place, set to a discharge pressure of 70 to 90 psi—not unrealistic—than only one of the combinations would work, the one equipped with the low pressure $^{15}/_{16}$-inch smoothbore tip, and only to deliver 150 gpm. Furthermore, note the excessive reaction force of many weapons combinations.

2-Inch Handline

The 2-inch handline is a relatively new tool for the American fire service. Faced with that age-old tactical dilemma of finding a reasonable balance between maneuverability and high volume flow, the 2-inch handline was developed. There are some fire departments that have adopted this size hose, as their primary handline for high-rise and standpipe operations. One of those is the Los Angeles City Fire Department (LAFD).

As with the 1¾-inch handline, the 2-inch handline has specific advantages and some disadvantages. The prepared company officer must first understand and become thoroughly familiar with the realistic flow associated with this weapon. With the 2-inch handline, 200 gpm is certainly a very realistic and achievable flow. However, once again, when operating off of a standpipe system, firefighters must be prepared for low-pressure situations, which may limit the ability to achieve optimum flows. As with the 1¾-inch handline, realistic flow from the 2-inch handline includes the same factors of friction loss, nozzle pressure, and nozzle reaction.

Let's take a look at the potential flow from various different combinations of weapons, starting with 150-feet of 2-inch attack line, supplied from a standpipe for the purposes of combating a fire in a high-rise building. We want to achieve a flow of 150 gpm, once again, that which is considered the minimum for structural firefighting in residential occupancies.

Hydraulic calculations for 2-inch hoseline

The following flows were calculated using the friction loss formula for 2-inch hose, from *Fire Service Hydraulics*, Second Edition, by Dick Sylvia.[2]

$$40(2Q^2 + \tfrac{1}{6}Q)C$$

Q = gpm/100

C = 0.1 (The friction loss coefficient for 2-inch hoseline)

150 feet, 2-inch, 150 gpm (100 psi automatic combination fog nozzle)

1. Approximately 28 psi friction loss in 150 feet of 2-inch hose flowing 150 gpm
2. 100 psi nozzle pressure in order to deliver 150 gpm
3. 76 pounds force nozzle reaction
4. Required standpipe outlet pressure = 128 psi

150 feet, 2-inch, 150 gpm (low pressure 75 psi, combination fog nozzle)

1. Approximately 28 psi friction loss in 150 feet of 2-inch hose flowing 150 gpm
2. 75 psi nozzle pressure in order to deliver 150 gpm
3. 66 pounds force nozzle reaction
4. Required standpipe outlet pressure = 103 psi

150 feet, 2-inch, 150 gpm (low pressure, 50 psi, $^{15}/_{16}$-inch smoothbore nozzle)

1. Approximately 28 psi friction loss in 150 feet of 2-inch hose flowing 150 gpm
2. 33 psi nozzle pressure in order to deliver 150 gpm
3. 46 pounds force nozzle reaction
4. Required standpipe outlet pressure = 61 psi

Next, we want to achieve a flow of 200 gpm, which is considered to be a realistic and achievable flow from 2-inch handline.

150 feet, 2-inch, 200 gpm (100 psi automatic combination fog nozzle)

- Approximately 50 psi friction loss in 150 feet of 2-inch hose flowing 200 gpm
- 100 psi nozzle pressure in order to deliver 200 gpm
- 101 pounds force nozzle reaction
- Required standpipe outlet pressure = 150 psi

150 feet, 2-inch, 200 gpm (low pressure 75 psi, combination fog nozzle)

- Approximately 50 psi friction loss in 150 feet of 2-inch hose flowing 200 gpm
- 75 psi nozzle pressure in order to deliver 200 gpm
- 87 pounds force nozzle reaction
- Required standpipe outlet pressure = 125 psi

150 feet, 2-inch, 200 gpm (low pressure, 50 psi, 1-inch smoothbore nozzle)

- Approximately 50 psi friction loss in 150 feet of 2-inch hose flowing 200 gpm
- 45 psi nozzle pressure in order to deliver 200 gpm
- 71 pounds force nozzle reaction
- Required standpipe outlet pressure = 95 psi

Next, we want to achieve a flow of 250 gpm, which is considered the minimum for structural firefighting in commercial occupancies.

150 feet, 2-inch, 250 gpm (100 psi automatic combination fog nozzle)

- Approximately 77 psi friction loss in 150 feet of 2-inch hose flowing 250 gpm
- 100 psi nozzle pressure in order to deliver 250 gpm
- 126 pounds force nozzle reaction
- Required standpipe outlet pressure = 177 psi

150 feet, 2-inch, 250 gpm (low pressure 75 psi, combination fog nozzle)

- Approximately 77 psi friction loss in 150 feet of 2-inch hose flowing 250 gpm
- 75 psi nozzle pressure in order to deliver 250 gpm
- 109 pounds force nozzle reaction
- Required standpipe outlet pressure = 152 psi

150 feet, 2-inch, 250 gpm (low pressure, 50 psi, 1⅛-inch smoothbore nozzle)

- Approximately 77 psi friction loss in 150 feet of 2-inch hose flowing 250 gpm
- 44 psi nozzle pressure in order to deliver 250 gpm
- 87 pounds force nozzle reaction
- Required standpipe outlet pressure = 121 psi

Of the previous different weapons combinations, most would not be capable of achieving the target flow when operating off a low-pressure standpipe. In fact, if a factory preset pressure reducing valve is in place, set to a discharge pressure of 70 to 90 psi—not unrealistic—the only combinations that would work are the ones equipped with the low pressure 15/16-inch smoothbore tip flowing 150 gpm, and the 1-inch smoothbore tip flowing 200 gpm. Once again, the reaction force of several weapons combinations is significant.

One can easily see that the performance of the 2-inch hoseline, in terms of flow, is better than the 1¾-inch hoseline. The friction loss alone in 2-inch hoseline is approximately half that of the 1¾-inch. By utilizing a low-pressure nozzle, fire departments like the LAFD are in some circumstances able to achieve a reasonable flow, while reducing nozzle reaction and ultimately making it easier for the attack team to advance and handle the line.

Advantages of 2-inch handline

The obvious advantage of the 2-inch handline is the higher flows that can be achieved at lower overall operating pressures. When compared to the 1¾-inch handline, the one disadvantage is a slightly less maneuverable line. I would consider the 2-inch handline to be the smallest recommended size attack line for fire departments to utilize during high-rise and standpipe operations, but only if it is backed up by 2½-inch attack line.

The advantages of the 2-inch handline and the operational success of many fire departments, including the LAFD, prove that this can be a very effective tool for many high-rise and standpipe operations. However, because of the nature of standpipe systems, and the potential for extremely low-pressure situations, fire departments will achieve the best overall operational success with one of the oldest tools in the American fire service. That tool is the 2½-inch handline.

2½-inch Handline

Almost as old as the American fire service itself, the 2½-inch handline, also known as the *big line*, is the lowest pressure, highest volume handline available to firefighters. There are no two ways about it, the *big line* is, and always will be, an engine company's greatest physical challenge. The obvious disadvantage of the 2½-inch handline is that it is less maneuverable than the smaller handlines. The clear advantage is that it is a high volume weapon that can achieve the high volume at very low pressures, making it very suitable for standpipe operations.

ADULTS acronym

The ADULTS acronym is used as a guideline for firefighters and company officers as to when and where the 2½-inch attack line should be used. A brief review of the A D U L T S acronym is:

- **A** – Advanced fire upon arrival
- **D** – Defensive operations
- **U** – Unable to determine the extent or location of the fire
- **L** – Large, uncompartmentalized areas
- **T** – Tons of water to cool the heat
- **S** – Standpipe operations

Most of the criteria contained in the A D U L T S acronym can be applied to operations at high-rise and standpipe-equipped buildings. Starting with *advanced fire upon arrival*, officers must proactively select this more powerful handline to combat those fires that are obviously in an advanced stage upon arrival of first due companies. However, with an emphasis on being proactive, a good fire officer will also make the weapon selection based on the probable and projected fire growth during the reflex time needed to get operating forces up to the fire floor and place an attack line in service. Although perhaps not advanced upon arrival, a fire that is on an upper floor of a high-rise building will have additional time to grow to an advanced stage before fire companies begin the attack.

In situations where we are unable to determine the extent or location of the fire, both of these are significant tactical concerns that point directly to the need for 2½-inch attack line. *Unable to determine the extent of a fire* means that, although there may not be a heavy smoke condition showing from the exterior of a large high-rise or standpipe-equipped building, and perhaps no fire showing at all, there still could be a serious, fast-moving fire deep within the building. Unknown extent should direct an engine company to stretch a big line, especially at fires in high-rise and standpipe-equipped buildings, so that regardless of the fire volume that they encounter, they will be equipped with a powerful weapon.

Most high-rise and standpipe-equipped buildings, especially those that are commercial occupancies, typically have floor areas that are *large and uncompartmentalized*. This leads to rapid fire extension, with the potential of extremely large fire areas, requiring a significant flow if the fire is to be stopped. There will likely be a very narrow window of opportunity, that if missed, may mean fire control will likely not be achieved. Proactive selection of 2½-inch attack line is of paramount importance.

Even in those situations where we have the advantage of compartmentalization, such as in residential occupancies, the proactive selection and utilization of a 2½-inch handline is still very important. Compartmentalization, although excellent for limiting fire spread, also allows for an extreme buildup of dangerous and debilitating heat conditions. The capability to deliver *tons of water* precisely—one ton of water every minute from a properly supplied 2½-inch attack line—is critical to cool this extremely high heat. Fire officers should not rely on compartmentalization to protect them and their crew members, but, rather, be prepared for a deadly and dangerous heat condition. This is no place for a low flow handline. To stop the fire, cool the atmosphere, and prevent a deadly flashover, using a 2½-inch handline is essential.

Standpipe operations is the last component in the A D U L T S acronym. The recommended use of 2½-inch handline for standpipe operations is once again based upon the low pressure typically associated with most standpipe systems, including those situations where a building has improperly set pressure-regulating devices. In addition, the reflex time associated with standpipe operations, and overall firefighting operations in large high-rise buildings is very significant. The fire that may appear from the street to be rather benign and easily controlled with a 1¾-inch handline, will, in most cases, be a completely different enemy 10, 15, or 20 minutes later, when the attack actually begins. Add to that an unforeseen wind condition forcefully driving the fire out of the apartment and into the public hallway, and the need for an extremely powerful, high volume, low pressure weapon becomes critical. If this equipment is still on the pumper down at street level, or worse yet, if those tools are not available to firefighters at all due to organizational complacency, fireground failure and possible tragedy might be the end result.

Even with the recommendations listed in the A D U L T S acronym, along with the endorsement of countless fire service professionals experienced in high-rise firefighting and standpipe operations, there is still a segment out there that just cannot comprehend it. They simply disregard the facts and figures, and place their entire weapons selection criteria on the speed and maneuverability associated with 1¾-inch attack line. Those are, in fact, very valuable benefits of the smaller weapon. However, all the speed and maneuverability in the world will not overcome a low-pressure situation and the overwhelming Btus that will likely be encountered at many high-rise operations. The proactive selection of a more powerful tool will, at worst, end up being too much, which is not a bad thing. It's better to have too much power nine out of 10 times, than to not have enough power just once. That one time could be someone's last time.

Hydraulic calculations for 2½-inch hoseline

Let's take a look at the potential flow from a combination of weapons, including 150 feet of 2½-inch attack line, supplied from a standpipe for the purposes of combating a fire in a high-rise building. We want to achieve a flow of 150 gpm, the minimum for structural firefighting in residential occupancies.

The following flows were calculated using the friction loss formula for 2½-inch hose, again from Mahoney's *Fire Department Hydraulics*.[3]

$2Q^2 + Q$

$Q = gpm/100$

150 feet, 2½-inch, 150 gpm (100 psi automatic combination fog nozzle)

1. Approximately 9 psi friction loss in 150 feet of 2½-inch hose flowing 150 gpm
2. 100 psi nozzle pressure in order to deliver 150 gpm
3. 76 pounds force nozzle reaction
4. Required standpipe outlet pressure = 109 psi

150 feet, 2½-inch, 150 gpm (low pressure 75 psi, combination fog nozzle)

1. Approximately 9 psi friction loss in 150 feet of 2½-inch hose flowing 150 gpm
2. 75 psi nozzle pressure in order to deliver 150 gpm
3. 66 pounds force nozzle reaction
4. Required standpipe outlet pressure = 84 psi

150 feet, 2½-inch, 150 gpm (low pressure, 50 psi, ¹⁵⁄₁₆-inch smoothbore nozzle)

1. Approximately 9 psi friction loss in 150 feet of 2½-inch hose flowing 150 gpm
2. 33 psi nozzle pressure in order to deliver 150 gpm
3. 46 pounds force nozzle reaction
4. Required standpipe outlet pressure = 42 psi

Next, we want to achieve a flow of 200 gpm, which is considered to be an easily achievable flow from 2½-inch handline.

150 feet, 2½-inch, 200 gpm (100 psi automatic combination fog nozzle)

1. Approximately 15 psi friction loss in 150 feet of 2½-inch hose flowing 200 gpm
2. 100 psi nozzle pressure in order to deliver 200 gpm
3. 101 pounds force nozzle reaction
4. Required standpipe outlet pressure = 115 psi

150 feet, 2½-inch, 200 gpm (low pressure 75 psi, combination fog nozzle)

1. Approximately 15 psi friction loss in 150 feet of 2½-inch hose flowing 200 gpm
2. 75 psi nozzle pressure in order to deliver 200 gpm
3. 87 pounds force nozzle reaction
4. Required standpipe outlet pressure = 90 psi

150 feet, 2½-inch, 200 gpm (low pressure, 50 psi, 1-inch smoothbore nozzle)

- Approximately 15 psi friction loss in 150 feet of 2½-inch hose flowing 200 gpm
- 45 psi nozzle pressure in order to deliver 200 gpm
- 71 pounds force nozzle reaction
- Required standpipe outlet pressure = 60 psi

Next, we want to achieve a flow of 250 gpm, which is considered the minimum for structural firefighting in commercial occupancies, and is a flow that can be realistically achieved (during low pressure situations) *only* by 2½-inch handline.

150 feet, 2½-inch, 250 gpm (100 psi automatic combination fog nozzle)

- Approximately 23 psi friction loss in 150 feet of 2½-inch hose flowing 250 gpm
- 100 psi nozzle pressure in order to deliver 250 gpm
- 126 pounds force nozzle reaction
- Required standpipe outlet pressure = 123 psi

150 feet, 2½-inch, 250 gpm (low pressure 75 psi, combination fog nozzle)

- Approximately 23 psi friction loss in 150 feet of 2½-inch hose flowing 250 gpm
- 75 psi nozzle pressure in order to deliver 250 gpm
- 109 pounds force nozzle reaction
- Required standpipe outlet pressure = 98 psi

150 feet, 2½-inch, 250 gpm (low pressure, 50 psi, 1⅛-inch smoothbore nozzle)

- Approximately 23 psi friction loss in 150 feet of 2½-inch hose flowing 250 gpm
- 44 psi nozzle pressure in order to deliver 250 gpm
- 87 pounds force nozzle reaction
- Required standpipe outlet pressure = 67 psi

Of the previous different weapons combinations, none of the ones utilizing the automatic combination fog nozzles would be capable of achieving the target flow when operating off a low-pressure standpipe. However, all of the weapons combinations equipped with low-pressure nozzles would achieve the target flows. Specifically, the weapons combinations equipped with the low pressure smoothbore tips have extremely low pressure requirements (at the standpipe hose valve outlet), at 42 psi with a flow of 150 gpm, 60 psi with a flow of 200 gpm, and 67 psi with a flow of 250 gpm. If a factory pre-set pressure reducing valve is in place, set to a discharge pressure of 70 to 90 psi—not unrealistic—all of the weapons equipped with smoothbore tips would easily achieve the target flows, and two of the weapons combinations equipped with low pressure fog would work to achieve the target flow, at 150 and 200 gpm, respectively. The reaction force for almost all of the weapons combinations equipped with the automatic combination fog is very excessive, whereas the reaction force for all of the smoothbore and most of the low pressure combination fog examples are very reasonable.

A good rule of thumb regarding standpipe operations is that any weapons combination that requires less than 100 psi standpipe outlet pressure to supply is realistic and achievable from most standpipe systems. With regard to nozzle reaction, anything greater than 100 pounds force will be very difficult and potentially dangerous to operate and control. Clearly, the combination of high volume, low pressure weapons, specifically those equipped with 2½-inch hoseline and smoothbore nozzles, consistently achieves the overall best flow at a minimum pressure, while still maintaining a reasonable nozzle reaction.

The formulas used to calculate reaction force for the previous weapons combinations are from the *Fire Stream Management Handbook* by David P. Fornell.[4] They are as follows:

Reaction force formula for smoothbore nozzles:

$$RF = 1.57 \times (BD \times BD) \times NP$$

- RF = Nozzle reaction in force/pounds
- BD = Smoothbore diameter in inches
- NP = Nozzle pressure

Reaction force formula for combination fog nozzles:

$$RF = gpm \times \sqrt{NP} \times .0505$$

- RF = Nozzle reaction in force/pounds
- gpm = Gallons per minute flowing
- NP = Nozzle pressure

At this point, those readers who have ever operated a 2½-inch handline may be thinking *easier said than done*. I agree. Operating and proactively advancing a big line is a very challenging proposition. However, I believe that there are several very specific factors that can truly make operational success with a 2½-inch handline a reality, as well as safe and effective, even with minimum manpower. I refer to these factors as the seven keys to operational success with 2½-inch handline.

Seven Keys to Success with the Big Line

There are several very important components associated with the use of 2½-inch handlines, which create a synergistic effect that will ultimately culminate in fireground success. I like to refer to these components as the *Keys to Success with the Big Line*.

1 User-friendly apparatus and equipment

Ideally, a user-friendly pumper apparatus is set up with at least one 2½-inch preconnected attack line and a large static hose bed with a substantial supply of 2½-inch hose—a minimum of 600-feet, depending on a needs assessment of the specific response area. The 2½-inch in this static bed should be set up in a reverse lay, with a nozzle attached to the lead length of hose. The first two or three sections can be set up in horseshoes for easy stretching of the line (fig. 8–1). In addition, three or four separate sections of 2½-inch hose should be set up for, and dedicated to, standpipe operations (fig. 8–2).

2 User-friendly standpipe hose packs

Fire departments that protect standpipe-equipped buildings, both highrise and low-rise, have a critical need to be equipped with user-friendly standpipe hose packs. In a two-part series on high-rise/standpipe hose packs—"High-Rise/Standpipe Hose Packs: A Primer," Part 1 and Part 2) in the July and August 1999 issues of *Fire Engineering* magazine[5]—

I listed some essential considerations when designing and assembling a standpipe hose pack to ensure that it remains user friendly:

A. The hose pack should be as lightweight and as compact as possible.

B. The hose pack should be relatively easy and comfortable to carry.

C. The hose pack should be designed for easy, fast, and efficient stretching on the fireground.

I have seen advertisements for various devices used to carry standpipe hose packs. One particular system is very large and appears to be quite cumbersome, resembling a body bag slung over a firefighter's shoulder. Other devices use a system of straps and buckles to hold several

Fig. 8–1 A static hose bed with a substantial supply of 2½-inch attack line is an essential tool for engine company success.

Fig. 8–2 A minimum of three lengths of 2½-inch hose should be preassembled and set up for high-rise standpipe operations.

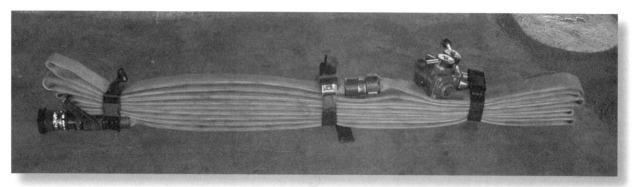

Fig. 8–3 This heavy, bulky hose pack is an example of an impractical and inefficient method to carry all of the standpipe hose and appliances.

sections of hose together. Many fire departments have purchased such devices and use them to carry all of their standpipe equipment, including three sections of hose in a very heavy and bulky one-size-fits-all hose pack (fig. 8–3).

The theme with regard to hose packs should be to "keep it simple." The overall high-rise/standpipe equipment package should not consist of one large, heavy, bundle of equipment. The entire equipment package should be designed as a modular system, with several, lightweight, manageable components that are transported to the point of operation, not by one firefighter, but by a team of firefighters. On arrival, the company officer in charge of the fire attack group determines how much hose needs to be stretched and what equipment should be used for that particular operation. Members of the fire attack group then proceed to quickly place it in service.

Operational efficiency and effectiveness are best achieved by dividing the labor of transporting the entire equipment package among several team members. Single-hose packs consisting of one 50-foot section (one length) of 2½-inch hose per team member is ideal. The single-hose pack should be as lightweight as possible. Lightweight hose is available on the market today. The DFD specifies a high-rise 2½-inch hose that weighs only 20 pounds per 50-foot length. Reducing the overall weight that one firefighter must carry up several flights of stairs by specifying lightweight hose and dividing up the labor is very effective (fig. 8–4).

One-piece hose packs, with two or three lengths of hose and with all the ancillary equipment attached, can weigh well in excess of 50 pounds, and that's using smaller diameter 1½-inch or 1¾-inch hose. Even the biggest, strongest, and most physically fit members will quickly become exhausted transporting such a heavy and cumbersome equipment package up several flights of stairs.

Fig. 8–4 Teamwork should be used to transport all of the necessary standpipe equipment items to the point of operation.

Properly stretching the attack line

Once the equipment has been brought up to the point of operation, typically the floor below the fire, members can determine the amount of hose needed to effectively reach the fire area; generally 150 feet is a good starting point in most high-rise and standpipe-equipped buildings. The hose must then be connected and properly stretched (fig. 8–5).

When using the recommended 2½-inch attack line, it is critically important that the entire attack line be fully and properly stretched out prior to calling for water. "Flaking" the hose out on the landing is not acceptable and will not work. Even a small pile of 2½-inch hose, perhaps only five or 10 linear feet that has not been properly stretched out will develop several serious kinks once the line is charged with water (Fig. 8–6).

Kinks are among the engine company's greatest enemies on the fireground and can lead to operational failure and firefighter injuries. Some kinks are a fact of life and will naturally occur during the advancement of the hoseline at most operations. Those kinks can usually be addressed and corrected easily by members of the attack team. No member, including members of other companies and command officers, should ever pass a kink without taking the time to eliminate it. Keep in mind that most kinks can be prevented by properly stretching the hoseline from the beginning.

In addition, when hose must be laid out up a flight of stairs, and specifically when the hose is stretched up past the fire-floor landing (stairwell stretch), members must secure the hose before water is called for and while the line is being charged. This is a simple process of holding onto the line or using your body weight by kneeling on it to hold it in place. If this isn't done and the hose is not secured, the weight of the water will pull the hoseline down the stairs when it is being charged, and all your hard work

Fig. 8–5 Taking the extra time to properly stretch the attack line prior to charging with water will pay significant dividends when the attack commences.

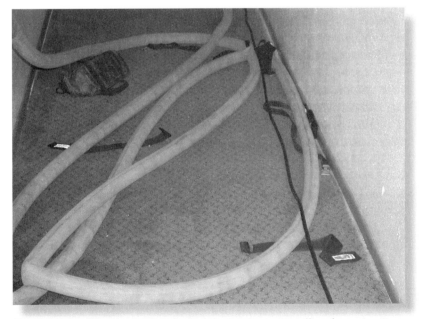

Fig. 8–6 If the line is not properly stretched out prior to calling for water, it will result in several kinks, and most likely, a failed operation.

will be for naught (fig. 8–7).

Remember, when stretching a 2½-inch handline, especially during high-rise and standpipe operations, the line *must* be fully and completely stretched out, with *no* piles of hose at any point, and members must physically secure the line before it is charged. Simple, basic fundamentals are the keys to success (fig. 8–8).

4 Proper operating pressures

Without exception, excessive and improper operating pressures are at the heart of every failed 2½-inch handline operation. Once again, because this tool is most applicable at our low-frequency events, a lack of experience and training can quickly lead to a failure because of excessive pressures. Simply put, with 2½-inch handline, less is more. It is a high-volume weapon, but, because of its size, there is much less friction loss than what we typically experience with 1¾-inch handlines. In fact, the friction loss in a big line is six times less than that of a 1¾-inch handline at equal flows.

Short stretches of 2½-inch hose (150 to 200 feet) off standpipe, typically require very low operating pressures. If the outlet pressure is more than two digits, the 2½-inch attack line is probably over-pressurized. For example, a typical standpipe stretch, 150 feet of 2½-inch hose, using a low-pressure smoothbore nozzle (1⅛-inch tip, 266 gpm at 50 psi), requires an operating pressure of only 75 psi. In fact, when using some of the best brands of new modern fire hose, a proper stretch, with no kinks and minimal turns, pressures can be reduced by 10% and still maintain effective flows. So at 65 to 70 psi standpipe outlet pressure, a 150-foot stretch of 2½-inch attack line can flow well over 250 gpm.

5 Low-pressure nozzles

One of the very best tools to help firefighters achieve a successful fireground operation when using a 2½-inch attack line is the low-pressure nozzle. In fact, I consider the use of low-pressure nozzles, specifically the smoothbore nozzle, to be an absolutely essential part of a successful big line operation. Those who insist on using combination fog nozzles, specifically, high-pressure and automatic-type nozzles, for operations with 2½-inch handlines are setting themselves up for probable failure.

With the required 100 psi for nozzle pressure, coupled with the friction loss in the line, the overall required pressure is in excess of 125 psi. This pressure makes the line much stiffer and more difficult to maneuver. Add to this the significant nozzle reaction, and something has to give. The nozzleman likely will have to partially close the

Fig. 8–7 Engine company members stretching the attack line up a flight of stairs past the fire floor landing, must be positioned to secure it prior to calling for water.

manufacturers produce excellent low-pressure combination fog nozzles with rated flows at 75 and 50 psi.

6 Unglamorous operating positions

The vast majority of firefighters would likely agree that one of the most popular assignments on the fireground is that of the nozzleman. It is truly the attack team's most glamorous and satisfying position. A good nozzleman protects the crew with proper water application and can see immediate results of his actions at most fires. Unfortunately, there is only one nozzle position, and we can't all be on the nozzle at once. Therein lies the problem: without the proper training and fireground discipline, some firefighters tend to sneak up on the line in an attempt to get a piece of the action.

Ideally, two engine companies with a minimum of six personnel (not counting engineers/pump operators) should be used to effectively advance and operate a 2½-inch attack line. The second engine company should be used to support the first engine company's advancement by positioning the members at critical friction points within the stretch. Those locations include corners, doors, and stairs (fig. 8–9).

The first engine company or the nozzle team should consist of, preferably, at least three members—the nozzleman, a backup, and a company officer leading the attack. The second engine company, or the support team, should consist of,

Fig. 8–8 This member is stretching out the remaining hose on the floor below before charging it with water.

bail to maintain control of the line, thus reducing the needed flow and adding to the excessive (static) pressure within the overall system. The firefighters attempting to operate this line will be left with the memory of a miserable experience and hence will become less likely than ever to select a 2½-inch handline for future fireground operations.

For those who, for whatever reason, cannot accept the facts that clearly distinguish the smoothbore nozzle from the combination fog as a superior weapon for interior structural firefighting, I strongly suggest that they at least consider using low-pressure combination fog nozzles. Once again, the goals are to lower the nozzle reaction without reducing flow and to lower the overall pressure needed to supply the handline and nozzle. Some very good nozzle

Fig. 8–9 This firefighter is helping advance the 2½-inch handline by pushing additional hose up onto the fire floor from the floor below.

preferably, at least three members. A firefighter should be positioned at the closest friction point behind the nozzle team and the company officer at the next friction point, so he can get the best overall picture of conditions and maintain best contact with both of his members. The remaining firefighter should be positioned at the next friction point and will likely have to operate in a fluid and dynamic manner for long stretches, moving forward and backward as necessary to free the line and pull more line in at the entry point (fig. 8–10).

answer is always very simple—use everything you would normally use at a real fire. As the saying goes, "you practice like you play."

Quality training involves being very creative. Company officers and training officers must constantly work to create new, interesting, and innovative ways to train firefighters, especially on repetitive evolutions and the fundamentals. When drilling with the big line, it is always best to make the drill as realistic as possible. Make every attempt to use actual buildings for stretching lines.

7 Training, training, and more training

Of the seven keys to success when using a 2½-inch attack line, the seventh item or key is by far the most important. Just as it's written—"training, training, and more training"—training never ends, and you're never finished preparing.

To become efficient and effective with a 2½-inch attack line, an engine company must schedule a reasonable amount of training into their daily routine. A short, but productive drill, preferably one hour, on the use of 2½-inch hose at least every third shift, more if possible, will yield results in a very short time. The frequency and consistency of the training are important, but so, too, is the quality of the drill. Pulling a preconnected 2½-inch line off the rig in front of the firehouse, and squirting some water on the grass while wearing gym shorts and a tee-shirt is certainly better than nothing, but it is far from a realistic and truly valuable drill. When you arrive at a fire, you're wearing full PPE and SCBA. For interior operations, you're down low on the floor, with your SCBA face pieces on, breathing air from the SCBA cylinder. During drills, some firefighters will ask, "Do you want us in full gear? Hoods too? Do I need my gloves?" The

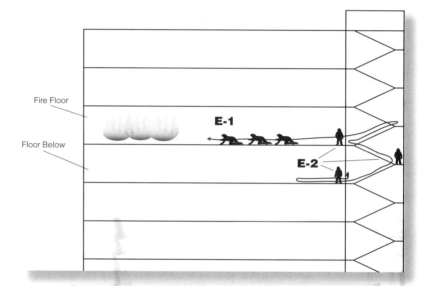

Fig. 8–10 This diagram illustrates the proper positions for a six-member, two-engine-company attack team.

Fig. 8–11 Stretching a line into the basement of the firehouse is a simple, yet very productive drill.

When I was the captain of DFD Engine Co. 3, there were a lot of abandoned buildings in my district. I would frequently take my crew out for drills in those buildings. In our dismally litigious society, it now takes an act of Congress for many fire departments to use private property for training. Once their lawyers talk to our city attorneys, it's all over. However, make every effort to find buildings scheduled for demolition, and be tenacious during the long process of getting approval, from the property owner, the fire department bosses, and any lawyers that get involved. Nothing is better than live training in acquired structures, especially if live fire is part of the curriculum.

If the opportunities for acquired buildings and training on private property are few and far between, remember, you have a great location you can use anytime. Pull the pumper out on the street, hook up to the hydrant in front of the firehouse, and stretch a big line into the basement of the firehouse (fig. 8–11). Another option is the playground at your local city park. On numerous occasions, I took my firefighters down to Curtis Park for a drill. When the playground was not occupied, we'd stretch the big line to one side of the playground, call for water, go on air, and move through the playground equipment, periodically flowing water. It probably looked odd, but what an excellent way to hone the basic skills associated with moving a big line through numerous friction points.

Whether you have a fire in a commercial building 100-feet away from the pumper, or a fire on the 50th floor of a high-rise, you can achieve operational success with 2½-inch handline. The seven keys to success are a good foundation to help you achieve that fireground success.

The High-Rise/Standpipe Hose Pack

For high-rise firefighting and standpipe operations, I recommend the use of 2½-inch hose. Once again, the engine company should be equipped with a minimum of three, preferably four separate lengths of 2½-inch hose, preassembled into separate hose packs. Keeping the previously listed goals in mind when assembling a hose pack, the following procedure can be used to build a reliable 2½-inch hose pack.

First, I recommend using as lightweight a hose as possible, but one that is durable and proven reliable during interior structural firefighting operations. The standpipe hose pack that I recommend is used by the DFD, and I'll refer to it as the *Denver Hose Pack*. This hose pack is assembled in a horseshoe configuration. Three lightweight, quick release straps are used to hold the hose pack together.

Assembling the hose pack begins with a measurement of 32 inches from the outside of the female coupling. It is a good idea to mark all of your high-rise hose at 32-inches with a dark, permanent marker.

The first bend is taken at 32 inches, and the hose pack is assembled in a horseshoe (fig. 8–12). A minimum of three members should be used to assemble the hose pack, with a focus on keeping the pack as tight as possible while folding it, making it easier to store, carry, and transport. The member located at the couplings should ensure that the hose is stopped short of the couplings, and each fold should be staggered, one short, one long (fig. 8–13). This

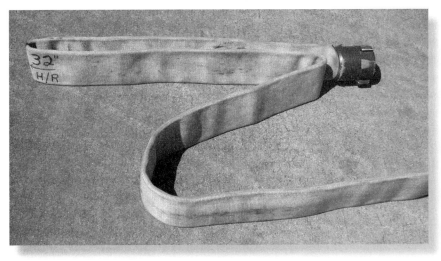

Fig. 8–12 The Denver Hose Pack is assembled in a horseshoe type configuration, starting with the first bend at the 32-inch mark.

procedure also helps keep the hose pack as tight and compact as possible.

Folding continues until the pack is completed. The final steps include connecting the couplings together to protect the male threads (fig. 8–14). In addition, any extra hose can be tucked back into the hose pack on the opposite side from the male coupling (fig. 8–15). The completed hose pack requires three straps to hold it together. Place these on both sides, as close to the couplings as possible, and one on either side near the top of the hose pack (fig. 8–16).

One of the three hose packs should be equipped with a nozzle in order to make operations faster when connecting the hose packs together and stretching for fireground operations. The nozzle should be placed bail down, with the tip just past the end of the hose fold (fig. 8–17).

Fig. 8–13 The hose folds should be staggered, one short, one long, which helps keep the hose pack tight and compact.

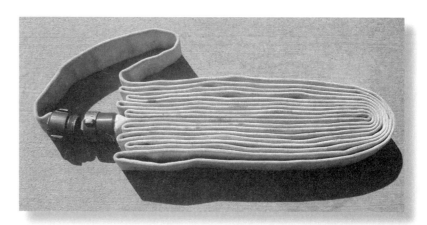

Fig. 8–14 The couplings should be connected together to protect the male threads.

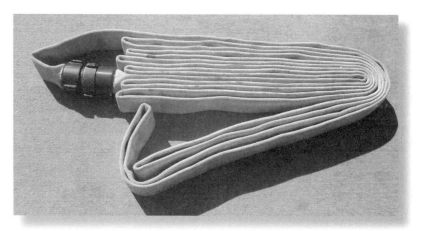

Fig. 8–15 Any extra hose should be tucked into the hose pack, on the opposite side from the male coupling.

Overwhelm the Enemy with Disproportionate Force

Let me close this chapter with a few final thoughts. At the Fire Department Instructors Conference in Indianapolis, my HOT Team member, Harry Lee Davis, Firefighter, FDNY Squad Co. 18 (Retired), while instructing attendees of our Standpipe Hands on Training program, cited former Secretary of State Colin Powell's military philosophy. In the now-infamous "Powell Doctrine," Colin Powell, then chairman of the Joint Chiefs of Staff, outlined his vision for efficient and decisive military action. Specifically, Powell stated in reference to military force, *"the force, when used, should be overwhelming and disproportionate to the force used by the enemy."*[6] I find this to be a very inspirational philosophy that can and should be applied to a fire department's operations when faced with the enemy that is fire. The concept of using overwhelming and disproportionate force can and should be applied by all fire departments, including even the smallest of fire departments, by training on the use of, and proactively selecting, the appropriate weapons for battle. The 2½-inch handline is just such a weapon that can truly overwhelm a formidable enemy. It should be used to combat a wide range of fires encountered by fire departments in big cities and small communities. It's an essential weapon for fires that occur in high-rise and standpipe-equipped buildings. With the proper attitude and application of the *seven keys to success*, the use of this weapon can be a reality for any fire department.

Summary

Several different hoseline sizes are available to firefighters. Each size has various advantages and disadvantages. Although 1¾-inch handline is more maneuverable, its use for high-rise firefighting and standpipe operations is not recommended due to high friction loss, low flow, and low pressures typically encountered with standpipe systems. The 2-inch handline should be the minimum sized attack line for operations off standpipes and high-rise firefighting, but only if 2½-inch hose is available for backup. The best weapon for high-rise firefighting and standpipe operations is the 2½-inch handline, due to its high volume flow at extremely low pressures. With a good training program, and application of specific techniques, fire departments can achieve operational success with the 2½-inch handline, even with minimum manpower.

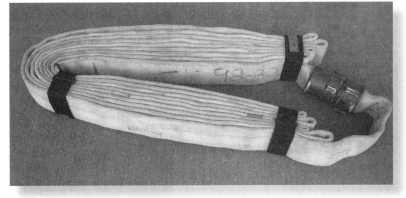

Fig. 8–16 Two of the lightweight Velcro straps used to hold the finished hose pack together should be placed as close to the end of the folds as possible, with a third strap placed on either side near the top.

Fig. 8–17 One of the hose packs should be a nozzle section, with the nozzle preattached, bail down.

References

1. Mahoney, Eugene F., 1980, *Fire Department Hydraulics*, Boston, Allyn and Bacon, Inc., p. 202.

2. Sylvia, Dick, 1970, *Fire Service Hydraulics*, Second Edition, Fire Engineering Books and Videos, a Division of Pennwell Publishing Company, p. 234.

3. Mahoney, Eugene F., 1980, *Fire Department Hydraulics*, Boston, Allyn and Bacon, Inc., p. 536.

4. Fornell, David P., 1991, *Fire Stream Management Handbook*, Fire Engineering Books and Videos, a Division of Pennwell Publishing Company, p. 223.

5. McGrail, David M., 1999, High-Rise/Standpipe Hose Packs: A Primer, Part 1 and Part 2, *Fire Engineering* Magazine, July and August Issues, 1999, A Division of Pennwell Publishing Company.

6. Davis, Harry Lee, 2005, Engine Company Standpipe Operations, Hands on Training (HOT), Powell Doctrine (Former Secretary of State, Colin Powell).

CHAPTER 9

Nozzle Selection

The type of nozzle to be placed at the business end of an attack hoseline is a critically important decision. The selection of what type nozzle to use should be based on a conscious and deliberate decision. It truly does matter, especially for high-rise firefighting and standpipe operations.

The Great Debate

For nearly 50 years, the American fire service has had three very distinct groups with regard to nozzle preference. The first group consists of individuals who prefer the use of a combination fog nozzle for most, if not all operations. Many members of this group are firefighters who came into the fire service when indirect attack and the use of combination fog nozzles was common practice. Other members of this group are individuals who were mentored by fog nozzle proponents, or work for fire departments that use fog nozzles exclusively. Most were introduced to and taught firefighting tactics using combination fog nozzles with little emphasis on the use of smoothbore nozzles.

Most firefighters from this group have embraced the use of combination fog nozzles having, in many cases, never actually utilized a smoothbore nozzle, specifically for interior structural firefighting. This may in fact be the only water delivery tool that most firefighters from this group have used throughout their entire career, and at this point, most have become so accustomed to using the combination fog nozzle, that they are very reluctant to even consider a change. I believe that many firefighters from this group were taught to utilize the combination fog nozzle in an improper manner, unfortunately, having passed these techniques along from one generation to the next.

The next group consists of those firefighters who prefer the use of a smoothbore nozzle for most operations, specifically for interior structural firefighting, and especially for high-rise firefighting and standpipe operations. The first advocates of the smoothbore nozzle were comprised of a significant number of men who where in the fire service before the introduction of indirect attack and the combination fog nozzle. Having experience in the use of the smoothbore nozzle gave them the knowledge and ability to compare the direct attack using a smoothbore nozzle to the newly introduced indirect attack using a combination fog nozzle. I learned the value of direct attack using a smoothbore nozzle from my father, who was part of this pre-fog generation.

The late William E. Clark was also a member of this pre-fog generation of firefighters. Long considered to be a true expert on fireground operations, Clark spent years teaching and writing about proper water delivery and application. Clark was selected to write the "Fire Company Operations" section of the *Fire Chief's Handbook*, Fifth Edition. Speaking to indirect attack and the use of water fog, Clark states in Chapter 17: "Ridiculous claims were made for it, but worse, they were accepted without question, except by a few of us who believed what we saw rather that what we were told." Clark goes on: "This revision of fire stream tactics [from fog back to smoothbore/solid stream] is the strongest trend of the nineties."[1] I believe that the revision is still occurring, stronger than ever before. If this is new to you, welcome aboard.

Clark, like so many others, never fully embraced the indirect attack and the wide spread indiscriminate use of water fog. I agree with Clark, and believe that there was a significant change with regard to fire stream tactics that took place in the nineties, and is continuing today. In fact, many of the newest generation of firefighters, having been exposed to a resurrected philosophy about water delivery, are now among a growing number of smoothbore nozzle proponents.

I have found that most fire departments, at least the progressive and forward-thinking ones, are now teaching their firefighters proper water application. Specifically, the water should be delivered in a straight stream pattern when using a combination fog nozzle, or in the much-preferred solid stream from a smoothbore nozzle for interior structural firefighting. It is important to note, the straight stream is a pattern selection from a combination nozzle, which creates a straight stream pattern, albeit still a broken stream. On the other hand, the stream created by a smoothbore nozzle is actually a solid stream of water.

I admire and respect the firefighters from both of these very distinct groups. Members of these two groups have strong opinions about one of the most important tools we use. And although their relationship may seem to be adversarial at times, when it comes to water delivery, they both have the same primary goal in mind, to extinguish the fire.

It is the third and final group that gives all of us the most trouble. That group is represented by the individuals who simply don't care, or think that it doesn't matter what type of nozzle is used. In some pathetic cases, they might not even know whether or not there is, in fact, a nozzle attached to the end of an attack line, having failed to even check it at the beginning of their shift.

The guys in this group will make ignorant statements like "I don't care what nozzle's on there, just as long as water comes out the other end of it." Or the guys who want to play Switzerland and remain neutral will say, "I'm not getting involved in that argument." Get off the fence, take a stand, and have an opinion, but base that opinion on facts, not fiction. "Because that's the way we've always done it" is another sad comment that speaks volumes about the mind-set of the individual who said it and his inability to learn, grow, and evolve within a very dynamic fire service.

My Personal Experience

Let me give you a little background on my evolution and history with regard to nozzle selection. I came into the fire service 25 years ago, and landed a job on a suburban Denver metro fire department. Because I was a fireman's kid, I thought I knew a whole lot more about being a fireman than I actually did. In fact, I was a very young, immature kid, without the first clue about firefighting.

My dad let me go out into the real world, without any interference, I believe, in an attempt to let me discover the world of firefighting on my own. Don't get me wrong, he was always there to assist, provide guidance, and answer any questions, but because I thought I knew more than I actually did, I unfortunately didn't ask for much help in the beginning.

In reality, it was all new, overwhelming, and a whole lot scarier than I imagined. My instructors in that first fire academy taught me how to deliver water using a combination fog nozzle. The bible was the *IFSTA 200*, which has evolved to become the *IFSTA Essentials Manual*. My first fire department used fog nozzles almost exclusively, and, in fact, my exposure to smoothbore nozzles was primarily from pictures in the *IFSTA Manual* and the occasional encounter with a few old smoothbore nozzles around the job that were rarely actually placed on the end of a hoseline and used for real firefighting.

Keep in mind, this was a time long before the current benefits of significant fire service evolution. We were equipped with ¾-length turnout boots, and Nomex hoods were unknown to most of us, except for a few pioneers who purchased their own and were subsequently made fun of for it, often times referred to as "race car drivers" when they wore their hoods. Needless to say, our personal protective equipment was limited.

I was taught to use a wide, 60-degree, fog pattern from a high pressure, 100 psi, 95 gpm, combination fog nozzle for interior structural firefighting. The idea was to make little water droplets, said to absorb more heat faster. I remember one of my academy instructors talking about the proper way to rotate the nozzle, all because we were on this side of the equator. Firefighters on the other side of the equator had to rotate their nozzles the other direction. I always wondered how you would have to rotate the nozzle if you had to fight a fire right on the equator.

Another thing that was taught—well, it was actually driven home forcefully—was the issue of water damage. I was taught to use this fog pattern primarily because it reduced water damage, and to always shut the nozzle down quickly in order to avoid any and all water damage. It was like a mortal sin if there was one drop of water on the floor that hadn't been converted to steam and used to extinguish the fire.

Furthermore, it was constantly emphasized that a fog pattern from a combination fog nozzle could actually provide protection for firefighters. I remember my first experience with the propane Christmas tree. It was absolutely exhilarating! We did the drill at night, and when my training instructors fired up that tree, it lit up the entire drill ground. The sight and sound of the fire roaring out of that tree was awesome.

We were taught to use two wide fog patterns, from two separate handlines, side by side to approach the tree, and thus provide the necessary "protection." When we got close enough, one man would squat down and shut off the valve controlling the flow of propane and the fire would go out. For a young kid, it was an absolutely incredible training experience. Unfortunately, after 25 years, I have still yet to respond to a propane Christmas tree on fire.

During that first fire academy, we did a few training burns and each time I was amazed at how much of a beating you took fighting these fires. After experiencing a few of these fires, I had a whole lot more respect for my father and all firefighters, thinking how I couldn't believe just how hot, miserable, and scary it was to attack an interior structural fire.

After I graduated from the fire academy, I found myself in the real world, responding to calls, and occasionally some good working fires. I tried to be tough, but something just didn't seem right. I was very reluctant to approach my father, thinking how disappointed he would be if I complained about the heat and my fear about being inside these fire buildings where the atmosphere just seemed way too hot.

I remember seeing guys just get hammered, coming out of buildings with skin hanging off their face and neck. Some guys sustained significant burns to their ears, sometimes even requiring hospitalization. Little did I know, I was about to get my own taste of this, an experience that would truly change my life.

It seems like it happened yesterday, but it occurred over 20 years ago. I was assigned to an engine company located in a fairly old and busy area of this suburban city. We got a fair amount of calls, including some good fires. There were some good guys on the crew and I learned a lot from most of them. Unfortunately, my officer was probably one of the worst I have ever worked for in my entire career. I remember him making fun of me when I would read my *Fire Engineering* magazine in the firehouse kitchen. He would always refer to it and my other fire service periodicals as "comic books." We did little or no company level training, so I found myself out on the apparatus floor by myself most days, trying to learn the job. The other men would occasionally help and teach me things, but without leadership at the top, we didn't do much as a group.

Oddly enough, every time we got a decent fire, this lieutenant could be found anywhere but inside the fire building. Still, being the new guy, I just assumed that was the way it was done. The officer stays outside and the men take the line in and put out the fire. Maybe he was practicing an early version of incident command. Perhaps he was a pioneer and I have underestimated him. But I really don't think so. Bottom line, he could certainly talk a big fire, but come to find out, that's all it was—just talk. He was my boss, and I was stuck with him for a few years. As they say, sometimes you learn more from the bad officers than you do from the good ones.

The day of my career-changing experience started out just like any other day.—roll call, housework, apparatus and equipment maintenance, a few runs, nothing major, and a trip to the store for groceries. It was a little after noon when we got the call. It came in as a fire under a bridge. Our company was the only unit dispatched to what was believed to be just some trash burning under a bridge.

When we got to the location, there was a fair amount of smoke showing from the street level. My experience was still very limited, and my only guidance came from my "favorite" lieutenant. I didn't don my SCBA pack, and my boots were of course not pulled up. My only protection was from my turnout coat and helmet. My lieutenant told the other fireman and me to stretch the booster line, and the engineer (pump operator) gave us water.

We started down a steep embankment toward the source of the smoke. It was difficult terrain, and as we got closer the smoke became much heavier. Once we got under the bridge, the smoke was very heavy, and you could definitely feel the heat. Because of the smoke, we couldn't see it at the time, but a group of homeless transients had built a shanty to live in under the bridge. The walls were plywood, with a dirt floor, and the ceiling was the concrete bridge structure. It was actually a good-sized structure, with an actual working door.

I was on the nozzle, and I did what I was taught; I opened the line up to a wide fog pattern and attempted to aggressively move forward and attack this fire. Reflecting on my training, I was attempting to "protect" myself and the backup firefighter as we approached. What actually happened was that I pushed the heat, fire, and smoke back into the structure, but it had no place to go. It immediately came back out the door and enveloped me. I hung in there, allowing inexperience and pride to override common sense and firefighter safety. Eventually I switched to a straight stream pattern, but the low flow from the booster line required continuous water application, until we eventually depleted the tank. As the line went dry, we finally came out from under the bridge, and the fire was still burning.

As you may have guessed, my lieutenant was standing up on the street. I was dazed. I remember feeling very dizzy (carbon monoxide will do that) and it was all I could do to climb back up to the street. My lieutenant had a look on his face like I had never seen before. Just shocked; eyes wide open, and speechless; for the first time. He looked real scared, but I had no idea why. He told me and the other firefighter to sit down. He began talking on his radio, asking for an ambulance and paramedics. I didn't know why, we didn't rescue anyone from the makeshift structure. The engineer ran over to us with the first aid kit and resuscitator. He had a real concerned look on his face too. He asked us what happened. I was still clueless. And then it hit me. I thought; am I hurt? As it turns out, I had sustained burns to my face, neck, and ears. I took in a lot of smoke and I remember coughing uncontrollably. It was at that point that I started getting a little scared myself.

As it turned out, I survived a serious fire event, (a near miss) probably due mostly to fire service luck. None of the injuries were serious and certainly not life threatening. What hurt more than anything else was my pride. I was especially ashamed of what my father would think when he found out. However, the entire experience became one of the most important events of my fire service career.

The mentoring begins

I saw my dad the next day. After making sure I was okay, he asked me a couple of questions. As it turns out, he wasn't ashamed or mad, but he knew that it was time to start teaching me a few things based on his experience. I described the incident and fire conditions to him. His first question for me was "What size line were you using?" His next question was "What type nozzle were you using?" I learned more from my dad in one afternoon than I had in my entire first three years on the job. He had over 30 years on the DFD at that time, and I was in awe of his knowledge, then and still today. I had no idea he knew so much (fig. 9–1).

As that afternoon became evening, I felt like the heaviest weight had just been lifted off my shoulders. I was relieved to hear from my dad that you don't have to take a beating every single time you fight a fire. He introduced me to the fine points of water delivery within the art of firefighting. Most importantly, he explained that there are other tools out there besides small handlines and combination fog nozzles.

During those first three years on the job, I had been going along without a clue. I was nothing more than an employee of a fire department. After that afternoon and evening spent with my father I became a professional firefighter, dedicated to excellence and the art of water delivery.

After testing for nearly five years, I achieved my life long goal and was appointed to the Denver Fire Department (DFD). My education and development on the art of water delivery would continue on the DFD, as I found a large segment of firefighters and officers using large attack lines and smoothbore nozzles. In fact, the culture of the DFD was entirely different from my previous department. That's not meant to insult or disparage the fine men who I worked with on my first fire department, but by and large, the firefighters I was now working with were much more aggressive, and not at all reluctant to utilize much more powerful tools to deliver water. Furthermore, I rarely heard about water damage anymore, the attitude was completely different.

Many of the firefighters from my father's generation had learned about big lines and smoothbore nozzles before the fog frenzy. They recognized fog as a useful tool for specific, but limited, operations. They held on to their beliefs with regard to smoothbore nozzles, based on their experience and countless tried-and-true operations. I was sitting on top of the world, now being mentored and taught by great firemen, who were, themselves, taught by my father.

Captain Jack Rogers, DFD retired, was a lieutenant at the time, and one of my drill instructors in the DFD academy. Having worked with my father for many years on Rescue Co. 1, he was passing on this powerful water delivery philosophy to me and the other recruits in my academy. I remember him being next to me during many of our live burn training drills. It was a completely different feeling delivering water in proper form and volume, with one of my mentors right next to me guiding me the whole way.

My comfort level and confidence inside fire buildings began to rapidly increase. I was now seeing firsthand that proper and aggressive water application, using the direct method of attack from a smoothbore nozzle, resulted in a faster knockdown on the fire, helped maintain good visibility, and, most importantly, decreased unwanted steam and the resulting exponential increase in temperature. Furthermore, my new mentors demonstrated that you can kill a lot of fire from a greater distance by using the power of a penetrating solid stream of water produced by a smoothbore nozzle, thus affording true protection and safety.

Although I was now going to many more fires in much larger more complex city as a member of the DFD, I was no longer experiencing the extreme temperature conditions I had grown accustomed to. It quickly became clear that the high volume, low pressure weapons used to attack these fires, coupled with the delivery of water in a solid form, was the key to this astounding discovery.

Beginning in my early years on the DFD, I became a serious student of water delivery. I read and studied every piece of information I could find with regard to water delivery. I also listened and learned vicariously from the experiences of countless other firefighters. I quickly learned how to differentiate between fact and fiction. As time went on, it became clear that many of those who were staunch proponents and supporters of combination

Figure 9–1 My number one fire service mentor for the past 25 years, my father, Pat McGrail, retired division chief, DFD

fog nozzles for interior structural firefighting and standpipe operations had never actually tried using a smoothbore nozzle. "I've been doing it this way for 20 years, and I'm not changing now." Just because someone has been doing something a certain way for 20 years, doesn't make it right. On the other hand, of those individuals who were once strong fog advocates, but had open minds, and were willing to give the smoothbore a try, the vast majority would rarely go back to applying water via a fog.

Andy Fredericks

The icing on the cake for me with regard to my water delivery education was the discovery of a man by the name of Andy Fredericks. It was 1995, and I was a captain on the DFD working on Engine Co. 3, a very busy inner city engine company. I felt that my knowledge and experience base with regard to engine company operations, handlines, and nozzles was strong; but there was always room for improvement.

I was quickly thumbing through the latest edition of *Fire Engineering* one day, when I came across an article titled "Return of the Solid Stream." Needless to say, I stopped at that page, sat down, and for the next half hour I read all of the information that I wished I could put down on paper and express to others as eloquently as this guy. Andy Fredericks was saying everything I believed, but could never quite express in such an awesome and professional manner. I thought, "who is this guy, and where is he from?"

Andrew A. Fredericks, from FDNY Engine Co. 48, the bio read. With some additional research, I quickly found out that FDNY Engine Co. 48 was a very busy Bronx engine company that worked many serious fires on a regular basis. Fredericks was a man with a strong formal education who could eloquently express his message and had the down-and-dirty firefighting background gained in the hallways of many a Bronx tenement to backup every word.

I was so moved by the article that I immediately got on the phone in an effort to express my appreciation to Andy Fredericks and, hopefully, learn more from him. I reached Bill Manning, who was the editor of the magazine at the time, and got a phone number. I called and spoke to Andy's wife, Michelle, at home, and she was kind enough to give me the telephone number to Engine Co. 48. I called Andy at work, and we talked for quite some time. It was a great conversation, and, at the end, I knew that I had a new friend and ally in the effort to continue moving the American fire service and my own fire department forward in regard to fire streams and fire attack. At that time, I had no idea how powerful Andy Fredericks's message would become and how many firefighters he would eventually reach. Our brother, the late, great Andy Fredericks, left us with countless articles and quality videos that thoroughly explain the value of the smoothbore nozzle, the direct versus the indirect attack, and the inappropriate application of water via the utilization of water fog.

Smoothbore versus Combination Fog

By now, you probably have a very clear understanding that I am a strong proponent of the smoothbore nozzle. But keep in mind, I am also a proponent of combination fog nozzles, for certain, specific applications. However, an appropriate use for or application of a combination fog nozzle is definitely *not* when you are operating off standpipe systems, and specifically not inside high-rise buildings. In fact, I firmly believe that fog nozzles should not be used for interior structural firefighting, period.

I am going to lay out the facts behind my philosophy, but most importantly, I base this opinion on my own personal experience. As previously stated, I suffered burn injuries early in my career when using combination fog nozzles. Since then, I have become an advocate of direct fire attack and the use of smoothbore nozzles. I have not suffered serious heat- or thermal-related injury since that unfortunate event some 22 years ago, even though I have been assigned to some of the busiest fire companies on the DFD.

Most proponents of the combination fog nozzle cite a couple of factors regarding their preference for this type nozzle. One of those is their desire to have the versatility to adjust the pattern from a straight stream to a cone shaped fog pattern, narrow or wide. Many in the fog camp also emphatically state that they would never actually open up to a fog pattern when operating inside a fire building, due to the potential to create steam. That comment always brings up the question: "Why are you taking the fog nozzle into the fire building to begin with?" Then come the standard answers, "just in case," or "because I might want to hydraulically ventilate," and of course, my personal favorite, "I might need the protection."

Protection Myth

Since it's my favorite, let's start with that one—protection. It will be a tremendous event in American fire service history when the entire fire service agrees, and fully understands, that what has been incorrectly referred to as "protection" for nearly 50 years, is not! A curtain of finely divided water droplets between you and a fire is not protection. The word *protection*, used in conjunction with a combination fog nozzle, must be replaced, by another word that starts with the letter *P*. That word is *pushing*. Yes, that's actually what's occurring when a combination fog nozzle is opened up to a cone-shaped pattern: the nozzle operator is pushing all the stuff in front of him away. That stuff is heat, fire, smoke, and other nasty products of combustion.

Now, pushing this harmful stuff away from you is not always a bad thing, as long as you have a good place to push it. For example, I always selected a good, high-volume, low-pressure, constant-gallonage—combination fog nozzle to attack a vehicle fire, that is, when it's outside in a ventilated atmosphere. I can attack the fire from a distance using the straight stream pattern, and after the initial knock down, I can approach the vehicle, opening up to a cone-shaped fog pattern, and I will push all the nasty stuff away from me. But I realize this is not protection, it's pushing. The protection came from my initial attack, putting water directly on the fire, and knocking the main body of fire down.

In my travels, teaching and speaking with other firefighters across the country, I find that when these facts are thoroughly explained to them, most are in complete agreement. However, many are still reluctant to break free from nearly 50 years of bastardized water application.

Those lost in this thick fog of bastardized water application will frequently cite this "need for protection" as their reasoning behind wanting to use a fog nozzle. The truly progressive thinkers in the American fire service now understand that this so-called "protection" is truly a myth. Capt. Dave Fornell gives a comprehensive explanation of this fog nozzle "protection myth" in his text, *Fire Stream Management Handbook*. As Fornell states, "All the wide pattern accomplishes is to give the firefighters a false sense of security."[2] Here are the facts. In order to achieve this so-called "protection," you must open the fog pattern up from a straight stream to a wide fog pattern. (Keep in mind, it is not a solid stream from a combination nozzle, but a broken stream in a straight pattern.) If you open up to a wide fog pattern, here's what happens. First, whatever fire, heat, and/or smoke conditions exist in front of you will be forcefully pushed away. That's not a bad thing if you are pushing that heat and smoke to an open atmosphere with no exposure concerns, such as while extinguishing a vehicle fire on the street.

However, in the typical, open floor, center core construction of a commercial high-rise building, you can push the bad stuff all the way around the center core and back on top of yourself, possible cutting off your escape route, or the escape route for other firefighters. Furthermore, that forceful pushing of heat, fire, and smoke will seek the path of least resistance, which could be any one of the countless hidden void spaces in most buildings (such as the plenum in a commercial high-rise building), thus, potentially extending your fire laterally and vertically, and making the overall fire conditions much worse.

So it's not *protection*, it's *pushing*. That is, pushing the fire somewhere else, which will most likely have to be dealt with again, later in the operation.

This wide-fog approach can have even more devastating results in a typical residential high-rise fire. I frequently hear firefighters talk about wanting the fog pattern to help them make a long, hot hallway filled with heat and smoke. A friend and I were discussing nozzle tactics one night at the firehouse. He's a good man, who resides comfortably in the fog camp. He stated that he wanted to have the fog nozzle on his standpipe pack so he could open it up in a hot hallway and provide protection for him and his crew. His scenario involved a fire inside a high-rise multiple dwelling. In as diplomatic a manner as possible, I proceeded to explain to my friend just how dangerous this tactic is, and how detrimental it could be to occupants of the building, including other firefighters, and that it could ultimately compromise the entire operation.

Ask yourself, how did the heat, fire, and smoke get into the public hallway in that high-rise multiple dwelling in the first place? It certainly didn't get there by accident. Experience has shown time and again that there is a high probability that an occupant from the fire apartment more than likely left their door open to the public hallway when they hastily fled the fire environment. As the fire continues to develop, eventually the fire alarm audible horns begin to sound and countless other complacent occupants open their doors to investigate and possibly attempt to escape via the public hallway. Many will retreat back into their apartments, unable to close their doors to the public hallway due to extreme heat and smoke conditions.

Once the engine company makes the fire floor, the hallway is well involved with fire. The perceived "protection" behind a powerful cone of small water droplets will not protect, but it will probably kill, and has! If we are coming down that hallway with a wide fog in front of us, we are once again, forcefully pushing everything away from us. Furthermore, none of these finely divided water droplets are likely reaching the burning solid fuels that are causing all that heat and smoke to begin with. So, rather than extinguishing the fire and dealing with the real problem (the disease), we are merely treating the symptom of that disease by forcefully pushing the heat, fire, and smoke being produced by those burning solid fuels into open, uninvolved apartments and on top of the people we are supposed to be helping (fig. 9–2).

✳ It might even be some of our own guys that we are harming, brother firefighters attempting to complete a primary search, when all of the sudden it feels like the world is coming down on top of them as they are dangerously chased out (if their lucky enough to get out) by the heat and smoke. Those firefighters who have been on the receiving end of this know exactly what I am talking about. It's no different than having some buffoon open up a nozzle, or worse yet, a master stream, to a wide fog pattern from the outside while the real brothers are taking a beating on the inside.

✳ Plain and simple, *protection* comes from killing the fire. If the fire is killed, quickly, you protect yourself and others. When a combination fog nozzle is opened up to a cone-shaped pattern, a very powerful venturi of air is created that forcefully pushes heat, fire, smoke, and other products of combustion ahead of the attack team. Outside, this is not a problem; but when this tactic is employed inside a structure, it can lead to serious problems.

I'll address the steam and visibility issues later, but right now, the force of the entrained air will violently push the heat, fire, and smoke forward, and any occupants attempting to escape will likely be killed. Any other apartments with doors open to the public hallway will also soon be involved with fire. Any occupants still inside these other apartments with open doors, if not protected,

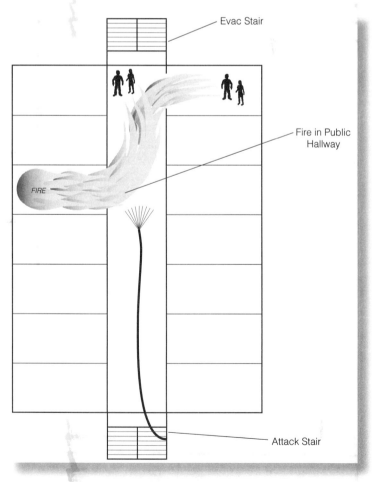

Fig. 9–2 A wide fog pattern will push heat, fire, and smoke into uninvolved locations and on top of occupants.

will likely be incinerated by this powerful current of superheated air. And, all the while, not a drop of water will likely reach the burning solid fuels. In his book, Fornell also states, "Additional air movement caused by improper use of fog streams can also accelerate burning, releasing more heat than if the fire were left to itself."[3]

The Line-of-Duty Death of Firefighter James Heenan

On January 1, 2001, a young firefighter by the name of James Heenan died of injuries sustained while operating at a structure fire in West Deptford Township, New Jersey. The New Jersey Division of Fire Safety conducted a comprehensive investigation into the incident. A report was issued on December 15, 2003, as part of their Firefighter Fatality and Serious Injury Report Series. In the report, investigation of the tragic incident revealed that there were three primary factors contributing to Firefighter Heenan's death. According to the report, "the introduction of fog fire streams into the hole in the floor and through the exterior basement windows pushed the fire and superheated gases back down upon FF Heenan, thus causing the burn injuries that ultimately led to his death."[4] The facts relating to the line-of-duty death of Firefighter James Heenan are extremely sad and tragic. Those include the fact that fire and superheated gases were *pushed* onto Firefighter Heenan, and there is no mention of *protection* from fog streams, because it didn't protect him.

Treat the Disease, Not Just the Symptom

This inappropriate use of water fog is a classic example of attempting to treat the symptom without actually attacking the real disease. Remember, until water of sufficient volume reaches the burning solid fuels, the fire will continue to burn, and the heat, gases, smoke, and other products of combustion will continue to be produced. A curtain of finely divided water droplets between you and a fire does nothing to attack and kill that fire. A powerful, solid stream of water, rotated vigorously and directed forward toward the probable seat of fire, will cool the upper atmosphere during advancement, preventing flashover, while delivering solid water droplets to burning solid fuels, in essence, a direct hit to the fire.

The power of a fog stream to move air has been compared to that of a modern positive pressure blower. There are circumstances when this tool can be used to give us a specific tactical advantage, such as when operating outside at a vehicle fire. Once again, the propane Christmas tree drill has been used for years to demonstrate the power of a fog pattern, but, unfortunately, the word *protection* has often been used to describe what's taking place. Once again, it's not protection, it's pushing. The protection only occurs when the fire is extinguished.

So, those proponents of the combination fog nozzle must be honest, and change their terminology, something as simple as, "I prefer to use a combination fog nozzle so I can push the fire away from me." Okay, that's an honest answer; not a very good tactic, but nevertheless, an honest answer. Just like the propane Christmas tree drill, what allows the firefighters to approach, operate the valve, and shut off the fuel supply, is the force of the water fog and entrained air pushing the heat and flame away from the approaching firefighters.

Firefighters have been burned, and seriously injured during the propane Christmas tree training exercise when improper placement of the fog streams allowed heat and flame to contact the approaching firefighters. So, ultimately, the fog stream is not providing protection, it is pushing, and the protection is only achieved after the fuel is shut off. The worst part of this training is that many instructors teach this in such a manner that the student firefighters will attempt to apply this protection myth inside buildings, during interior structural firefighting.

Fog Application in a Center Core High-Rise

The use of a combination fog nozzle for high-rise and standpipe operations is very dangerous. This is especially true in high-rise buildings constructed with a center core design. In essence, a cone shaped pattern used to initiate attack and provide the *protection* for advancing firefighters will only serve to forcefully drive all the heat, fire, smoke, and other products of combustion all the way around the center core and back on top of advancing firefighters. This could easily injure or kill firefighters by driving the fire directly on top of them, or, at the very least, by cutting off their escape route. This has been referred to by many as the *donut effect* (fig. 9–3).

The Effect of Wind

Sometimes, the wind conditions at many high-rise fires are an extremely significant and dangerous factor. Severe wind can be encountered at any level, but is particularly strong at the upper levels. At most of our more frequent fires, such as a typical single family dwelling fire, when fire breaks through a window, it typically vents to the outside. This is not always the case at high-rise fires. A forceful wind, driving toward the fire side of a high-rise building, can literally force the fire back inside the building and make conditions exponentially worse.

FDNY Battalion Chief Jerry Tracy instructs nationally on high-rise operations. He has discussed the lessons learned from a high-rise fire in Manhattan at the apartment of Lionel Hampton. The severe wind conditions at this fire required the use of several engine companies operating multiple 2½-inch handlines for a period of over one-half hour just to make it down the public hallway to the fire apartment. Jerry tells me that while presenting this information at a conference, a fire officer from another state commented to him that had they been using combination fog nozzles, it would have given firefighters the necessary protection to make the fire apartment more easily. Jerry was diplomatic and attempted to exorcise the demons from this brother firefighter's head.

The force of wind driving into the upper floors of a high-rise building cannot be underestimated. Those who have not encountered it simply cannot understand how powerful it truly is; it's Mother Nature at work. She makes the rules, not us. Placing a fog pattern in front of firefighters attempting to fight their way toward the seat of a serious fire, under extreme wind conditions, could literally steam the firefighters to death.

I remember my first encounter with a strong wind condition at a high-rise fire. The fire was on the seventh floor of a 20-story high-rise, multiple dwelling. The building had an interior atrium from the lobby to the roof. I was a firefighter on Rescue Co. 1 at the time of the fire. When we arrived in the building lobby, we could see heavy fire, in the form of a blue flame pushing out into the open atrium from what was apartment #702. We didn't realize it at the time, but it was an extremely hot fire being fed by a strong wind from the north side, and exacerbated by a very strong stack effect inside the atrium.

Several years ago I attended a training class at our Safety and Training Division. A retired chief officer from a large city was the instructor. During his dissertation, he recommended the use of a fog nozzle on a wide fog pattern operated on the floor above the fire in a high-rise to prevent auto exposure. I was shocked buy this ridiculous statement. It was obvious to me that he had never really tried this, because he was still alive.

Those who have worked a few serious high-rise fires are familiar with the unusual and unpredictable air movements and wind currents. A strong stack effect and heavy wind conditions (not unusual, especially higher up in a high-rise building), and you could literally kill firefighters by attempting such a dangerous maneuver. Furthermore, just how

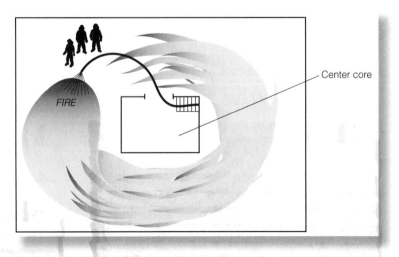

Fig. 9–3 The donut effect

many individual hoselines and fog nozzles would be needed on the parameter of a 10-, 15-, or 20-thousand-square-foot floor to stop a serious case of auto exposure? I am still performing exorcisms on a few of the sad souls who actually believed some of this guy's propaganda.

Lloyd Layman

The history of combination fog nozzles dates back to the 1800s, when the first fog nozzle was actually invented. However, the manufacture and use of fog nozzles was virtually nonexistent until the early 1950s, when a man by the name of Lloyd Layman began to pioneer the use of water fog as an extinguishing agent. Layman was the Chief of the Parkersburg, West Virginia, Fire Department. He also gained firefighting experience in the military as a member of the United States Coast Guard.

Layman's Coast Guard experience included shipboard fires, many of which occurred in the lower cargo holds, and were virtually impossible to extinguish. This was during the 1940s and 1950s, long before the widespread use of quality personal protective equipment (PPE) and self contained breathing apparatus (SCBA). Sending firefighters down deep into the lower levels of a ship to extinguish a fire would guarantee certain death.

Layman did his homework, and developed what was the first use of water fog for a so-called indirect fire attack. Just as we know it today, the indirect attack involves using water fog, by introducing it from the outside of a fire compartment, shutting it down, closing up the compartment, and allowing the water to covert to steam, and, thus, smother the fire. Layman's discovery and development were very significant, and led to an entirely new way to fight some fires.

Layman documented his findings, and even went so far as to even write a couple of books on the topic. The books, *Attacking and Extinguishing Interior Fires*, published in 1952, and *Fire Fighting Tactics*, published in 1953, are extremely important in the quest to understand the history of water fog.

When Layman first developed the use of water fog for indirect attack, he concluded and documented several facts. Layman was emphatic that several critical factors be closely adhered to with regard to indirect attack using water fog.

In order to have a safe and successful operation, when using water fog for an indirect attack, there must be:

1) Extremely high heat conditions, in order to successfully convert the water fog to steam, and thus initiate the process of extinguishing the fire by smothering it.

2) No ventilation of the fire compartment, in order to keep the steam inside, allowing it to smother the fire and provide extinguishment.

3) No occupants inside the fire compartment. Because of the creation of steam and the exponential increase in (*wet heat*) temperature, human survival within such an environment would be highly doubtful.

Therefore, as Layman said, there can be no occupants inside the fire compartment or environment where the indirect attack is to take place.

Steam produced inside a fire environment during the extinguishment process has led to countless injuries to civilians and firefighters alike. Something Layman warned us not to do over 50 years ago has been taking place for years and has resulted in many injuries, and, in some cases, death. The misuse and inappropriate application of water fog is today referred to by many as the *bastardization* of Lloyd Laymen's theories, and many injuries and some deaths can be attributed to water delivery techniques that were never advocated by the father of the indirect attack.

The insurance industry

So, how did we get to this point, using water fog in a manner that the pioneer of its use never really advocated? A significant portion of the fog frenzy that inundated the American fire service in the 1960s and beyond must be attributed to Lloyd Layman himself, but not entirely. Certainly, many individuals in positions of leadership within the American fire service and the American insurance industry are culpable parties to the widespread misuse of water fog.

It was during the *war years* (the 1960s and 1970s), with tremendous fire activity, when most urban fire departments were responding to and operating at one fire after another. The large number of fires led to numerous insurance claims for fire damage during the war years. However, there were also a very significant number of claims made for water damage. In fact, at many fires, the cost to repair the damage caused by water was far greater than the cost to repair fire damage.

Anyone who has ever made an insurance claim knows all too well that the insurance company is not nearly as happy to give out money for a damage claim as they are to take it when your premium is due. Face it, they're in business to make money, not lose it. The insurance industry is filled with plenty of intelligent people. In their ongoing quest to make a profit and satisfy a board of directors and their share holders, insurance companies, like many for-profit organizations, are continually thinking of ways to cut costs.

For the insurance industry, it led them to study firefighting operations, tactics, and water delivery during the war years. It was the insurance industry that lobbied and pushed the American fire service to implement proactive methods to prevent water damage during structural firefighting operations. That's right, some guy in a three-piece suit, sitting around a cherry wood conference table in an air-conditioned building, decided how you should deliver water inside the dirty, dangerous, dynamic, and uncontrolled environment of an interior structural fire. They studied and saw Layman's ideas as an opportunity to reduce water damage and save their money. After all, one of the many misleading characteristics of the indirect attack is its so-called efficient use of water.

The fire service

However, once again, the insurance industry is not totally to blame for this movement. We must also look within the fire service itself, and you can quickly see that we too must share a huge portion of the blame. We took what the insurance industry proposed, and bought into it, hook, line, and sinker. We disregarded one of the fundamental rules that Layman himself emphasized when he first introduced indirect attack, that is, that there should be no occupants in the area or compartment where the indirect attack takes place.

Remember, once we decide to operate in an offensive mode, we firefighters become occupants of the building. The issue of water damage, and preventing it at all costs, unfortunately, became a central theme of water delivery in the American fire service, overriding common sense and the need to protect occupants of the fire building, including firefighters.

Water damage

It absolutely infuriates me when I hear of a training officer, company officer, or especially a chief officer berating a young firefighter over the issue of water damage, either before, during, or after a fire. Some of these young firefighters become so intimidated and concerned about the reprimand they might be subjected to, that they are literally afraid to flow water, perhaps even when it's most critical to do so. What a terrible shame. In the quest to prevent water damage at all costs, we are injuring and killing the members of our own profession.

The entire fire service must accept and send out a loud and clear message that water damage, is in fact, a post-fire control issue, and should not be on the nozzleman's mind during the attack of an uncontrolled fire. Yes, we must teach proper nozzle tactics, opening up, and shutting down in a timely manner, but we must also teach that killing the fire is the goal. Not killing it in a sensitive, politically correct, or gentle manner, but killing it, and killing it now! Once the fire has been extinguished, and extension of the fire has been stopped, along with thorough primary search and the confirmation that there are no other fires in the building, then, and only then, does water damage become a fireground issue.

We must remember the fireground priorities that have been established and adopted across most of the American fire service. They are:

1. **Life safety.** That is the number one priority, life safety, and not just life safety for the civilian occupants of a building, but also for the firefighters who become occupants of that building. Firefighter safety is at the top of the list, long before water damage should even be considered.

2. **Incident stabilization.** This is a fancy term for stopping uncontrolled nature, which is the fire. Until we stabilize things, stop the fire, stop the extension, search for and rescue or remove any occupants, and search for and extinguish any other fires, we have not stabilized the incident.

3. **Property conservation.** The issue of water damage falls under this third and last priority on the fireground. Let me repeat that; it's the *last* priority! Yes, it is our professional responsibility to provide good service, and that we will do. But we must do it in a proper

order, with a high priority placed on firefighter safety and survival. Many like to use the term *customer service*. Remember, if the *customer service* providers become injured or killed in the process of attempting to provide good *customer service*, then customer service will not be achieved at all.

The extremely serious nature of just how absurd our obsession with water damage had gotten was clearly illustrated in the third edition of the *IFSTA Essentials Manual*. A young firefighter pointed this out to me a several years ago. Located in the section of this manual on salvage, filling the top one third of an entire page, was an artist's rendition of several firefighters operating at a structural fire. It clearly depicted a fire burning in a cockloft area with direct flame contact on structural components depicting a lightweight wood truss assembly. Three firefighters were operating on the roof, apparently attempting vertical ventilation directly over a burning lightweight wood truss. In addition, five more firefighters are seen operating inside the structure, under the burning lightweight wood truss. Of that group, three firefighters seem to be part of an attack team, with two members on a hoseline (not flowing water), and another member apparently poised to pull ceiling with a pike pole. The other two members in the illustration are apparently part of a salvage group and are shown preparing to throw a salvage cover in order to protect two rolltop desks.

Needless to say, I was furious that such an inappropriate and highly dangerous operation was depicted in a nationally recognized training manual. Tens of thousands of firefighters would be exposed to this over the period of this manuals life span, and, unfortunately, some might even actually apply such a procedure. That procedure places the value of two inanimate objects—rolltop desks—above the lives of eight firefighters, all for the purpose of preventing some water damage. It clearly reverses the fireground priorities, placing property conservation at the top of the list, followed by incident stabilization, with life safety as the last priority.

For me, this served as a clear indication of a symptom of a much greater problem, a cancer, a disease, that could be directly correlated with an overall philosophy related to inappropriate fireground procedures. The good people at IFSTA apparently recognized the absurdity of this illustration as well, and I am happy to report that it was not included in the fourth edition of the *Essentials Manual*.

Proper Nozzle Selection

As addressed in previous chapters, the specifics of the One Meridian Plaza fire involved improperly set pressure reducing valves (PRVs) at standpipe outlets which created a very low pressure. Coupled with this, the Philadelphia Fire Department was using 1¾-inch attack line with automatic combination fog nozzles as their weapons to combat this fire. Ultimately, all of these factors contributed to an inability to control the fire, and, thus, it rapidly spread and quickly became far too large to control.

In his book, Capt. Dave Fornell illustrates and describes the results of his research regarding the performance of different types of nozzles for standpipe operations. Fornell first shows the low pressure performance of an automatic combination fog nozzle, similar to the type used by the PFD at One Meridian Plaza. At approximately 40 to 45 psi outlet pressure, the flow meter read zero (0), indicating a flow of less than 50 gpm. With regard to the type of automatic combination nozzle used for the test, Fornell further states that "it is designed for optimum performance when input pressures are above 75 to 85 psi."[5] In the second illustration Fornell utilizes a $^{15}/_{16}$-inch smoothbore/solid stream tip. With the same 40 to 45psi outlet pressure, the flow meter reads 110 gpm. Both tests were conducted using 100 feet of 1¾-inch hose.

One must keep in mind that the 110 gpm discharged from the $^{15}/_{16}$-inch smoothbore, although not the optimum water delivery (optimum for the $^{15}/_{16}$-inch smoothbore tip is 185 gpm @ 50psi nozzle pressure), was significantly greater than the discharge from the automatic combination fog nozzle. Ultimately, one must remember that proper weapon selection involves not only the proper nozzle but also the hoseline size. As Fornell states, "Large diameter hose, either 2-in. or 2-1/2-in., should be used when operating from standpipes."[6]

Hands-on training (HOT)

For the past several years I have been involved with standpipe hands-on training (HOT). Nothing drives home a point better than letting firefighters see and feel the results for themselves. I have been fortunate to have a great team for many of the standpipe hands-on training drills. Together, we have conducted standpipe training at several different venues across the country. We always include Fornell's research in our drills, and actually replicate the One Meridian Plaza scenario. Furthermore, we compare the low pressure performance of the 1¾-inch hoseline to that of the 2½-inch hoseline, which is the recommended tool of choice for standpipe operations.

In our hands-on training, we use 150 feet (three lengths) of 2½-inch hose, with a 1⅛-inch smoothbore/solid stream nozzle. With the same 40 to 45 psi outlet pressure hooked up to the outlet on the floor below, we are consistently able to deliver over 200 gpm, with a very effective fire stream. We compare this to 150 feet of 1¾-inch hoseline with an automatic combination fog nozzle. And, just like Fornell, our results prove that less than 50 gpm is delivered, with an ineffective fire stream of about 10 to 15 feet. We estimate the actual flow to be perhaps 25 to 30 gpm at best (fig. 9–4).

It would certainly be nice to consider myself and the members of my standpipe hands-on training team as pioneers that have discovered something new. The fact is that this is nothing new at all. We are quite simply the ambassadors of a message that has always been based on scientific fact and the recommendation of many fire service authorities, including the National Fire Protection Association (NFPA). That's right! As I said in Chapter 4, the NFPA makes a specific recommendation regarding the use of hose and nozzles for standpipe operations.

Specifically, NFPA #14 reads, "It is very important that fire departments choose an appropriate nozzle type for their standpipe firefighting operations. Combination fog, constant pressure, (automatic) type spray nozzles should not be used for standpipe operations because many of this type require a minimum of 100 psi of pressure at the nozzle inlet to produce a reasonably effective fire stream. In standpipe operations, hose friction loss might prevent the delivery of 100 psi to the nozzle. In high-rise standpipe systems with pressure reducing hose valves, the fire department has *little or no control* over hose valve outlet pressure."[7]

"The 2½-in. smoothbore with a 1⅛-inch tip produces a usable stream (250 gpm) at 50 psi inlet pressure requiring 65 psi at the valve outlet with 100 ft. of 2½-inch hose."[8]

After the tragic One Meridian Plaza fire, NIOSH produced and distributed an Alert Bulletin regarding standpipe systems, PRVs, and the use of automatic combination fog nozzles, etc. That happened almost 15 years ago. Today, in the American fire service, countless fire departments are still equipped with the same 1¾-inch high rise/standpipe hose packs, with an automatic combination fog nozzle attached to the end—the same weapons combination that the National Institute for Occupational Safety and Health (NIOSH) and the NFPA warned us *not* to use over 15 years ago, after three of our brother firefighters were killed at One Meridian Plaza.

It is very disturbing to think we have a very good chance of seeing more firefighters killed, in the same manner as has occurred in the past, doing something we simply should not be doing. Specifically, many fire departments have equipped their firefighters with tools and equipment that were not designed for high rise/standpipe operations. Keeping the names of our brothers off the granite stone is a significant motivation behind the work involved in writing this book, and in providing countless hours of engine company standpipe operations hands-on training to firefighters across the American fire service.

Fig. 9–4 At 40 to 45 psi, simulating One Meridian Plaza standpipe outlet pressure, a flow of approximately 25 to 30 gpm from 150 feet of 1¾-inch hose with an automatic combination fog nozzle.

There are many great firefighters out there who are attempting to eliminate this dangerous equipment combination from their respective departments' high-rise standpipe hose packs. You continually face the negative, condescending attitudes, and cynical comments from those who simply don't get it. They will frequently say something ignorant like, "Hey, this isn't Philadelphia." While they continue to demonstrate their insecurity, you must remain calm and arm yourself with knowledge and facts.

Many people will say that the Philadelphia fire was a worst case scenario, and "that can't happen here." Well, maybe you won't encounter something as severe as what occurred in Philadelphia, but for any fire service professional to make a comment like that, or underestimate the potential of what could possibly happen anywhere, especially after 9/11, is nothing short of negligence.

Fig. 9–5 Dual pressure automatic combination fog nozzles should be flow tested at a wide range of low pressures, typical of standpipe operations, in both the high and low pressure settings in order to clearly demonstrate how dangerous they are in low-pressure situations and why they should not be used for high-rise firefighting and standpipe operations.

Everyone, including the complacent types, must understand that many buildings with standpipe systems were likely designed, specified, and installed according to the minimum standards of your fire and building codes. As said earlier in the Chapter 4, in many jurisdictions across the American fire service, minimum standards means 65 psi at the topmost standpipe outlet for countless existing buildings, and 100 psi at the topmost standpipe outlet for newer buildings and those built today. We must be prepared to account for friction loss, nozzle pressure, and head pressure due to elevation. What if you only have 65 psi, or maybe 100 psi at the outlet? You do the math. It better be a small fire if you're equipped with small weapons!

One particular nozzle manufacturer has developed and produced a very popular new style automatic combination fog nozzle that can be switched from a high pressure mode to a low pressure mode. Although this nozzle may have some merit for certain applications, high-rise and standpipe operations are definitely not among them.

As with any type of nozzle, fire departments that own and utilize these dual pressure type automatic combination fog nozzles should flow test them at a wide range of low pressures. I have conducted numerous flow tests utilizing this particular type nozzle and have found that at 40 to 45 psi (simulating the One Meridian Plaza scenario), when you switch from a high pressure mode to a low pressure mode, the hose line literally goes limp, and the flow wouldn't be sufficient to water a small patch of grass (fig. 9–5). At 65 psi (the fire code standard for many buildings) there is an attractive looking stream, but the flow is still below 50 gpm. At the 100 psi (the NFPA recommendation for new standpipe installations) the stream is once again, good looking, but the flows are consistently well below 100 gpm.

Although many different nozzles have been used for this test from many different fire departments across the country, they were all the same dual pressure type from the same manufacturer. Flows varied depending on the age of the nozzle used, and how it had been maintained. The flows were consistently below 100 gpm, however, throughout that range of low pressures.

Keep in mind, 100 gpm should be considered a minimum flow for fighting a small vehicle fire. It is a far from acceptable flow for interior structural

Table 9–1 Flow test results from various weapon selection combinations at FDIC West in 2005.

	1-3/4" Hose 150-Feet	2" Hose 150-Feet	2-1/2" Hose 150-Feet
Standpipe Outlet Discharge Pressure			
40 PSI @ The Outlet	TFT Auto Fog Dual Pressure	TFT Fog (LAFD) 200GPM @ 75PSI	TFT Auto Fog Dual Pressure
Started @ HP (Blue)	HP 0 GPM	137 GPM (TFT Fog LAFD)	HP 70 GPM
Switch From HP to LP	LP 0 GPM		LP 71 GPM
Started @ LP (Red)	LP 0 GPM		LP 71 GPM
	SB 15/16" 123 GPM	SB 1" 160 GPM	SB 1-1/8" 222 GPM
		SB 1-1/8" 184 GPM	SB 1-1/4" 278 GPM
65 PSI @ The Outlet			
Started @ HP (Blue)	HP 0 GPM	167 GPM (TFT Fog LAFD)	HP 85 GPM
Switch From HP to LP	LP 0 GPM		LP 91 GPM
Started @ LP (Red)	LP 98 GPM		LP 237 GPM
	SB 15/16" 148 GPM	SB 1" 192 GPM	SB 1-1/8" 265 GPM
		SB 1-1/8" 219 GPM	SB 1-1/4" 315 GPM
100 PSI @ The Outlet			
Started @ HP (Blue)	HP 80 GPM	203 GPM (TFT Fog LAFD)	HP 163 GPM
Switch From HP to LP	LP 96 GPM		LP 197 GPM
Started @ LP (Red)	LP 137 GPM		LP 351 GPM
	SB 15/16" 177 GPM	SB 1" 230 GPM	SB 1-1/8" 329 GPM
		SB 1-1/8" 260 GPM	SB 1-1/4" 376 GPM
Nozzle Weight	TFT DP 1-3/4" 7.5 #	TFT Fog LAFD 6 #	TFT DP 2-1/2" 9.5 #
TFT = Task Force Tip DP = Dual Pressure HP = High-Pressure (Blue) LP = Low Pressure (Red) SB = Smooth Bore			Akron 2.5" x 1.5" Shutoff w/ 1-1/8" Tip 3#

firefighting, especially for fires that occur in the large, open floor areas of many high-rise buildings.

Flow tests at FDIC West, 2005

With the assistance of my FDIC engine company standpipe operations hands on training (HOT) team, I conducted extensive flow tests at FDIC West in 2005. During that test, we compared three different sizes of hose—1¾-inch, 2-inch, and 2½-inch—and several different types of nozzles, including smoothbore, low pressure combination fog, and high pressure, dual pressure automatic combination fog.

The three different sizes of hose were compared side by side, using three lengths for each size, or 150 feet each. A standpipe inline pressure gauge was used to simulate standpipe outlet pressures of 40 psi (One Meridian Plaza), 65 psi (NFPA #14 pre-1993), and 100 psi (NFPA #14, post 1993), and pressure was controlled using a large, 2½-inch × 2½-inch gated wye. The gated wye was supplied from an LAFD pumper, which was in turn hooked up to a municipal fire hydrant at the LAFD training facility. An inline flow meter was placed ahead of the pressure gauge and gated wye. It was placed in this location for convenience, so that it would not have to be moved with each separate flow test. This gave a flow reading in gpm for each separate flow test.

The results of the flow tests were not at all surprising. Flows from the smoothbore nozzles on each hoseline were greatest, and flows from the 2½-inch hose were the greatest. What may be a surprise to some is just how low the flows were from the various weapons combinations, but especially the 1¾-inch hoseline, and when the high pressure automatic combination nozzles were used (table 9–1).

Steam and disruption of thermal layering

Most firefighters are aware of the fact that a nasty by-product of the inappropriate application of water fog is steam, accompanied by a violent disruption of the thermal layering. Yes, we all know it. Just as Layman said some 50 years ago, and as Fredericks tried to explain to the American fire service during his unfortunately short life, if you utilize water fog for interior structural firefighting, you are, in essence, attempting an indirect attack.

For years, firefighters have been taught that the magic number is 212° Fahrenheit, the temperature at which water turns to steam. Furthermore, we've all been taught that the expansion ratio is 1,700 to 1, or that 1 gallon of water will expand 1,700 times when converted to steam. However, those numbers are a bit misleading.

The temperatures encountered by firefighters in the modern fire environment, specifically interior structural fires, especially those that occur in large, uncompartmentalized, high-rise buildings, are not going to be a mere 212° Fahrenheit. In many interior structural fires, specifically in situations where the fire is fully developed, in the second or free burning stage, ceiling temperatures can be well in excess of 1,000° Fahrenheit. I'm not sure what the exact expansion rate of water to steam is at those extremely high temperatures, but it is certainly exponentially greater than 1,700 times. Because of this, it stands to reason that the effect on human beings inside such an atmosphere would be immediately debilitating.

Once again, many fog nozzle proponents emphatically state that they would never use the combination fog nozzle on a fog pattern during interior, structural firefighting operations. If you're going to use a straight stream anyway, why not go with a smoothbore? This lowers the nozzle pressure, which in turn reduces the nozzle reaction, yet the GPM increases. Less pressure, less work, more water, and the water is delivered in the form we want it in to begin with. Furthermore, in the low visibility of heavy smoke conditions typical at most serious interior structural fires, it is often impossible to actually see what type of pattern is being produced and delivered from a combination fog nozzle. In complete darkness, the smoothbore nozzle has two settings, open or closed. No mystery as to what pattern is being produced in total darkness.

Standpipe system debris

I have encountered several debris incidents at fires in standpipe-equipped buildings over the years, and I have heard countless horror stories from other firefighters in the DFD and brothers from all across the country who have had similar experiences. A brother firefighter in the DFD, Captain Thor Hansen of DFD Engine Co. 13, who was at the time assigned to our Fire Prevention Bureau, gave me a very powerful piece of evidence. During his work, he encountered a standpipe system in a Denver high-rise building with a buildup of debris that would be the envy of most cardiologists. Captain Hansen was kind enough to give me a short section of pipe from that standpipe system, which has served as a valuable visual aid for several years (fig. 9–6). At a high-rise/standpipe training class presented to the Oakland (CA) Fire Department, an Oakland firefighter told me that the proper term was *tuberculated* pipe. I certainly wouldn't want to hear that from my doctor.

Debris passing: Fog versus smoothbore

Another argument that we all must face revolves around the fact that there are many new low pressure, constant-gallonage combination nozzles being produced by various manufacturers. Many of these are very good pieces of equipment, and represent a tremendous improvement over the good old 100 psi combination fog nozzle. There are 75 psi and 50 psi low pressure fog nozzles available today.

Fig. 9–6 This short piece of pipe, almost completely clogged with rust and scale, came from an active standpipe system inside a Denver high-rise.

Yes, the 50 psi version can operate at the same pressure as the recommended smoothbore nozzle. However, if you ever encounter a debris problem within your standpipe, and you will, there is a very high probability that it will eventually clog up a typical combination fog nozzle, even in the flush mode. Yes, even a smoothbore nozzle could become clogged up with larger pieces of debris, but you will recognize it much more easily, and have a better chance of correcting it. Furthermore, it is much less likely to damage a smoothbore nozzle than a combination fog.

Try the M&M's® test. Simply pour a small package of peanut M&M's® pt in the female end of your nozzle, and attach your hoseline. Charge the line with water, open the nozzle, and evaluate the performance. With the smoothbore, you will have a chocolate treat thrown some 100 feet away from the nozzle team. When using any type of combination fog, even on the flush position, you will at best have a distorted stream, with a dangerously low GPM rate. Your standpipes will not have M&M's® in them, but most of them will likely contain large chunks of rust, scale, debris, sediment, and almost anything else you could imagine.

Nozzle weight

Another important consideration is weight. Use the tallest building in your district as the example. Create a scenario with a fire on the top floor. The elevators are either out of service, or unsafe to use. Climb the stairs with all your stuff (full PPE, SCBA, spare air cylinder, hose, nozzles, appliances, and forcible entry tool). If you're going more than 10 flights, you'll wish you could lose some of that unnecessary weight. Guess what? You can! But be proactive, do it before the fire occurs.

A standard smoothbore/solid stream nozzle tip weighs less than 1 pound. Most combination fog nozzles, including many that are attached to high-rise standpipe hose packs, are in the 5- to 6- pound range, with some over 10 pounds. It doesn't seem like much, but carry it up 10 or 20 flights of stairs and you will likely become a believer (fig. 9–7). Dropping a pound here and a pound there, or specifically dropping five, six, or 10 pounds at once can make a big difference toward reducing the physical demand placed on your firefighters. We must be able to get up there as quickly and efficiently as possible, and most importantly, we must still have something left to fight the fire.

Cost, durability, and maintenance

The issues of cost, durability, and maintenance are all significant concerns. Combination fog nozzles can and do fail. It's a fact of life when you are dealing with equipment that involves moving parts. If you have a good maintenance program, that will help. However, you simply don't have to worry about the smoothbore, as there are no moving parts in that tip, and therefore no maintenance required.

How about durability? The fireground is not a gentle place, neither is the fire apparatus. Equipment on board the pumper takes a significant beating just coming in and out of the compartment. How's the old saying go? Give two ball bearings to a firefighter, and within a very short period of time, one will get broken and the other lost. This is not meant to insult or disparage the group to which I belong, but firefighting equipment needs to be extremely tough—tough enough to survive consistent abuse and still be operable the night we need it on the forty-second floor.

Dropping a combination fog nozzle tip accidentally down a flight of stairs while preparing to initiate attack on a high-rise fire will likely damage this piece of equipment, and in many cases may render it useless. You could drop a smoothbore tip from the fifth floor of a drill tower, and it would survive with only a few scratches. After such abuse, put these two nozzles to work and compare the performance.

And than there's the cost. Are the bean counters cutting your budget too? Of course they are! A standard smoothbore nozzle tip costs less than $100.00, verses a typical combination fog nozzle tip at $400.00 to $800.00 for the more expensive automatic nozzles. Think of all the stuff you could buy with the savings.

Figure 9–7 The much preferred smooth-bore nozzle weighs about 2.5 lbs, whereas the automatic combination fog nozzle weighs almost 10 lbs.

Summary

With all this information and our own tragic history, there will still be a segment of the American fire service that simply won't give up their fog nozzle, even during the low-pressure, debris-laden standpipe operations. For firefighters involved with standpipe operations, debris in the system and low operating pressures are a fact of life. They are real-world problems that you will likely encounter during most standpipe operations. The utilization of a low-pressure nozzle will help to address the issue of the low operating pressure. The smoothbore/solid stream nozzle, with an operating pressure of 50 psi is the lowest pressure nozzle available that will also pass debris better than any other nozzle type. This is the obvious tool of choice for standpipe operations, because it addresses both the low pressure and the debris issues. Once again, there are low pressure combination nozzles available today, but they will clog up with even the smallest pieces of debris.

Debris, low pressure, improperly set PRVs, and equipment weight are just a few of the numerous indisputable water delivery facts associated with high-rise firefighting and standpipe operations. However, even with these facts, and several tragic events in fire service history, there will still be those individuals resistant to leave their beloved fog nozzle behind. Most of those individuals haven't yet encountered a tragic event, debris, or low pressure situation to gain the firsthand experience illustrating the negative aspects of fog nozzles. Therefore, since it hasn't happened to them yet, they figure, why change? There are things I haven't personally experienced yet, either, like big planes crashing into high-rise buildings, but I know it can and has happened, and in this day and age, I'd better be prepared for it in my city. The truly dedicated fire service professional is constantly on the look out for *Firenuggets*—bits of new information about our profession—even if it happens to be a vicarious nugget of experience gained by listening to, or learning from a brother firefighter.

For many of the tens of thousands of firefighters raised on fog nozzles, it frequently seems to be very difficult to teach some of these old dogs a new trick. Nevertheless, there are a lot more believers out there today than 10 years ago.

As for your high-rise standpipe hose packs and equipment, stay away from the highly marketed break-apart nozzles. When these nozzles first came out several years ago, I thought they were a good idea. It was a compromise, providing two tools in one. In fact, we purchased several of these on the DFD for our high-rise standpipe hose packs. With experience and flow testing I have found that the break-apart nozzle quickly becomes a large, long, heavy fog nozzle with a 7/8-inch or 15/16-inch reducer between the fog tip and the ball valve shutoff. Most departments rarely break off the fog tip, and the small slug tip can reduce the overall flow when using the fog tip.

You may have noticed that several nozzle manufacturers have been working for the past decade to make their fog nozzles more like smoothbores. They have given them fancy names, and big prices to go along with them. Believe me, they see the fire stream movement as well, and they are not at all happy about many in the American fire service purchasing more of the cheap, durable, maintenance-free nozzles that last a lifetime.

Fig. 9–8 The recommended nozzle for high-rise firefighting and standpipe operations consists of a lightweight, 2 ½-inch x 1 ½-inch ball valve shutoff, with a lightweight 1 1/8-inch smoothbore tip—a lightweight, low pressure, high volume, debris passing weapon.

Most of the people in the nozzle industry are good hard-working individuals, but remember, they are also businessmen. The bottom line for them is profit, or loss. We will continue to purchase their best fog nozzles for some of the work we do. It just so happens that their best product for high-rise firefighting, standpipe operations, and any interior structural firefighting is also their cheapest product.

For those of us raised on the fog nozzle, the wide, 60-degree fog pattern was replaced in recent years by the *30-degree power cone*. Another real cool name! Unfortunately, that name was thought up and developed by several individuals wearing ties, sitting around a fancy cherry wood conference table, in the comfort of an air-conditioned building. Although many are good, well-intentioned people, the marketing department of a nozzle manufacturer is no match for a group of firefighters pushing down a hallway on their bellies. If I want a power cone, I will assemble the companies in my district, and we will meet up at the local Dairy Queen. The only "power cone" I'm going to be using is one coated with chocolate.

For those who are fog guys, and that's it, not going to change, ever, no way, please at least heed this advice. You must steer clear of the automatics and invest in low pressure, constant gallonage, fog nozzles for your high-rise and standpipe operations. It is critical to flow test them before your fire, and at low pressures—100, 65, and 40 psi are some of the potential pressures you might encounter. You must know how much water you are *not* going to flow before your fire occurs.

Last, but certainly not least, have a Plan B. Your fog nozzle will clog up with debris, someday, somewhere, and you aren't going to fix it in the heat of battle up on floor 27. Have a spare nozzle and extra hose. If you are using 1¾-inch hose with your fog nozzle, buy some 2½-inch to stop that fire that is probably going to get away from you some day.

Please remember this: the information and recommendations in this chapter are related specifically to interior structural firefighting, especially high-rise and standpipe operations. There are countless exterior fireground operations that require the use of a good combination fog nozzle. Don't get rid of all your combination fog nozzles. It is an important and valuable tool for many operations. However, it is a dangerous and deadly choice for standpipe operations.

Also remember, what makes the smoothbore nozzle a great tool for high-rise and standpipe operations also makes it an excellent choice for all of our other interior structural firefighting operations as well (fig. 9–8). I certainly encourage all of you to continue to purchase and use good quality combination fog nozzles, but only for their intended purpose. It is also my hope that each of you will give the smoothbore nozzle a try, that is, if you haven't already done so. And for safety sake, I pray and hope you will use it for all of your high-rise standpipe operations.

References

1. Clark, William E., *The Fire Chief's Handbook*, Fifth Edition, Chapter 17, Fire Company Operations, Fire Engineering Books and Videos, A Division of Pennwell Publishing Company, 1995, p. 637.

2. Fornell, David P., *Fire Stream Management Handbook*, Fire Engineering Books and Videos, A Division of Pennwell Publishing Company, 1991, p. 125.

3. Ibid, p. 54.

4. New Jersey Division of Fire Safety, Firefighter Fatality and Serious Injury Report Series, FF. James Heenan, December 15, 2003.

5. Fornell, David P., *Fire Stream Management Handbook*, Fire Engineering Books and Videos, A Division of Pennwell Publishing Company, 1991, p. 192.

6. Ibid.

7. National Fire Protection Association (NFPA) 14, Standard for the Installation of Standpipe and Hose Systems, A.7.8, Quincy, Ma. 2003. p. 25-26.

CHAPTER 10

The Standpipe Equipment Kit

In addition to hoses and nozzles, there are several other water delivery appliances, tools, and equipment that are essential to fireground success for the engine company when operating off standpipe systems and fighting fires in high-rise and standpipe-equipped buildings. Most of this equipment can be included in a portable *standpipe equipment kit*.

The standpipe equipment kit consists of a large, lightweight tool bag, with several different pieces of lightweight equipment carried inside (fig. 10–1). A complete standpipe equipment kit, with all the necessary tools and appliances needed for standpipe operations, should be standard equipment for every engine company that operates at high-rise and/or standpipe-equipped buildings.

When assembling a standpipe equipment kit, fire departments should strive to achieve a balance between having all of the essential equipment and still keeping the kit lightweight and manageable. The standpipe equipment kit for any engine company will likely contain many of the same, essential items. However, some specific tools and equipment needed by some fire departments or specific engine companies might not be necessary for others. The following list includes all of the essential items, along with some optional equipment that may or may not be necessary,

Fig. 10–1. A typical standpipe equipment kit with all the associated tools and appliances.

depending on the needs of a specific engine company, based on experience, pre-fire planning, and response district needs:

1. 2½-inch × 2½-inch, standpipe inline pressure gauge
2. 2½-inch × 2½-inch, 60-degree, elbow (2)
3. 2½-inch × 1½-inch, ball valve shutoff with a 1⅛-inch smooth-bore/solid stream tip (primary nozzle, attached to nozzle length and hose pack)
4. 2½-inch × 1½-inch, ball valve shutoff with smooth-bore/solid stream stacked tips, 1¼-inch × 1⅛-inch × 1-inch (spare, backup nozzle)
5. 1½-inch × 2½-inch increaser (2)
6. Adapters (area specific, such as 3-inch to 2½-inch reducer, or pipe thread-to-national-standard-thread)
7. Spanner wrenches (2)
8. 18-inch pipe wrench
9. Wire brush
10. Spare operating hand wheel (for standpipe hose valves)
11. Hand wheel wrench
12. Wood door wedges (door chocks) (15–20)
13. Door strap/search markers (15–20)
14. Grease pencils, markers, chalk (carried in rigid box)

Standpipe In-Line Pressure Gauge

The standpipe in-line pressure gauge (SPG) is probably the most important of the water delivery appliances, second only to the nozzle. The SPG was first used by the FDNY and has now been widely adopted by fire departments across the American fire service.

Historically, when operating off standpipe systems, fire department engine companies had little or no knowledge of the actual pressure at the outlet. Of course, common sense, judgment, and experience were used to evaluate the effectiveness of the stream and to determine whether or not there was sufficient pressure. However, this was not always a reliable method and, in some cases, led to an unsuccessful operation.

Quite simply, the SPG gives the attack team information remote from the pump panel of the pumper. In essence, you actually have a pump operator on the floor below at the standpipe outlet that gives the attack team the necessary finesse to deliver water in the most effective and efficient manner possible. When resources will allow, one firefighter can and should be assigned to the standpipe outlet to monitor the discharge pressure via the SPG. This firefighter can literally operate as a remote pump operator, increasing or lowering pressure by using the hand wheel to make adjustments to the standpipe hose valve, given an appropriate operating pressure within the standpipe itself. Furthermore, this firefighter can communicate with the officer in charge of the attack team and the engineer/pump operator in the street, with regard to pressure adjustments.

The SPG that is specified and used by fire departments should be a very durable piece of equipment, designed to withstand the abuse subjected to it. Fire departments should specify the SPG, which has the actual gauge recessed into a protective casing mounted on the body of the appliance. Several different nozzle companies manufacture quality standpipe gauges, and with the increased popularity of this appliance, most are continually attempting to improve on it. SPGs that are designed and equipped with an unprotected gauge extending out, away from the appliance body, are not very durable, and are prone to damage. This type, although ideal for flow testing and fire stream evaluation, should not be used for standpipe operations.

The SPG should be carried inside the standpipe equipment kit for protection, but it is still susceptible to damage. Specifically, the lens on the gauge is made of glass, and can be damaged and scratched. Over time, sitting on the apparatus with other tools inside the kit vibrating against the gauge, the lens can become so scratched up that it is impossible to read the gauge. A proactive measure is to protect the lens, and a simple plastic cap works very well.

Several years ago, I discovered that using the plastic cap from a can of cake frosting works perfectly to cover the gauge on many types of SPGs and would thus serve to protect it from damage, such as scratches. This may seem to be a minor issue, but it's the little things that make a big difference. So, next time you make a chocolate cake for dessert at the firehouse, save the cap from the frosting container, and put it to good use on your SPG, protecting the lens.

Testing the SPG

It has been said, false information is worse than no information at all. This couldn't be more accurate, especially on the fireground. The SPG is designed to give information regarding the pressure at a standpipe outlet; however, if the pressure reading on the gauge is not accurate, it could negatively affect the entire operation. Therefore, the SPG, along with all the other appliances and tools in the standpipe kit, should be checked regularly, and any necessary maintenance performed in a proactive manner.

Engine companies should remove the SPG from the kit once a week for inspection, and once a month for flow testing. During the test, the SPG should be attached to an outlet on the pumper.

Hose, preferably the high-rise hose, should be stretched from the discharge outlet and the nozzle from the high-rise/standpipe kit should be used. The hoseline is charged and flow tested to the appropriate pressure, at which time the reading on the SPG is compared to the readings of the discharge and compound gauges on the pumper's pump panel. The readings from all three gauges should match, which would verify accuracy (fig. 10–2). Anytime a SPG is not operating properly, it should be repaired or recalibrated, and a spare gauge should be provided to the engine company.

Labeling the SPG

It is also a good idea to label the SPG with predetermined pressures that will correspond with common hose stretches at standpipe operations. For example, a 150-foot stretch is very common at many standpipe operations. Sometimes a 200-foot stretch is necessary, and occasionally a 250-foot

Fig. 10–2. The standpipe inline pressure gauge should be flow tested at least monthly to verify accuracy.

stretch. These being the most common lengths that engine companies will likely encounter, a label affixed to the SPG with the proper pressure for each layout is very beneficial (fig. 10–3). Those pressures, based on hydraulic calculations, and verified with a flow meter are:

150 feet: 75 psi
200 feet: 85 psi
250 feet: 95 psi

Note: 5 psi per floor can be added as necessary for head pressure due to elevation, as the attack line should be supplied from a standpipe outlet at least one floor below the fire floor.

Static versus flow pressure

Firefighters must understand *static* versus *flow pressure* in order to properly regulate the attack line pressure at the standpipe outlet hose valve. Initially, when the hoseline is first charged with water, there will be a typically high pressure reading on the gauge, perhaps well over 100 psi. This is a static pressure. After the nozzle is opened up, the pressure will drop on the gauge, at which time the firefighter assigned to make adjustments can do so.

Without training and experience, firefighters might attempt to adjust the high static pressure by closing the valve several turns. There will be no change on the gauge, but when the attack team opens the nozzle, they might have a dangerously poor stream at a very low pressure. The firefighter at the standpipe hose valve outlet will then need to rapidly open the valve back up. Once again, adjustments can only be made based on flow pressure, after the nozzle is opened and water is flowing. It is recommended that the firefighter assigned to the standpipe hose valve outlet, initially open the valve approximately three quarters of

Fig. 10–3. Labels affixed to the standpipe inline pressure gauge provide predetermined pressures for the three most common hose layouts.

Fig. 10–4. Two elbows, inline, attached to a standpipe outlet facing straight up. This will prevent severe kinking of the hoseline.

the way, and leave it in this position, which will allow for rapid adjustments in pressure, either up or down.

60-Degree Lightweight Angle Elbows

Several years ago, the DFD added two, 60-degree, lightweight angle elbows to our standpipe kits. Like most of the equipment in the kit, the elbows were added to solve a specific problem encountered by fire companies. Recall from Chapter 4 that most standpipe-equipped buildings in Denver, unfortunately, have the standpipe outlets located inside cabinets out on the floor. Several years ago I responded to a fire that occurred in a lower downtown Denver residential high-rise building. In the heat of battle, engine companies had a difficult time with the attack line kinking at the outlet. This was because the standpipe outlets were installed with the 2½-inch outlet facing straight up from the bottom of the cabinet. After the fire, a comprehensive investigation showed that the installation met the requirement of the code, that is, "there must be an accessible 2½-inch outlet." However, it doesn't specify that it should be pointing in a convenient direction (something other than pointed at the ceiling), just that it be accessible.

Needless to say, there was no way the owners of this building, or as it turns out, the owners of countless others in our city with the same issue, were going to spend their own money to solve the problem. After all, someone from the city, either the fire department or the building department, maybe even both, signed off on this as being acceptable when the buildings were built. Most likely, the inspector who approved it had probably never been involved in a high-rise fire or standpipe operation. Once again, firefighters had to come up with a solution.

The short-term solution was to use the elbows located on the pumper's discharge outlets. However, this was only a short-term fix, as these elbows are heavy chrome elbows—too heavy for inclusion in a standpipe kit. Eventually the fire department purchased the lightweight, 60-degree elbows that are currently carried in the kit. The DFD includes two of these elbows in each engine company's standpipe kit.

The purpose of the elbows is to simply eliminate the potential for the hose to kink, specifically when the hose is attached to a standpipe outlet that faces straight up in a cabinet. In this situation, two elbows must be attached in series in order to prevent a severe kink at the outlet (fig. 10–4). For fire departments that don't encounter this type of problem, I still recommend that at least one elbow is included in your standpipe kit, to deal with outlets in tight quarters that might cause small to medium-sized kinks to occur near the outlet (fig. 10–5). Make sure to specify a lightweight elbow, and include it in the maintenance program. Regularly inspect the threads, especially on the male end, the female threads, and swivel, along with the gasket.

Fig. 10–5. One elbow may be necessary to prevent minor to moderate kinking for outlets located in tight spaces.

Spare Nozzle

The standpipe kit should include one spare nozzle. I recommend the same nozzle as the primary nozzle, which consists of a 2½-inch to 1½-inch lightweight ball valve shutoff, with a 1⅛-inch smooth-bore tip, or, preferably, smooth-bore stacked tips, sizes 1¼-inch × 1⅛-inch × 1-inch (fig. 10–6). This spare nozzle serves the purpose of being immediately available in the event that the primary nozzle malfunctions, or more likely, in the event that the primary attack line has to be extended and another nozzle is needed.

I do not recommend carrying any type of combination fog nozzle or tip inside the kit due to weight, operational problems, and the potential for debris to clog this nozzle (see chapter 9). However, fire departments that feel strongly about the inclusion of a combination nozzle should include both a separate ball valve shutoff, and a nozzle tip. That way, when the fog tip becomes clogged up with debris and/or malfunctions, and it will, you can twist it off and replace it with the always reliable smooth-bore nozzle tip, which has no moving parts; therefore it will not malfunction and it is less likely to clog with debris. Hopefully, when the malfunction of the fog tip occurs, your attack team will be in a safe enough position to protect themselves while removing the problem and replacing it with the solution.

Fire departments that choose to utilize combination fog nozzles for standpipe operations, regardless of the strong recommendations against it, should make every effort to specify a low pressure tip that is as lightweight and compact as possible. I cannot emphasize enough that I consider automatic combination fog nozzles to be extremely dangerous tools when used with standpipes and for high-rise operations. If you are entrenched in the fog camp, and not planning on leaving anytime soon, at least go with a combination nozzle that is non-automatic, low pressure, small, and lightweight.

1½-Inch to 2½-Inch Increaser

The 1½-inch to 2½-inch increaser, which I'll refer to simply as the *increaser*, is an unusual adapter that, at first glance, meets with some skeptical looks (fig. 10–7). This relatively inexpensive piece of equipment can significantly expand an engine company's operational efficiency and effectiveness. There are many valuable applications for this appliance, the most notable of which happen during standpipe operations, specifically while operating on the upper floor of a high-rise, remote from the equipment on board the pumper, which might be 10, 20, 30, or more floors below on the street.

Extending a hoseline

The increaser is most valuable when a 2½-inch attack line needs to be extended. Using the increaser allows for an attack line to be extended without having to shutdown the water at the standpipe outlet. Generally, 150 feet of attack line is a good starting point inside most buildings for standpipe operations, based on typical fire and building code requirements, with regard to the distance between standpipe outlets. A stretch of 100 feet from the floor below will often come up short, and a stretch of 200 feet or more often makes hose management in the attack stairwell and on the floor below more difficult and cumbersome.

Fig. 10–6. The standpipe kit should include a spare nozzle with smooth-bore stacked tips

However, the standard 150-feet starting point can sometimes also come up short of the final objective, especially if the attack team must fight its way through heavy heat, fire, and smoke to make the seat of the fire. Extending 2½-inch hoseline off a charged 2½-inch attack line is accomplished by removing the nozzle tip, attaching the increaser to the 1½-inch threads on the ball valve shutoff, and then attaching the extra length or lengths of hose to the 2½-inch male connection of the increaser. This allows for a handline to be extended without shutting down at the standpipe outlet.

A very good example of an effective use of the increaser to extend the line is for a fire in a high-rise multiple dwelling. A typical scenario might be that, upon arrival at the fire floor, the attack team encounters heavy fire that has extended into the public hallway from the fire apartment. Unfortunately, the attack team was forced to use a stairwell that was much farther from the fire apartment than the other stairwell, because the closer stairwell was a fire/smoke tower and several occupants were already using it for evacuation.

The attack team stretched 150 feet of hose and initiated their attack in the hallway knocking down a heavy body of fire. Unfortunately, they came up about 15 feet short of the fire apartment. Using the reach and penetration of their solid steam, the attack team is able to knock down all of the visible fire in the public hallway and the fire just inside the entry way to the fire apartment. Truck company members are able to get to the fire apartment door, do a brief sweep of the entry area and close the apartment door, blocking the fire from extending back out into the public hallway.

At this point, with the door to the fire apartment controlled, the officer in charge of the attack team is able to safely order for an extension of the primary attack line. This operation will, in some cases, be faster and more efficient than waiting for a second attack line to be stretched and placed into position, especially when this line must come from two floors below, and will likely require at least four, maybe five lengths (sections) of hose, for a total of 250 feet. Keep in mind, with appropriate manpower on scene and available, a second, or backup, line will still be stretched by the next two engine companies, but an extension of the first, primary attack handline, will, in many cases, get water on the fire faster.

Members of the second due engine company, who were assisting the first due engine with the stretch and advancement of the primary attack line, have proactively brought to the entrance of the fire floor an additional length (section) of hose, in the form of a 50-foot, 2½-inch standpipe hose pack, with a nozzle and the increaser preattached.

Fig. 10–7. A 1½-inch to 2½-inch increaser

It is very important to preattach the nozzle and the increaser to this length, and to do so outside of the fire area, preferably on the floor below. Furthermore, leave the hose pack straps in place. This ensures that all of the necessary pieces of equipment needed to complete the extension will arrive on the fire floor, in one complete and compact package, and eliminates the possibility of dropping one or more pieces of the equipment package and potentially losing them in the zero visibility conditions typical of a smoke-charged fire floor. Seconds after it was requested, the additional hose arrives in the hallway next to the attack team.

Engine companies who have trained together frequently will hone their operational skills and learn to use this very efficient method to quickly extend the handline. The hose pack is placed on the hallway floor next to the attack team, positioned with the female coupling of the hose pack next to the attack team (fig. 10–8). From this position, members of the second engine company remove the hose pack straps, but keep them available (placing them in turnout coat pockets), as at least one will be needed to complete this operation. One member of the second engine company grasps and securely holds on to the female coupling of this hose length in one hand and the nozzle in his other hand. Another member of the second engine company, after receiving a verbal command to do so from the first firefighter, grasps the hose length at the midpoint, and stretches this back down the hallway, away from the fire and nozzle team (fig. 10–9).

This hose length is now fully stretched out in a horseshoe-type configuration (fig. 10–10). The firefighter holding the female coupling and nozzle instructs the nozzleman of the attack team to remove the nozzle tip, which should be placed in his pocket after removal. The firefighter holding the female coupling will remove the increaser and give it to the nozzleman, who will than screw it onto the exposed threads of the nozzle (fig. 10–11). The nozzleman will work with the firefighter holding the female coupling to attach the additional length of hose to the increaser (fig. 10–12).

Once the additional hose has been securely attached and the nozzle team is in place, the additional hose can be charged with water. This is done by simply opening the ball valve shutoff of the original nozzle. After this is completed, the nozzle team should bleed the line and ensure that there is adequate pressure and an effective fire stream. With adequate pressure in the standpipe system, operating pressure to the attack team can be increased as necessary by a firefighter operating from the floor below, while the pressure is monitored via the SPG. The nozzle team can now advance to the fire apartment door and prepare to make entry and attack the fire.

Fig. 10–8. A Denver hose pack, with a smooth-bore nozzle and increaser pre-attached, placed in position with the female coupling next to the original attack line in preparation for extending the line

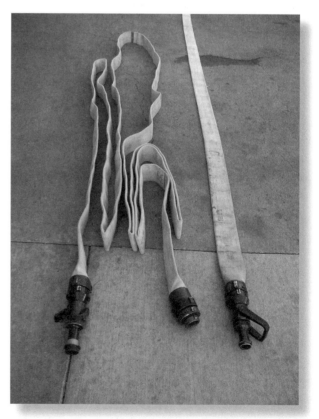

Fig. 10–9. The hose pack is put into operation by having one member securely hold the female coupling and the nozzle, while another member identifies the center point marked by a 1-inch line, and then stretches the hose pack out, fully, away from the attack team

Fig. 10–10). Additional hose length fully stretched out in a horseshoe type configuration, next to the original hoseline

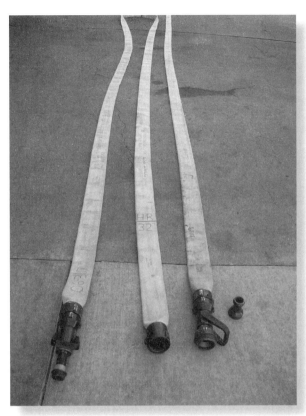

Fig. 10–11. The nozzle tip is removed by the nozzleman and he should place this in his turnout coat pocket. Increaser is removed from the hose pack, and the nozzleman places the increaser on the exposed 1½-inch threads of the original nozzle.

In less than one minute, two well-trained engine companies will have the additional 50 feet of hose attached, charged with water, and stretched to the door of the fire apartment. The truck company firefighters will open the door to the fire apartment and the engine company will begin knocking down the fire inside. The extra 50 feet of hose is, in most cases, more than enough to make the fire apartment door, advance into the apartment, and cover the entire apartment, completing extinguishment. If necessary, two lengths of hose (100-ft), or two hose packs, can be used in order to extend the hoseline by 100 feet.

This might sound too simple, but it can literally be done that fast. Obviously there is a need to determine tactically whether or not a hoseline should be extended or if a second attack line must be stretched immediately. This is based on the volume of fire, whether or not the fire in the public hallway can be controlled, and if the door to the

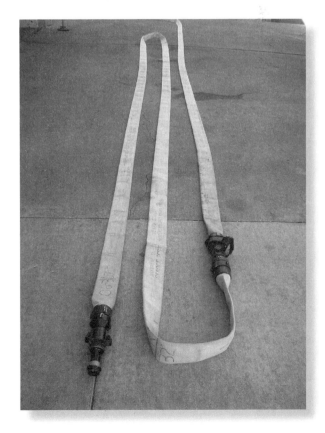

Fig. 10–12. Members work together to connect the additional hose length to the increaser and original attack line.

fire apartment can be closed. Equally important to this entire evolution is that the truck company firefighters are diligent in maintaining the integrity of the door to the fire apartment when forcing entry so that it can be closed if necessary. The main point of all this is to emphasize that a very small, lightweight, inexpensive piece of equipment—the increaser—allows for another, very valuable, tactical option that can be used on the fireground.

It is important to point out that in order for the increaser to be a viable option, fire departments must utilize nozzles that allow for the tip to be detached, exposing a 1½-inch thread. Some refer to these as break-apart nozzles, but in this case the nozzle is without the slug tip. Once again, the recommended nozzle is the lightweight, 2½-inch × 1½-inch ball valve shutoff with a detachable smooth-bore tip. When the tip is removed, the well-trained firefighter can quickly and easily attach the increaser to the 1½-inch threads of the ball valve shutoff. The hose is then attached and the new nozzle length is then charged using the ball valve shutoff. This procedure and the increaser cannot be used if the nozzle on the primary attack line is a one piece nozzle without a detachable tip.

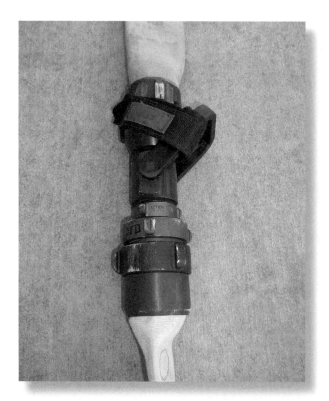

Fig. 10–13. A hose pack strap should be kept available and used to secure the nozzle bail to keep it from inadvertently being closed.

One final thought on the procedure to extend the hoseline. I recommended keeping the straps that are used to hold the hose pack together in a convenient location. A minimum of one strap can be used to hold the bail of the nozzle open, and thus keep it from being inadvertently shut if it is struck as an operating member crawls by. Simply run the strap around the handle of the bail, around the hoseline, and then back on itself and secure it tightly (fig. 10–13). In addition, pushing the hose against the wall will make it less of a target, preventing the bail from being struck and inadvertently closed.

This entire operation must be practiced by all engine company members on a regular basis, as its efficient completion in zero visibility smoke conditions is easier said than done. An effective training drill would include completing this evolution in zero visibility as well.

Creating a second 2½-inch outlet

Another valuable use for the 1½-inch to 2½-inch increaser is to provide a second 2½-inch outlet on Class III standpipe systems. Keep in mind that many Class III standpipe systems are equipped with two outlets, a 2½-inch outlet for fire department use, and a 1½-inch outlet for occupant use (as opposed to those Class III systems with only one 2½-inch outlet and a 2½-inch to 1½-inch reducer attached to meet the criteria for a Class III system). In many cities such as Denver, the occupant hose has been removed from the 1½-inch outlet, leaving only the outlet. In some situations, fire departments can use this 1½-inch outlet to great advantage.

By using the 1½-inch to 2½-inch increaser, and attaching it directly to the 1½-inch outlet, you have in essence created a second 2½-inch outlet. Some firefighters have a tendency to overanalyze this and focus on the perceived excessive friction loss that might occur in the smaller 1½-inch piping. However, the only assessment that must take place involves determining what size riser is supplying the 1½-inch outlet. Generally, on most Class III systems, the same large diameter riser supplies both the 2½-inch and 1½-inch outlets. Therefore, even with a minor amount of friction loss in the few feet of small pipe connecting the riser to the 1½-inch outlet, it can still be used to supply an adequate volume of water to a second 2½-inch attack line. It might not work in every situation, but I have had excellent results using this procedure many times. Remember, the operation cannot even be considered unless you're equipped with the appropriate appliance, which is a 1½-inch to 2½-inch increaser.

Some firefighters might be tempted to utilize the 1½-inch to 2½-inch increaser on a Class II standpipe system. The Class II system (see chapter 4) is, once again, for occupant use only, typically supplied by domestic water only, and in most cases without any means for the fire department to support it. Flows of only 100 gpm or less can be expected. Although the 1½-inch to 2½-inch increaser can easily be attached to a Class II standpipe system outlet, there will not be a sufficient volume of water to supply a 2½-inch attack line.

Ultimately, I certainly recommend that fire departments carry at least two of these adapters in their standpipe kit. All DFD engine companies are issued four 1½-inch to 2½-inch increasers. Two are carried in the standpipe kit, a second one is used in conjunction with a Bresnan distributor (cellar nozzle), and a third is kept in the nozzle compartment as a spare and to be utilized whenever necessary.

Other Adapters

Every fire department is confronted by a wide range of unique and specific problems within various buildings in their jurisdiction. Standpipe systems can have a wide variety of problems, including the need for special adapters just to hook up to an outlet. Depending on the city, jurisdiction, or specific building, a fire department may have to place one or more adapters in their standpipe kit.

In Denver, there are a number of high-rise and standpipe equipped buildings that require a special adapter be placed on the standpipe outlet, due to a slight variation in the thread size. The San Francisco Fire Department (SFFD) and Oakland Fire Department (OFD) use a 3-inch to 2½-inch reducer, as the standpipe outlets in many bay area buildings are equipped with a 3-inch outlet. The FDNY carries an adapter that is designed to go from pipe thread to national standard hose (NST) thread.

Ultimately, each individual fire department must investigate and preplan the buildings in their jurisdiction. It must be determined before a fire occurs whether the hose valve threads on a specific standpipe system are compatible with the fire department hose threads. Unfortunately, all too many times, the discovery of an unusual or noncompatible thread is made during the heat of battle. This was the case at the Indianapolis Athletic Club (IAC) fire, which occurred in Indianapolis in 1992. Two firefighters lost their lives operating at this fire. The operation was complicated when Indianapolis Fire Department hose thread was found to be not compatible with the thread on the IAC standpipe hose valve.

Such incidents underscore the importance of good pre-fire planning, but especially of taking every opportunity to check standpipe systems and hose valves. One opportunity that presents itself on a regular basis at many buildings is the countless fire alarm system activations, elevator emergencies, and EMS calls, where fire companies can take a few minutes (after the incident is under control) to check standpipe hose valves on their way out of the building.

Adapters, if any, are carried in a fire department's standpipe kit, and will more than likely be specific to that city or jurisdiction. As with all of the other appliances and hand tools in the standpipe kit, any adapters should also be as small, lightweight, and compact as possible. Stay away from the brass, and go with the aluminum. Keeping your standpipe kit as lightweight and compact as possible helps make a more efficient logistical operation.

Some fire departments opt to use a gated wye in conjunction with their standpipe operations. The gated wye can be attached directly to some standpipe outlets, given enough room, but, oftentimes, the gated wye is used with a short length of hose, usually a 3-inch hose, 10 to 15 feet in length. This short length is sometimes referred to as a *pigtail*. This equipment is designed to allow for two hose lines to be stretched from one standpipe outlet.

Although I consider this to be a good idea for high-rise and standpipe operations, it does have some limitations. Most notably is the weight of this equipment. Specifically, if fire departments are using the recommended 2½-inch attack line, the required wye would be a 2½-inch × 2½-inch gated wye, which is a large and heavy appliance. Add to that the short length of 3-inch hose, and you have a significant extra weight.

Keep in mind that the primary goal is to place one very powerful handline in service quickly, in an attempt to stop the fire during the typically short window of opportunity at the front end of a high-rise event. I believe the use of a gated wye for initial attack operations clouds that primary goal by focusing on two handlines before the first one is in place and operating. In addition, I believe the logistics associated with the gated wye are not best for initial operations. Specifically the extra weight during transport, the fact that there will now be three shutoffs involved with the layout, plus complications with the use of the standpipe in-line pressure gauge and a cumbersome hose layout, make this unsuitable for initial operations. In addition, there is a significant hose management issue associated with stretching two separate 2½-inch hose lines from a gated wye on the floor below.

I do believe, however, that this equipment should be included in an engine company's arsenal for extended operations. So, the use of a gated wye can be very valuable for overall high-rise and standpipe operations, when a specific tactical need for this tool is identified. For speed and efficiency, though, I suggest that the initial attack focus on one 2½-inch handline, used in conjunction with the previously mentioned equipment, specifically the standpipe inline pressure gauge.

Most importantly, I highly discourage the use of a 2½-inch × 1½-inch gated wye, as this tool focuses on placing two small handlines in operation, without consideration or the ability to place the recommended 2½-inch handline in service. This would mean more weight, a more cumbersome hose layout, and, ultimately, much less overall water delivery from two small handlines being supplied by a typical low pressure standpipe system.

Hand Tools

In addition to the nozzles and water delivery appliances that must be part of an effective and complete standpipe kit, there are also several very important hand tools. Without the recommended hand tools, it might be impossible to even attach the water delivery appliances to the standpipe outlet, or perhaps even turn on the hose valve.

1 Spanner wrenches

The spanner wrenches are for the most part self-explanatory to any firefighter. These very basic, yet essential, hand tools must be readily available during standpipe operations. Each standpipe kit should contain a minimum of two spanner wrenches.

Once again, lightweight must be the central theme. Although the extra weight of an old brass spanner wrench might not seem all that significant by itself, it is the cumulative effect of several small items that add up to several extra pounds.

When you must climb 10, 15, or 20 flights of stairs on your way to a staging location, you will be very happy to have proactively eliminated any unnecessary weight well ahead of time. The two spanner wrenches should be standard size spanners made of lightweight aluminum. Specifying lightweight folding spanner wrenches for the standpipe kit can be very beneficial; however, ensure that they are strong and durable enough to do the job.

Obviously, the main function of a spanner wrench is to loosen or tighten hose couplings. They can also be used to remove tight standpipe hose valve caps and some types of pressure restricting devices (PRDs). The handle of a spanner wrench can be used to probe the inside of a hose valve outlet to check for debris or other types of PRDs, such as an orifice plate.

2 Pipe wrench

Another essential hand tool that must be included in the standpipe kit is a pipe wrench. A large, preferably a minimum 18-inch, pipe wrench is ideal. This is a large tool, so keeping it compact and lightweight is not that easy. Specifying a lightweight aluminum wrench is essential. Also, a wrench with a jaw that will open up to as great a width as possible is very important.

The primary use of the pipe wrench is for opening and closing the hose valve when the hand wheel is missing. Sometimes, even with the hand wheel in place, the valve might be difficult, if not impossible to open. The pipe wrench can usually provide the necessary leverage to open a stubborn valve. In addition, a pipe wrench with a wide enough jaw can sometimes work to loosen a stuck valve cap or remove extremely tight PRDs.

3 Spare hand wheel

Some fire departments include a spare hand wheel for the hose valve inside their standpipe kit. This is based on past experience in buildings with a history of vandalism and theft. However, because there may be numerous buildings with different systems and hose valves, one hand wheel might not fit all of the hose valves in the various different buildings in an engine company's response district. Therefore, fire departments, and specifically individual fire companies, must once again survey the buildings in their district and collect pre-fire information.

Similar buildings or several buildings within the same complex will be most likely to have the same type standpipe systems and hose valves. If hand wheels are typically missing in those buildings, placing one of those type hand wheels in the standpipe kit can be beneficial. Once again, it's important to be practical. You simply can't put five or ten spare hand wheels in the kit to accommodate every potential building in your district. Don't hesitate to include one or two hand wheels in your kit for frequently encountered problems. But remember, keeping with the theme of lightweight and compact, one adjustable pipe wrench will fit on all of the valves that require multiple different hand wheels.

4 Hand wheel wrench

Many fire departments include a special hand wheel wrench in their standpipe kits. This tool is used when firefighters encounter an extremely tight hose valve. By placing the hand wheel wrench over the outside of the hand wheel, extensions on the wrench engage the hand wheel when the wrench is rotated and provide additional leverage to open the hose valve. I believe this is an excellent tool that should be included among the standard equipment

Fig. 10–14. Several wooden door wedges (door chocks) placed neatly along the bottom of the standpipe equipment kit tool bag.

carried in a standpipe kit. Once again, keep this wrench as lightweight as possible, especially if yours is homemade.

5 Wire brush

A wire brush is another valuable tool that should be included in the standpipe kit. The primary purpose of the wire brush is for cleaning the male threads on hose valve outlets. Oftentimes, especially in buildings with a history of vandalism and theft, a stolen or missing hose valve outlet cap can lead to dirty or damaged threads.

When firefighters attempt to attach an appliance to a hose valve outlet and it doesn't thread on, there may be paint, dirt, or debris on the threads, or the threads may have some damage or a minor burr on them. Vigorously cleaning off the threads with the wire brush can remove paint, dirt, and debris, and may also remove small burrs or other minor damage on the thread, allowing attachment of fire department appliances.

6 Wooden wedges (Door chocks)

Wooden wedges, also known as *door chocks*, require little explanation to the seasoned engine company firefighter. However, it is important to emphasize the need to have a good supply of door chocks on hand for operations in high-rise and standpipe equipped buildings. Although the mentally and physically prepared firefighters will have several chocks on their person, kept in turnout coat pockets and secured to their helmet with a rubber band, there may be a need to chock open many more doors than the number of chocks carried by individual firefighters.

I recommend a good supply of ten to fifteen chocks be included in the standpipe kit. Chocks are lightweight and the recommended number will not add a significant amount of weight to the kit. However, if thrown into the kit in a haphazard manner, they will take up a lot of space. Therefore, it is best to carefully place the chocks into the kit, lining the bottom of the kit with the chocks. This serves two purposes: first, it maximizes the space inside the kit, and second, it reinforces and strengthens the bottom of the kit. Reorganizing the equipment in the kit once a week during a weekly inspection is an excellent and proactive measure to help achieve operational success (fig. 10–14).

There are several other innovative methods used by firefighters to chock doors open, including a fabricated metal door chock that can be hung over a door hinge. Regardless of the tool used, the important thing is to chock open any door that a hoseline must be stretched through. This is particularly important during the initial stretch with a dry line. A door that closes on a dry line can be a dangerous situation.

Door strap/search markers

Additional equipment carried in the high-rise/standpipe kit includes rubber door strap/search markers (fig. 10–15). It is recommended to carry 15 to 20 of these inside the kit. Carrying them in groups of five, held together by medical tape, is effective and efficient. The specifics associated with the use of the door strap/search markers will be discussed in chapter 14.

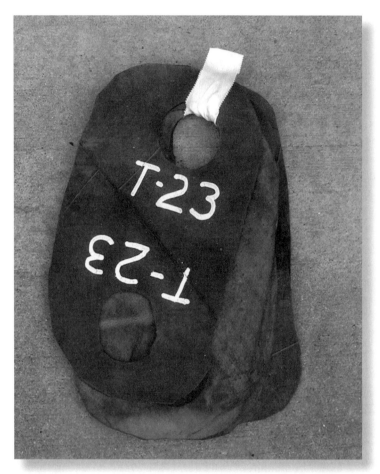

Fig. 10–15. Several door strap/search markers, held together with medical tape

Fig. 10–16. Attaching various appliances together inside the standpipe equipment kit will protect the male threads during storage and transport.

8 Colored markers, grease pencils, and chalk

Good communication on the fireground is central to operational success and safety, and that communication is not only verbal, face to face, and via radio, but also written. Various different types of writing utensils, such as colored markers, grease pencils, and chalk, are used to write messages on the various different surfaces typically found inside a high-rise or standpipe-equipped building. In addition, division supervisors can use these markers to write down pertinent information and to keep track of operating units within their respective divisions for fireground accountability.

It is a good idea to include a small, rigid box, such as a small Tupperware-type container, to carry all of the writing utensils. This will protect them from damage and keep them together and organized in one place. The specific uses of these writing utensils will be addressed in chapter 14 and chapter 17.

One last thought regarding the standpipe kit. It is a very good idea to carefully organize the tools and equipment inside the kit. Every attempt should be made to protect appliances, specifically the male threads. Attaching some appliances together and attaching adapters to appliances during storage in the standpipe kit is an excellent way to protect male threads (fig. 10–16).

Summary

The standpipe equipment kit is an essential part of an engine company's arsenal of high-rise firefighting and standpipe equipment. The standpipe kit should include all of the essential tools and appliances, but keep it simple. Each piece of equipment should be as lightweight and compact as possible, so as to minimize the overall weight and size of the kit. Most importantly, the appliances, tools, and other equipment inside the kit should be inventoried daily, inspected weekly, and thoroughly tested at least once a month.

CHAPTER 11

Engine Company Operations

The *engine company* is the primary firefighting unit of all fire departments. The basic, yet critically essential operations performed by the engine company, will literally define the outcome of every fire incident. Classic phases such as "the fire goes as the first handline goes" emphasize the importance of the engine company function. It has also been said that more lives are saved by a properly placed handline than by any other fireground strategy or tactic.

For several years, I taught a firefighting strategy and tactics class at a local community college. Each semester, my students were introduced to John Norman's book, *The Fire Officer's Handbook of Tactics*. In his book, Norman highlights the general principles of firefighting, with number one being the fact that *rescue* should take precedence over all other concerns. However, just as Norman explains in his book, and as I always reiterated, rescue may be best accomplished by stretching and properly placing a handline in service, in order to stop the fire.

These same basic tenets and principles can also be applied to engine company fireground operations in high-rise buildings, and anytime the engine company is stretching a line off a standpipe system. The high-rise operation is one of the most demanding and complicated fireground operations that most firefighters will ever face. The wide range and sheer volume of tasks that must be accomplished can quickly overwhelm even the largest of fire departments. Before you become completely overwhelmed, it is essential to remain focused on the basics, not the least of which is getting water on the fire—quickly!

Priority One: Stopping the Fire

Over the past several years, there has been an increased focus on many of the fireground support functions at high-rise operations, especially those associated with firefighter accountability. In an attempt to enhance firefighter safety and accountability, some fire departments have taken to assembling a complete command system (support system), including an extensive accountability system, prior to stretching hoselines and attempting to get water on the fire.

High-rise operational components such as *lobby control* are all essential, but none of them put out the fire. In fact, all fireground functions are important, but none is more important than that which will ultimately stop the fire. Therefore, the number one focus must be on getting that all important first handline in service off the standpipe in an attempt to quickly stop the fire. Remember, once the fire has been brought under control, all of the other fireground functions, from search operations to ventilation, become much more manageable. Furthermore, a firefighter's safety and ultimate survival is much easier to ensure once the fire has been stopped.

This philosophy is in no way meant to discount the importance of the numerous support functions and accountability necessary at a high-rise event, but rather to identify and establish a priority for each. The first arriving companies, preferably at least the first two engine companies and the first truck company, with a minimum of six, but preferably nine to 12, firefighters, should comprise the initial fire attack/search group.

Each of the countless support positions, including that of lobby control, can, in fact, be initiated by one company officer and or member, and then can be expanded as the incident grows and additional resources arrive on scene. To delay fire suppression in order to form a comprehensive support system requires many resources and does not address the primary problem; the fire.

Engine Company Functions

Regardless of the building size or occupancy type, the basic engine company fireground functions are essentially the same. From a single-family dwelling fire, to a fire on the thirtieth floor of a commercial high-rise building, the basic engine company functions, for all practical purposes, remain constant. Certainly, the central problems of exposures and logistics (identified in chapter 5) encountered at all serious high-rise fires, complicate these basic functions, but, nevertheless, the engine company is truly responsible for accomplishing the same mission at any fire building, that is, to locate, confine, and extinguish the fire.

Preconnected hoselines and tank water can make our most frequent operations at single-family dwelling fires seem routine and, in most cases, the basic engine company functions are usually accomplished quickly at these type fires. However, the reflex time, including the logistics of transporting necessary equipment to the point of operation, and the utilization of what may be a questionable standpipe system, can seriously complicate the typical high-rise/standpipe operation. Throw out all of the additional problems typically encountered at fires in high-rise and standpipe-equipped buildings, and the engine company functions are essentially the same at the high-rise operation as they are at any other structural fire. Specifically, an engine company is responsible for the following tasks during a fire.

Water supply

At every structural fire, the engine company is responsible for water supply. Specifically, the engine company must make every effort to establish and maintain a sustained water supply. In urban and suburban communities, this is usually accomplished by accessing and tying into a municipal fire hydrant. This will likely be the case in most, if not all, communities that have high-rise and standpipe-equipped buildings. Although there are typically several components associated with water supply in high-rise and standpipe-equipped buildings, including, in some cases, an on-site water supply such as gravity tanks, the engine company is ultimately responsible for ensuring that a sustained water supply is established. This typically involves tying into the building's fire department siamese connection (FDC) and pumping water into the building's standpipe system.

There are many aspects associated with water supply for fires at high-rise buildings, and other low-rise standpipe-equipped buildings. This topic is addressed comprehensively in chapter 12. Suffice it to say, the first function of the engine company at any structural fire event, including high-rise and standpipe operations, is to establish a sustained water supply.

Proper weapon selection

After establishing a sustained water supply, the engine company can than begin to determine what weapon should be selected for the firefight. Proper weapon selection for a given fire can actually begin long before the fire even occurs. Information contained in a good, comprehensive pre-fire plan will help guide firefighters on the selection of the appropriate weapon to combat the fire. Items such as building construction, type of occupancy, size of the building including the area and height, and what type of auxiliary appliances exist within the building, specifically whether the building has a standpipe system, are all part of the weapon selection size-up.

Most firefighters, fire officers, and chief officers who are experienced in high-rise and standpipe operations, and those in the fire service considered to be experts in the field of high-rise and standpipe operations, recommend the use of 2½-inch handline for operations off standpipes, especially to combat fires in high-rise buildings. There are countless reasons behind this recommendation, including the low pressure design of standpipes and the large non-compartmentalized areas typical of most high-rise buildings, especially commercial high-rises.

In addition, the reflex time necessary to place an effective weapon in service to begin attacking a fire in a high-rise building is significant, especially when that fire is 10, 20, 30, or more floors above the street. As that reflex time accumulates, the fire is left unchecked, and with a sufficient supply of fuel and oxygen will likely continue to grow, in most cases doubling in size every 30 seconds to one minute. Ultimately, a fire in a large, non-compartmentalized space of a high-rise building, with a significant modern fuel load will grow exponentially if not controlled quickly.

Far too many fire departments are equipped with only 1¾-inch hose for high-rise and standpipe operations. Some will boast that they have successfully stopped fires in high-rise and other standpipe-equipped buildings before with this weapon. I am sure there are some cases where that was true; however, if you were to closely examine these examples of successful 1¾-inch hose use during high-rise fires and standpipe operations, there would likely be many common elements.

First, the fires in question were likely small fires, located in compartmentalized areas, such as a one-room fire in a single apartment of a multiple dwelling, and thus, a fire that required a minimal fire flow to control, based on the size of the fire and the area of involvement. Second, the standpipe systems that supplied these high-pressure medium sized handlines were obviously working properly, supplying water of a sufficient volume and pressure necessary to support the greater friction loss associated with 1¾-inch hose. Last, there were likely no pressure-regulating devices encountered at these fires, and if there were, the devices were obviously set to provide enough pressure to properly supply a 1¾-inch hoseline.

Ultimately, 1¾-inch hose, which is an extremely valuable tool for most of our high-frequency events, might actually work as an effective weapon in some high-rise/standpipe situations. However, our own tragic history has proven that 1¾-inch hose can be a dangerous and deadly weapon to choose for standpipe operations, especially for fires in commercial high-rise buildings, and specifically in standpipe-equipped buildings that utilize pressure-regulating devices at the hose valves.

Once again, I recommend the use of 2½-inch hoseline for all operations off standpipes, but especially those operations in large, commercial high-rise buildings, and any high-rise or standpipe-equipped building that utilizes pressure-regulating devices at the hose valves. A comprehensive discussion of proper weapon selection and the use of 2½-inch handline compared to 1¾-inch handline can be found in chapter 8.

Many fire departments will claim that a lack of proper manpower levels prevents them from using 2½-inch handline. There is no question about it, most fire departments are severely undermanned, and 2½-inch hose used as an attack line requires more manpower to safely and successfully operate. However, if a smaller, more maneuverable handline is chosen specifically for its maneuverability, but firefighters are unable to control the fire with the lower flow, maneuverability becomes a moot point. It's all about GPMs overcoming Btus.

Two engine companies, one handline

The only way most fire departments will be able to achieve success with 2½-inch hose as an attack line is for them to utilize their limited manpower more efficiently. That is, using the limited manpower to place one powerful handline in service to fight a fire, rather than one or several smaller handlines. When a 2½-inch handline must be stretched and operated as an offensive weapon, specifically inside a fire building, with all the associated turns and friction points, additional manpower must be utilized. Fire departments that effectively utilize 2½-inch hose typically use two engine companies to operate and advance one handline.

For fire departments forced to make three-man companies work, pairing engine companies together for operations involving 2½-inch handline is essential. These two engine companies, depending on the assignment of pump operators, can build a team of four to five firefighters for their initial attack, when one pump operator is assigned to water supply outside. This is not ideal, but if these firefighters are well trained, physically and mentally prepared, with can-do attitudes, they can be very effective with a 2½-inch handline.

Many of the larger cities across the American fire service, including the Denver Fire Department (DFD), have minimum manpower levels of at least four per engine company. In the situation of a four-man engine company, pairing two engine companies together can yield a team of six to seven firefighters, with at least one pump operator initially supplying the system. A team of this size can be very effective moving and operating a *big line*, that is, if they are well trained.

In those unfortunate cases of three and two man staffing levels per engine company, assembling an attack team from at least three or possibly more companies may be necessary. Remember, these limited resources will be much more effective working together as a team, with one powerful, low pressure, high volume, 2½-inch attack line, than with two or perhaps more high-pressure, low-volume, 1½-inch or 1¾-inch lines.

Stretching the handline

With the components of water supply and proper weapon selection completed, the engine company can now focus on stretching the handline. From this point, I will focus specifically on the recommended weapon of choice for high-rise and standpipe operations, that being the 2½-inch handline. Obviously, techniques presented and utilized with this high volume weapon can also be applied to 1¾-inch and 2-inch handlines. Engine companies that train on and develop good skills with the use of 2½-inch handline will be able to effectively operate any of the smaller sized handlines when the use of those weapons is appropriate, specifically at our non-high-rise/standpipe, high-frequency operations. If you can do the big stuff well, you can certainly do the little stuff well.

Before the handline can be stretched, an accurate location of the fire must be determined. Locating the fire is actually part of the basic engine company functions (*locate, confine, extinguish*). However, for many fire departments, especially the larger ones, the task of locating the fire rests with the truck company. Locating the fire can also be completed by two or more members of a fire attack team or group, while the other members of the team stand by at a standpipe connection on the floor below. Once the location of the fire has been verified, the engine company can begin their stretch.

A good engine company is disciplined and will not simply start stretching their attack line blindly, without knowing the actual location of the fire. There are a number of important decisions that must be made, including which stairwell to utilize for attack. It is only after taking the necessary time to establish the exact location of the fire that this and many other decisions can be made, and made with accuracy. Don't start stretching until you know where you're going, and what the best, safest, and most efficient way is to get there.

Hookup on the floor below. It is critical that the hoseline stretch begin from as safe a location as possible, preferably away from and below the fire area. Hooking up on the floor below is a very commonsense and basic tenet of high-rise and standpipe operations. Unfortunately, the violation of this critically important rule has been the first domino to fall on a tragic path ending in firefighter line-of-duty injuries and deaths.

The temptation is significant. Firefighters arrive on a reported fire floor and encounter only a very light haze or smoke condition, or perhaps just a slight odor of smoke. The standpipe connection on the fire floor is close by, making it very convenient, especially in buildings that have standpipe connections located inside cabinets on the floor. I have seen it before. The line gets stretched out on the fire floor, hooked up to this convenient standpipe connection, and charged. Everything appears to be fine, until the door to the fire apartment is forced open. Then, the hallway quickly fills with heavy dark smoke, and visibility goes from good to zero in seconds. Fire service luck has allowed this procedure to run its course and end successfully on far too many occasions. Even when we do something that is clearly and fundamentally wrong, if we get away with it, we tend to do it again and again. However, there are numerous case studies of fires that resulted in firefighter injuries and line of duty deaths where our luck ran out. As Battalion Chief Ted Corporandy, San Francisco Fire Department (retired), says, "good luck often times reinforces bad habits" (fig. 11–1).

Due to the dynamic nature of fireground operations, sometimes things simply don't go as planned. Once that apartment door is forced open, there may be a loss of water pressure, perhaps due to a burst section of hose, one or several kinks in the stretch, or an unchocked door that closes over the hoseline. Maybe the attack team encounters a volume of fire that overwhelms them, coupled with a powerful wind condition pushing heat, fire, and smoke directly back at them. Suddenly, members must retreat. Low air warning alarms begin to sound and that once perfectly clear hallway is now like being in the middle of the Black Sea. Maybe the truck company failed to maintain the integrity of the door to the fire apartment. What seemed so routine has now become a struggle simply to get out alive.

Firefighters will attempt to utilize basic survival procedures, including the attempt to feel for hoseline couplings to identify which direction will lead them out, away from the fire. Even if firefighters are able to reverse direction, and follow the hoseline out, the hoseline, unfortunately, won't guide them all the way out to safety. Instead it leads to a dead end at the wall, just below the standpipe cabinet (fig. 11–2). Where is the door to the stairs from here? If luck runs out, one or perhaps several disoriented firefighters will lose their lives in a hallway that was, only moments ago, a clear, benign, and seemingly safe location. Hook up on the floor below!

In Denver, most of the high-rise buildings are equipped with a Class I or

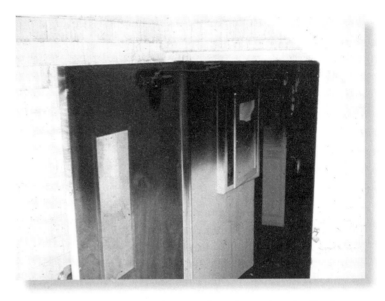

Fig. 11–1. This standpipe cabinet/outlet is located on the fire floor. Although more convenient logistically speaking, it would have been a dangerous and potentially deadly choice to use.

Fig. 11–2. When a hoseline is hooked up to the standpipe cabinet/outlet on the fire floor, it does not provide firefighters with an umbilical cord to safety, as it will not lead firefighters to safety in an emergency, but rather to the wall on the fire floor, which is a dead end, especially in smoke conditions.

III standpipe located within a wall cabinet on each floor, rather than a much-preferred location in the stairwells. This design was adopted several years ago as the fire and building code requirement for high-rise buildings and low-rise standpipe-equipped buildings. The intention behind this, I believe, was to allow for a fire attack to commence without having to keep exit doors leading to the stairwells open, thus reducing the smoke contamination inside the stairwells. Perhaps it was a good idea in theory, but obviously a concept thought of and designed by someone other than experienced firefighters. This design places all of its emphasis on an attempt to achieve life safety for the occupants in the building, with no regard for firefighter safety, and thus places firefighters out on the fire floor making connections. Ultimately, that which compromises firefighter safety and survival will eventually compromise the safety and survivability of civilians, as well. If we can't stop the fire and/or if we get into trouble attempting to do so, then we can't help them.

The need to maintain strict discipline and always begin the stretch by hooking up to the standpipe connection at a location below the fire floor, typically one floor below, is of paramount importance. At all other structural fire operations, we begin from a safe location, outside the fire area. Whether that happens to be a preconnect stretch to attack a dwelling fire, or a stretch from a static hose bed at a strip mall fire, it all begins in a safe environment away from the fire. That location might be on the front porch of a dwelling, or at the main entrance to a commercial building. At these fires, retreating firefighters have the ability to follow the hoseline, and it will guide them to a safe location outside the building. The same concept holds true when we hook up on the floor below. We don't take our preconnect into a dwelling fire and make our connections inside, thus, it makes no sense to place ourselves in potential harm's way by stretching and connecting our line on the fire floor of a high-rise or standpipe-equipped building.

The stretch must come from the floor below. If this is not possible, such as when there is a fire below grade or on the first floor, the engine company must then stretch from the pumper. In addition, numerous buildings in many cities have horizontal standpipes, such as warehouses, large malls, schools, auditoriums, sports venues, and so on (fig. 11–3). The design of many buildings equipped with a horizontal standpipe does not allow for hookup to the standpipe from below a fire, hence the term, *horizontal standpipe*.

A standpipe connection deep inside a large mall located relatively close to a fire can lure firefighters into a potentially deadly trap. Being lost in the hallway of a high-rise multiple dwelling is bad enough. Imagine losing your bearings inside a large, open mall or a 50-acre warehouse. That's exactly what can potentially happen when you hook up to a standpipe connection that is located within, or one that can quickly become part of the fire environment.

A long stretch from the pumper located hundreds of feet away outside the building, is not a very appealing alternative. However, having an umbilical cord that leads out and away from the fire is absolutely critical. The use of a horizontal standpipe must be avoided, unless hookup can be completed in an area that will be unaffected by heat, fire, and smoke conditions. If a horizontal standpipe is inadvertently used by the first arriving

Fig. 11–3. A horizontal standpipe hose valve inside a large shopping mall does not provide a margin of safety to firefighters, as the connection has to be made on the same level as the fire.

engine company or attack team, then the incident commander must make every effort to quickly place a backup handline in position coming from outside the fire building. This is also true anytime an attack line is hooked up on a fire floor, specifically inside a high-rise or standpipe-equipped building, where the fire occurs on the first floor or below grade, and the initial attack line is incorrectly stretched from the same level as the fire.

Connecting the hose lengths. From the floor below, prior to stretching the attacking line, the hose lengths should be connected together in a neat, organized fashion. The Denver hose pack is designed so that, when the hose lengths are properly laid out and connected together, the hose packs will pay out smoothly, without twisting or binding up.

This procedure begins with establishing the number of hose lengths necessary to safely reach the fire area, initiate attack, and advance to the seat of the fire. Once again, generally, 150-feet or three lengths is a good starting point inside most standpipe equipped buildings. However, in some situations, 100-feet will be more than enough, and in other buildings 200–250-feet or more may be necessary.

With the needed length of attack line established, firefighters place the necessary hose packs on the floor below, preferably out in the public hallway on the floor below, rather than inside the stairwell. The stairwell will be a very busy place, congested with firefighters and possibly evacuating occupants. From this location inside the floor below, members place the appropriate number of hose packs on the floor, with each one placed in the same position, with the female couplings facing one direction, and the male couplings facing the opposite direction toward the fire (fig. 11–4). One hose pack should have a nozzle pre-attached, and it should be the lead length, located on the side with the male couplings. From this lay out position, the straps holding the hose packs together are removed, and the hose packs are connected together (fig. 11–5). The hose stretch can now begin.

Apartment stretch. It is always much easier to stretch a dry hoseline rather than a charged one, especially in the case of the 2½-inch handline. However, smoke, heat, and actual fire conditions will dictate whether or not

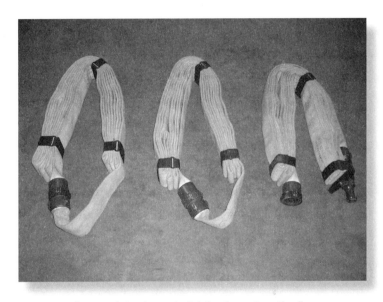

Fig. 11–4 The standpipe hose stretch begins out on the floor below by placing the hose packs in the same position, with female couplings going one direction, and male couplings going the opposite direction

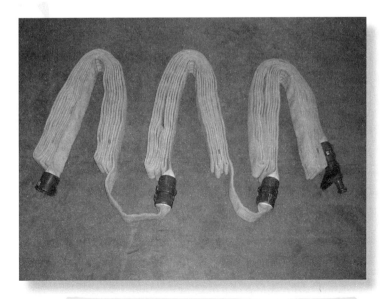

Fig. 11–5 With the appropriate number of hose packs in place, the straps are removed, the lengths are connected together, and the stretch can begin

this can be done, while maintaining the ultimate goal of firefighter safety. A fire that is still confined to a single apartment in a high-rise multiple dwelling is a great example of a situation where the attack line can, and should be, stretched dry to the apartment door. I refer to this as the *apartment stretch*. Of course, the apartment door must be closed and intact, and an evaluation of conditions in the public hallway must be made prior to the stretch, specifically with regard to any smoke and or heat conditions.

The apartment stretch begins on the floor below, after an appropriate attack stair has been chosen by the officer in charge of the fire attack group. After determining the appropriate length of the attack line and connecting the hose lengths together, members begin to stretch toward the fire apartment. The lead length, or nozzle section, is taken all the way up to a *"drop point"* in the public hallway adjacent to the fire apartment door. From this *drop point*, members identify the midpoint of the lead length, and fully stretch the lead length out away from the fire apartment, in the direction away from the supply. The lead length ends up in a U-type configuration, and the nozzle should end up about four feet from the apartment door (fig. 11–6). Final adjustments to the lead length of hose can and should be made to ensure that at least the first ten feet of attack line will be coming from the swing or hinge side of the fire apartment door. This provides for immediate water delivery into the fire apartment (if necessary) as the door is opened, and usually makes for a smoother advance into the fire apartment as members will not have to severely bend the attack line to make the corner.

All other lengths of hose in this layout should be fully stretched out in the attack stair and on the floor below, leaving no piles of hose prior to charging the line with water. This procedure will prevent kinking and subsequent loss of water volume and pressure. Due to the size of most typical apartments, the lead length of hose (50-feet) will be sufficient to reach all areas of the involved apartment. Therefore, the only friction point as the hose is advanced into the fire apartment will be at the entrance to the fire apartment, in the public hallway. This stretch, advance, and attack can be efficiently completed with minimal manpower. If necessary, for larger apartments, the second engine company can help advance additional hoseline from the floor below via the attack stair.

The same fundamentals can also be applied to commercial high-rise buildings. For a fire that is confined to a compartmentalized area, such as an office or utility room, with light smoke conditions and no heat outside of the fire area, a dry (apartment) stretch up to a location just outside the point of operation would be appropriate and most efficient, but only if the drop ceiling is opened up by truck company members to check the plenum space for fire overhead prior to the engine company's stretch.

Stairwell stretch. On the other hand, when the fire is not confined to the apartment of origin, and, in fact, has control over even a small portion of the public hallway, the attack line will have to be charged in the stairwell, prior to advancing out onto the fire floor. I refer to this as the *stairwell stretch* (fig. 11–7). A heavy smoke condition also dictates the need for a charged handline prior to advancing out onto the fire floor. This makes for a more

Fig. 11–6. An apartment stretch

difficult operation, but the safety of having water immediately available, should conditions change, far outweighs the convenience of stretching dry.

Ultimately, because of the large open areas that are typical of most commercial high-rise buildings, the possibility of a larger, fast moving, and unconfined fire is highly probable. A serious, non-compartmentalized fire that has control over a significant portion of a floor and or the public hallway, accompanied by heavy smoke and heat conditions, requires that a charged line be stretched from the entry point (stairwell stretch) which will be at the door to the fire floor from the attack stair.

In this type of situation, there are specific procedures and techniques that can and should be applied in order for this operation to be effective and efficient. Advancing a charged 2½-inch handline up a flight of stairs will be an extremely demanding operation. The engine company can gain a significant advantage by using gravity to assist with their initial advancement. Taking the extra time to properly stretch the handline before charging it will pay significant dividends in the long run.

Specifically, the engine company can stretch up past the fire floor to the next landing, be that a half landing in buildings with U-return stairs, or the landing for the floor above in buildings with a straight run of scissor stairs. Firefighters should stretch as much line as possible up past the fire floor, to the landing above, and then back down, with the nozzle placed about two to four steps above the fire floor landing. The line should be fully stretched out, using as much area on the landing above as possible.

In commercial buildings with a straight run of scissor stairs, this method will place approximately one length (one section) of hose, or 50 feet, above the attack team. Therefore, the attack team will be assisted by gravity for the first 50 feet of their advancement out onto the fire floor. In residential buildings the amount of hose that can be stretched above will be slightly less, at 20 to 40 feet, depending on the stair configuration.

There are some key points to remember when stretching the hoseline above the fire floor. First, if serious fire conditions exist, the door to the fire floor must be controlled and kept closed before firefighters can operate in the stairwell above the fire floor. Attempting to stretch the line above with heavy smoke, heat, and/or fire pouring into the stairwell is extremely dangerous, and could result in a firefighter injury or death. A firefighter must be assigned to stay at the door and ensure that it is kept closed until the officer in charge of the fire attack gives the order to open it.

Second, if firefighters are able stretch up to the landing above, once the stretch is completed, a firefighter must remain at the landing above, securing the line by pinning it to the floor, prior to and, most importantly, while the line is being charged. If this is not done, when the line is charged, the weight of the water filling the hose will forcefully pull the line down the stairs, gravity assisted, unless one or more firefighters is positioned above, securing the line as it is being charged.

I have seen firefighters work hard to complete a good stretch in a stairwell, only to end up with a pile of hose and numerous kinks on the fire floor landing because they didn't properly secure the line before calling for water. Once the line is completely charged, the weight of the water in the line will hold it in place on the stairs. However, if it

Fig. 11–7. This is a stairwell stretch which involves stretching a dry line inside the stairwell and up past the fire floor. The line is charged prior to advancement out onto the fire floor.

isn't secured during charging, it will end up as a pile of charged hoseline, with multiple kinks. There will be a low volume, ineffective stream, and advancement out onto the fire floor under heavy smoke and heat conditions will be extremely dangerous and probably impossible. Another thing to keep in mind is where the nozzle is placed. It is recommended to stretch the hose above, so that the nozzle ends up on the stairs, just above the fire floor landing (fig. 11–8). Sometimes firefighters have a tendency to bring the nozzle right up to the door and flake several feet of hose out on the fire floor landing. Keep in mind that the floor landings in most stairwells are not very large areas, especially in residential buildings. All firefighters know that when a hoseline is charged, it stretches out several feet. Rather than being pinned up against the stairwell door when the line is charged, and hastily attempting to recover by pulling several feet of charged hose back up the stairs, simply begin with a comfortable distance between the nozzle of the uncharged line and the stairwell door. It is much easier to pull a few more feet of charged hoseline *down* than *up*.

For this stairwell stretch, the remaining hose should be stretched out onto the floor below. If that area happens to be a hallway, then the hose will have to be stretched out in a *T* pattern (fig. 11–9). If there is a room or open area directly opposite the stairwell on the floor below, it is best to stretch the additional line straight out onto the floor below and into that room or open area from the stairwell (fig. 11–10). This makes advancement easier by minimizing the friction points that can bind a handline up. The main thing to remember when stretching 2½-inch handline is to take time to make time. Stretch the line out fully. There should be no piles of hose anywhere in the stretch. Remember, even a small pile of hose represents future kinks.

As for the stairwell stretch, keep in mind that some stairwells are so small, that stretching up past the fire floor landing becomes impractical, and should be avoided. In extremely narrow stairwells with small landings, attempting to stretch up past the fire floor landing usually results in severe kinking and a dangerous and cumbersome advancement out onto the fire floor. In these situations, the attack

Fig. 11–8. Nozzleman standing by to receive water, positioned a few steps up above the fire floor landing

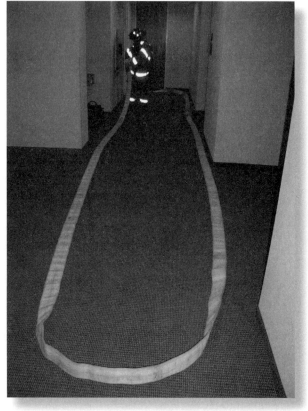

Fig. 11–9. Extra hose stretched out onto the floor below in a T-shaped pattern and down the hallway laterally in both directions.

team still initiates fire attack from the stairwell, but all of their line will be coming up from the floor below. This is not as easy and convenient as when we can stretch the line out on the stairs and landing above the fire floor, but in some cases, that's not an option.

Advancing the handline

It is not all that physically demanding to stretch a dry line out and position it for operation. That is actually more of a mental task than physical. However, it is a critically important task that must be completed carefully and judiciously. Success of the entire operation depends on completing a proper hose stretch at the beginning. The properly stretched hoseline will make for a much easier, more effective, and efficient advancement of the attack line once it's charged.

If the line was properly stretched, including stretching some hose up past the fire floor to the landing above, the initial advancement will be very smooth and relatively easy for the first 25 to 50 feet. One firefighter positioned at the fire floor landing can easily pull the hose down from above, gravity assisted, and feed line to the nozzle team (fig. 11–11). However, once all the hoseline from above has been advanced out, onto the fire floor, the real work begins. Now the line must be brought up from below, fighting gravity the whole way. This is where that second engine company comes into play.

The extra firefighters are placed at friction points, the so-called unglamorous positions. Yes, it's very unglamorous work, but absolutely critical. That second engine company, and those support firefighters, must maintain strict discipline. Although it would be much more exciting to crawl up to, and join the nozzle team, they must hold back and provide the physical support necessary to move the attack line. Two, four, six, or even eight firefighters located at the nozzle could pull on the hoseline all day long and not move it an inch, if there is no one assisting at the friction points. Firefighters must be positioned at corners, at the bottom, top, and middle of the stairs, and on the floor below to assist with and to lighten up on the

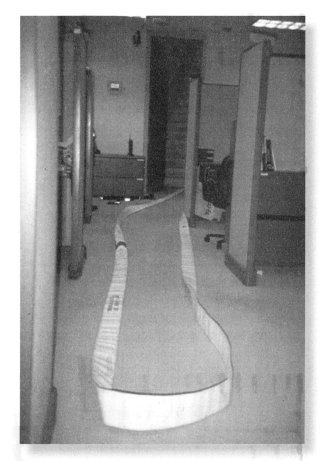

Fig. 11–10. Ideally, extra hose should be stretched out onto the floor below, straight away from the stairwell entrance, into a room, if one exists.

Fig. 11–11. One member positioned at the entrance door to the fire floor can easily pull hose down from above and out onto the fire floor.

line (fig. 11–12). If they're not properly positioned at those locations, the line will not move!

Now, there's a lot more to proper hoseline advancement than just heave-ho, pushing and pulling, and grunting and groaning. The good engine company firefighter is an unrecognized specialist who must utilize finesse. The nozzle team yelling, "more line, more line" will get just that, more line. Sometimes a whole lot more. And sometimes a lot more line than they really want or need, and line that is being pushed toward them at a pace much faster than they really would like to go.

When the nozzle team yells out for more line, the well-intentioned and very ambitious firefighter answers by aggressively pushing more line forcefully toward the nozzle team. Once again, the line may actually be pushing the nozzle team forward, perhaps even throwing them off balance. In addition, several feet of hose with no place to go will end up like a pile of spaghetti inside the fire area. Firefighters attempting to exit the smoke-filled fire area, especially during an emergency egress, will attempt to follow the hoseline out. Too much line forcefully pushed into the fire area means that firefighters attempting to make a hasty retreat might end up going in circles, attempting to follow piles of hose, until they become disoriented and lose their bearings. Remember, when feeding hose to the nozzle team, give them what they need, but not much more than that.

Finesse—that's the key to success. Whenever my FDIC hands-on training (HOT) team instructs firefighters on standpipe operations, we attempt to teach finesse. Rather than just simply yelling out "more line," firefighters should actually try and quantify their request. Ask for what you need, for example, "Give me five more feet," or, "Give me 10 more feet." This request is more specific, and gives the firefighters in the support positions a much better idea of what they have to do. A firefighter can gauge a quantifiable distance or amount of hose passing through his hands. If the nozzle team asks for five more feet, the support firefighters will move the line forward estimating the amount requested, and then they will stop, making for a smoother, more controlled operation.

Fig. 11–12. Firefighter positioned at the bottom of the stairs, assisting with the 2½-inch hoseline advance by feeding line up from the floor below.

At the same time, the nozzle team must be patient. Often times, a nozzle team calls for more line and expects instant results. They sometimes fail to remember that it's hard work for the support firefighters to move the line and overcome the many friction points, especially if there are only a few firefighters assigned to this task. The inexperienced nozzle team can quickly become very frustrated when the line doesn't immediately move upon their request. They yell louder and pull harder all the while the support firefighters are working as fast and hard as they can to give the nozzle team more hoseline. The nozzle team should make their request, and then be prepared to wait a few moments while that request is fulfilled.

It is also important to keep in mind that in the dark, noisy, and dynamic environment typical of an interior structural fire, verbal communications between team members who are spread out several feet might be impossible. Making the verbal requests over the radio is one option, but it too is not without potential problems. In the heat of battle, radio communication is never perfect. Ultimately, the support firefighters should remain fluid and be prepared to move forward on the line up to the next operating member at a friction point, or the nozzle team to directly communicate and then return back to the necessary position to support advancement of the line. This is a continual process throughout the advancement of the handline and fire attack.

The nozzle team

The nozzle team operating a 2½-inch handline must consist of at least two firefighters, but, preferably, three or more firefighters, plus a company officer. Fire departments with four-man companies will be able to provide two firefighters and an officer on the nozzle team (the engineer/pump operator is outside), when a second engine company is supporting their operation in the unglamorous positions. The nozzle firefighter should be given the necessary support by the backup man to counter the nozzle reaction, so that he can concentrate most of his efforts on operating the nozzle and directing the stream.

✻ For a fully developed fire in the free-burning phase, or a high heat fire in the growth stage that is approaching flashover, the nozzle firefighter must first focus his attention on cooling the upper area of the fire environment. Specifically, this means directing the stream, preferably a solid stream of water from a smoothbore nozzle, up toward the ceiling, sweeping the stream back and forth to cover as much of the area as possible. The powerful, solid stream from a large 2½-inch attack line will also serve to penetrate the plenum space by forcefully dislodging drop ceiling tiles typically found in commercial high-rise buildings (fig. 11–13). In the absence of sufficient truck company personnel, or if extreme heat conditions don't allow truck company firefighters to position themselves to pull ceiling tiles ahead of the engine company, the nozzle team can complete this task using their stream. The same procedure should apply to operations in residential high-rise buildings, even though the construction will likely be different, typically without a plenum space. The key is to cool the upper area of the fire environment to prevent flashover.

✻ The nozzle team should continue to proactively cool the upper areas, as high heat conditions are encountered during advancement toward the seat of the fire. This will allow for continued forward advancement, and prevent flashover. Once actual visible fire is encountered, or in extremely low visibility if high heat indicates fire conditions, a vigorous application of water by the nozzleman directing the stream to all areas of potential involvement will have the greatest impact on the fire. As conditions warrant, the nozzleman can shutdown the stream to allow for evaluation of conditions.

✻ Prior to any forward advancement, in addition to directing the stream toward the ceiling level for cooling the upper atmosphere, the nozzleman should also direct the stream at the floor and sweep it from side to side, forward and backward. This step will accomplish several important goals. First, by sweeping the floor prior to advancement, the water will cool the floor and allow for a more tenable advancement. In addition, many pieces of dangerous debris that are typically found in an interior structural fire environment will be pushed out of the way, once again, allowing for a safer advance of the attack team and hoseline.

✻ Last, the well-trained nozzleman can listen to the noise of the powerful stream and gather valuable information regarding what lies ahead. A noisy stream might indicate that the water is forcefully striking a solid floor, wall, desk, or other furniture. A quiet hollow stream could indicate a hole in the floor, perhaps not typical during most high-rise operations, but very possible during operations in high-rise buildings under construction. Also, in a typical, commercial high-rise building, a hollow-sounding stream could indicate to the nozzleman an opening such as an access stair,

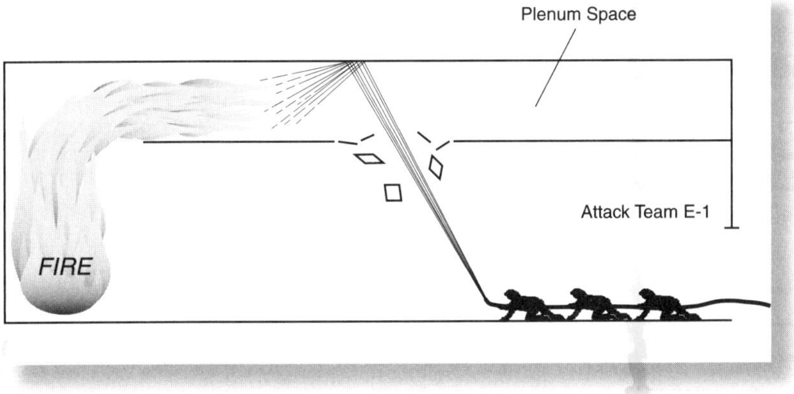

Fig. 11–13. The nozzle team can use the powerful stream from a 2½-inch handline with a smooth-bore nozzle to forcefully dislodge drop ceiling tiles, thus exposing the plenum space.

elevator hoist-way, or broken window. Large windows, sometimes located close to the floor, present a significant safety concern to operating firefighters, especially in zero visibility conditions. Recognizing the sound of a hollow stream could prevent a tragic fall out of the large, unobstructed opening of a broken window.

✳ As with any interior firefight, if visible fire is sufficiently darkened down, the company officer should direct the nozzleman to shutdown the stream and allow for a period to evaluate conditions. Furthermore, with the stream shutdown, less noise will give the entire nozzle team an opportunity to carefully listen for various sounds that can assist with operations in low or no visibility.

Dangerous Procedures

At this point, I must address some very dangerous and disturbing procedures that are unfortunately being introduced and recommended by some individuals. Many of us have heard the term *three-dimentional fire attack*, and about the use of the stream to *pencil* the upper atmosphere with short bursts of water in the form of a broken stream. I am very concerned and scared for the youth of the fire service, and specifically those who are taught and encouraged to use these bizarre methods of water delivery.

✳ The young, misled firefighter who is penciling the upper area with short bursts of water, as he attempts to advance deeper and deeper into a fire building, is on a collision course with disaster and potential tragedy. We must teach our youth to respect the fire, not play games with it. Cooling the upper atmosphere should be done with a forceful, powerful stream of sufficient GPM. A solid stream of water from a smoothbore nozzle is recommended as the best, most powerful, and safest method to cool the upper atmosphere during advancement toward the seat of fire. If a combination nozzle is used, and if you're lucky enough to not have it clog with debris from the standpipe, a straight stream pattern is recommended.

✳ Short bursts of water, especially in the form of water fog from a combination nozzle will do nothing to address the real problem—the burning solid fuel. Furthermore, in the high heat, unventilated atmosphere typical of an interior structural fire, especially those which occur in high-rise buildings, this penciling with water fog will only create an atmosphere of steam and exponentially greater "wet" heat conditions.

✳ Ultimately, the fire attack team must address the disease, not just the symptom, that is, water must be properly delivered to the burning solid fuels in order for the heat conditions to subside. By cooling the upper atmosphere with a powerful solid stream, large droplets of water will in most cases reach burning solid fuels, and even if the attack team can't actually see it, they might hear it. An occasional splash of water fog into the upper atmosphere is guaranteed not to reach any burning solid fuels, as it will be prematurely evaporated into the heat currents above, leaving the dangerous by-product, steam.

✳ Be extremely wary of concepts that include words and terms like *theory* and *new age*. Theory is defined as speculation, or a guess. New age stuff includes political correctness and bastardized ways of operating. Most importantly, avoid at all costs procedures that are not accepted and utilized by those who spend the greatest amount of time actually fighting fires in real buildings. Take the time to ask an experienced firefighter, who you would trust with your life, what he thinks. Would he do it? I'll bet his answer is, definitely not!

The First Handline, First

There are numerous other considerations with regard to successful engine company operations at high-rise fires and during any standpipe operation. A serious fire in any building will likely require the use of multiple handlines. Stretching and placing a backup line in position as quickly as possible is of paramount importance. However, with that said, nothing will cripple an operation faster than focusing on a backup or secondary handline prior to ensuring the success of the first handline. In other words, as important as the backup line is, at all operations, those responsible for subsequent lines must first identify if the first handline is in fact accomplishing the operational goals of confinement and extinguishment. It is particularly important when using 2½-inch handline that the second and later arriving engine companies assist those companies already operating, before committing to their own handline. This operational tenet cannot be emphasized enough.

The Backup Line

For standpipe operations, and specifically operations in high-rise buildings, any subsequent handlines stretched after the first may be positioned and operated in a wide variety of locations, depending on the needs and problems at a given fire. At most operations, the second or backup line is positioned directly behind the first line, following the path toward the seat of fire. However, the circumstances will dictate specifically where the second line should be placed.

For example, at a typical residential high-rise fire, with a common public hallway, the second handline can, and in most cases should, be stretched directly behind the first attack line up to the point of entry, typically the fire apartment door (fig. 11–14). In most cases, that's an excellent location for the backup team to position themselves to protect the egress for the members operating the first line. Given a large volume of fire that cannot be controlled by the first line alone, the second line can be ordered to advance into the apartment and assist with control of the fire.

Keep in mind, the typical apartment unit in most residential high-rise and low-rise buildings is relatively small, often times less than 1,000 square feet. Therefore, it becomes important to keep the number of firefighters operating inside the fire apartment to a reasonable number. A good rule of thumb is one engine company for fire attack, and one truck company for search and rescue, or a fire attack/search group of six to eight firefighters. I am a huge proponent of having plenty of resources for fireground operations, but those resources must be utilized and controlled appropriately. We don't need, and certainly don't want, the entire first alarm assignment inside the fire apartment. Without good command and control and preestablished comprehensive operating procedures, that's exactly what might happen.

On the other hand, a fire involving a portion of a floor at a commercial high-rise building might require a slightly different approach for the backup line. The vast majority of commercial high-rise buildings are constructed using a center core. That center core can create some very dangerous and complicated problems for operating firefighters, not the least of which is fire wrapping around the center core and back behind attack teams, potentially cutting off their egress. Because of this, a second or backup handline may need to be positioned prior to advancing the first handline. In this scenario, the backup line serves to protect the egress point for the primary attack team (fig. 11–15). In the event the fire wraps around the center core, this backup line is positioned to stop the fire and keep it from cutting off the egress path of the primary attack team.

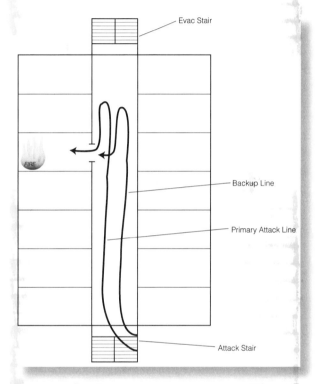

Fig. 11–14. The first or primary attack handline is stretched and operating inside the apartment. The second or backup handline is stretched to the entry point, and positioned here to protect the egress. It may be advanced into the fire apartment if necessary.

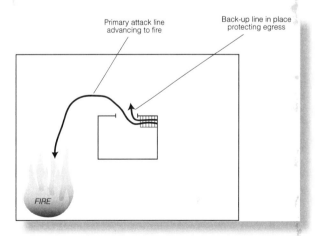

Fig. 11–15. In a commercial high-rise with a center core, a backup line might need to be positioned at the entry point, prior to advancement of the primary attack line.

It is important to note, once again, that the use of a combination fog nozzle set to a cone-shaped pattern will greatly exacerbate the potential of pushing fire back around the center core directly onto operating firefighters (fig. 11–16). The extremely powerful force of a combination nozzle can create a wind current equal to or greater than that of a positive pressure blower. This is one of many justifications against the use of combination fog nozzles for high-rise operations, and, for that matter, any interior structural fire attack.

An attack can possibly be employed by using the primary attack line and the backup line simultaneously. The lines are advanced out onto the fire floor in separate directions. In a center core configuration, positive advancement of both lines toward the seat of fire can act to confine the fire with a coordinated attack from opposite directions (fig. 11–17). Obviously there is a concern regarding the use of opposing streams whenever multiple attack lines are approaching from opposite directions. However, the danger associated with opposing streams is, and always has been specifically related to those operations where combination fog nozzles are set to cone-shaped patterns. The entrained air created by a fog stream is a powerful force serving to push heat, fire, smoke, and other products of combustion toward the team attacking from the opposite direction. A properly executed and coordinated attack from opposite directions, using the appropriate and recommended tool, specifically the smoothbore nozzle, will yield positive results, as the solid stream of water delivered by the smoothbore will not push fire, but will extinguish it, and, thus, not push heat, fire, and smoke toward firefighters operating opposite that team.

Attack teams must be prepared for extremely strong wind currents, forcefully pushing back toward the attack team. This occurs when windows on the windward side of the building are breeched by the fire, and can be greatly aggravated by a powerful stack effect. When this occurs, operating forces may have to employ an attack with multiple handlines, such as two separate handlines operating side by side (fig. 11–18). This can be an extremely punishing operation that will require significant resources, including several engine companies located in staging for relief.

Some proponents of the combination fog nozzle have suggested its use for situations such as these. The numerous dangers of combination fog nozzles for interior operations, specifically those in high-rise buildings, should be enough evidence to eliminate this tool from the arsenal used for high-rise firefighting and standpipe operations. If it were to be employed in an effort to push against a powerful wind current, firefighters could sustain severe or fatal injuries.

Man and machine are simply no match for Mother Nature. An extremely powerful wind condition, coupled with the a strong stack effect within a high-rise building, can create wind currents with enough power to drive heat, fire, and smoke

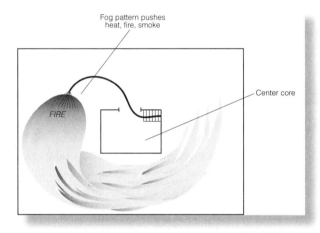

Fig. 11–16. A combination fog nozzle set to a cone-shaped pattern will forcefully push heat, fire, smoke, and other products of combustion back around the center core, cutting off firefighter egress.

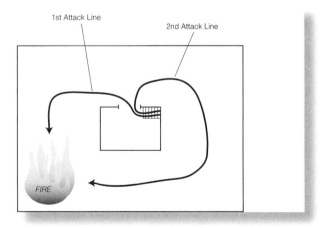

Fig. 11–17. Fire attack can and may have to be accomplished using two lines coming from opposite directions. The problem of opposing streams is not an issue when using the recommended smooth-bore nozzle.

back toward and on top of attacking firefighters with hurricane force. The powerful wind-driven current of heat cannot be stopped by a curtain of water from a fog stream. Rather, this current of heat will quickly penetrate the fine water droplets within the fog stream and thus envelop the attack team members in a debilitating atmosphere of steam.

✳ Firefighters must keep in mind that a cone-shaped pattern created by a combination fog nozzle is not going to protect them, but rather, at best, it may serve to push products of combustion away from them, thus potentially creating a feeling of protection. Actually, this would be best defined as a false sense of security. However, when the forces of nature from the opposite direction are much more powerful than the fog stream, such as what may be encountered during high-rise operations, any attempt to create protection by pushing the products of combustion away will quickly fail. The water from the fog stream will be driven back toward the attack team in the form of steam with enough force to create a thermal shock and instantly incapacitate a human body, including a firefighter in full personal protective equipment.

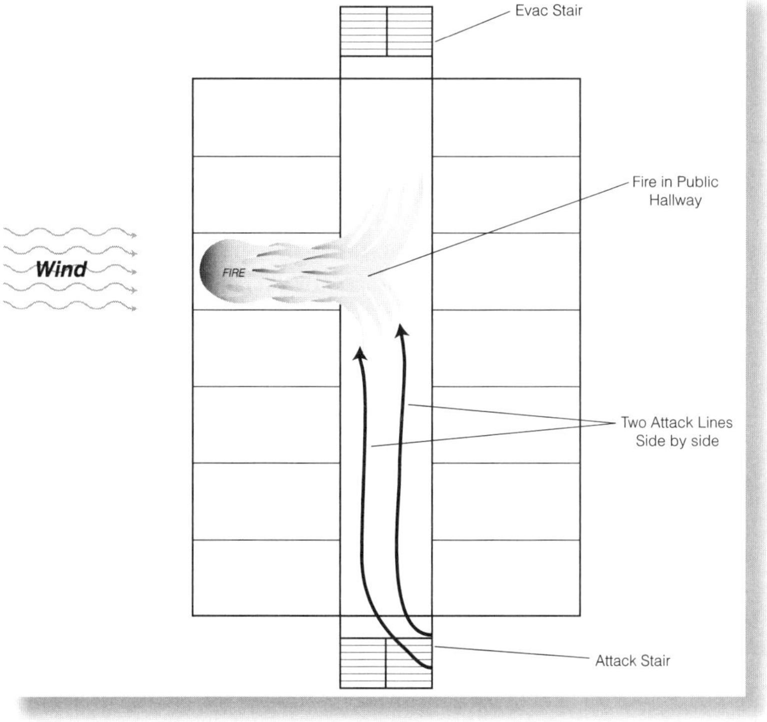

The only possible way to overcome these powerful forces is to use a very powerful weapon, such as the 2½-inch handline, or portable master stream appliance with a smoothbore nozzle. The powerful, solid stream of water has a much greater chance of penetrating the wind currents typical of high-rise operations, and thus allow for solid droplets of water to reach burning solid fuels. Remember, these burning solid fuels represent the real problem, the *disease* if you will. The heat created by these burning solid fuels is a symptom of this disease. If the battle is to be won, the attack team must treat the disease, not just the symptom. As my friend and Brother Harry Lee Davis, Firefighter, FDNY Squad Co. 18 (retired), teaches our students at FDIC, "It is best to have a long distance relationship with the fire."

Using Master Stream Appliances Offensively

Whenever we think of master stream appliances, I think we have a tendency to focus specifically on defensive operations. However, as addressed in previous chapters, the option of using master stream appliances for defensive operations is limited, based on logistical constraints, specifically the height of the fire. Unless the fire happens to be within reach of our aerial mounted master stream devices, or if we can utilize other buildings as platforms to deliver the water, exterior master stream use is not going to be possible. However, we must be proactive and utilize our tools as appropriate, even if that means utilization of some tools in what might be considered an "unorthodox" manner.

As we can sometimes get over-reliant on the use of 1¾-inch handline, we can also become locked into thinking that handlines (even 2½-inch) are our only option for interior structural fire attack in large buildings. At fires in large low-rise, standpipe-equipped buildings,

Fig. 11–18. Two attack lines working side by side, and advancing toward the seat of the fire.

Fig. 11–19 Placed across a doorway, a metal bar or other strong forcible entry tool of sufficient length can be used to effectively counter the nozzle reaction of a portable master stream appliance.

Fig. 11–20 A portable master stream appliance can be used as an effective offensive weapon for high-rise and other large area standpipe firefighting operations, supplied by a standpipe system

such as large warehouses, and in serious high-rise fire situations, we must be well trained and prepared to use master stream appliances, as well. Yes, the use of master stream appliances, for some specific interior operations, is not only possible, but is a very realistic and valuable tactical option.

Several years ago, my friend and mentor Jerry Tracy, Battalion Chief, FDNY, conducted tests at a Manhattan high-rise under construction. Using a portable master stream appliance, Jerry had companies supply it from the standpipe, and was able to deliver a high volume of water from an upper floor, to the outside open air. Prior to supplying this very powerful weapon, Jerry had members of FDNY Rescue Co. 1 use a "Ram Set" to shoot a lag into the concrete floor. This served as an anchor point to which the master stream was chained in order to counter the significant nozzle reaction.

Another good friend and mentor, Phil Miller, Lieutenant, DFD Engine Co. 3, has been developing high-volume water delivery techniques for the past several years. One of those involves the use of a master stream appliance for interior operations, supplied by a standpipe system. Lt. Miller places the appliance as close to the fire area as possible, and just inside a doorway. This could be the doorway to the fire floor from the attack stair, or any other doorway that provides direct access to the fire area, while protecting operating members. The appliance is placed in position, with a heavy, straight, forcible entry bar (long enough to span a typical doorway), placed across the doorway behind the master stream appliance. The rear feet on the master stream base are placed against the straight bar (fig. 11–19). When the weapon is supplied with water, the nozzle reaction force is transmitted to the bar, and onto the doorjamb/wall/building. This is more than sufficient to counter the nozzle reaction, even at high flows, including 1000 gpm with a 2-inch smoothbore tip (fig. 11–20).

Lt. Miller has developed several specific procedures associated with this operation that truly makes it an offensive tool. Namely, a jumbo shutoff (2½-inch x 2½-inch ball valve shutoff, with a 2-inch waterway) is used to control the water delivery at the master stream itself. In addition, hose clamps are placed several feet behind the master stream, which, once again, allows for control of the water. These are both very important safety measures, as well. Lt. Miller also teaches members to lay the supply hose out in a specific manner, with loops between the supply and the device, so that the hose can be clamped near the supply. This allows for an attack team to actually pickup the uncharged hose lines and the master stream, enabling them to advance the weapon forward several feet, and position it at a new anchor point.

I am very proud to have Lt. Miller serving as a member of my FDIC Engine Company Standpipe Hands on Training (HOT) Team. For the past several years, assisted by other team members, Lt, Miller has introduced this procedure to countless members of the fire service. It will be used by some of them during their careers, and will likely make the difference at a serious fire in a high-rise or a large, standpipe-equipped building.

There are questions that always arise from those who have not actually seen this evolution or done it. Some are skeptical, but once again, it is very doable. First is the issue of water supply. This weapon, like the 2½-inch attack line is a high volume, low pressure weapon. So a typical low pressure standpipe system can certainly achieve the necessary pressure. Remember the requirements of NFPA 14, those being 500 gpm at 65 psi (pre-1993), and 500 gpm at 100 psi (post-1993) required pressure and volume for the top most, or most remote outlet within a standpipe system. Based on those figures, the delivery of 500 gpm, using smaller diameter smoothbore tips, is very realistic.

Tests conducted by the author with Lt. Miller demonstrate that standpipe systems that can be safely and successfully augmented by fire department pumpers allow for a very high volume water delivery, using the portable master stream appliance. Flows in excess of 1000 gpm, using the 2-inch smoothbore tip have been achieved. However, to maintain maximum safety, manufactures of most of these devices recommend that the maximum tip size, when using these as portable master streams, should be the 1¾-inch smoothbore tip, flowing 814 gpm at 80 psi nozzle pressure. Furthermore, unless the device is securely anchored, it should not be lowered below 32°, as this could cause the device to forcefully move backward, due to the significant nozzle reaction. Using Lt. Miller's method to anchor the device, we have never had the device move on us, even at flows in excess of 1000 gpm. For training purposes, and, if possible, during actual fireground operations, establishing a secondary anchor point, and using a chain to secure the device, will provide an added margin of safety.

The master stream devices used for training, testing, and at actual fires are the large devices that serve as a deck gun on the pumper that can be disconnected and placed on a portable base. It is important to note that several good nozzle manufacturers have developed and produced a new generation of lightweight, portable master stream appliances that can be very effective, especially for high-rise operations. However, these devices are limited to 500 gpm, based on manufacturer's recommendations.

Remember, we are trying to pulverize a very large, fast moving fire. Unless you want to make a rainbow, stay away from the combination fog tips on this, or any other master stream appliance *firefighting* operation. We are trying to deliver an extremely high volume of water a significant distance and have it arrive intact, in order to cool burning, solid fuels. The solid stream from a smoothbore will arrive where it's needed, and will cool the fuel.

Strategically, utilizing a master stream appliance is based on an extreme situation, where a large volume of fire is burning out of control. The large warehouse is a low-rise example. In the high-rise arena, a technique called *"flowing the floor"* has been introduced to serve as a somewhat defensive measure at extreme fires. Once an entire fire floor is burning, specifically an un-compartmentalized fire floor in a commercial high-rise building, that fire might involve 10, 15, 20,000 or more square feet. A fire area that large is well beyond the capabilities of manual firefighting techniques, specifically handlines. In fact, it will be beyond the capabilities of master stream appliances as well. However, using the master stream proactively, by literally soaking the fuel on the floor or floors above while allowing the fire on the fire floor to consume fuel and burn down, might serve to prevent most vertical extension, and thus slow the overall fire down. This strategy might allow operating members to stop vertical extension altogether, and eventually make it possible to extinguish a fire on the fire floor once it reaches a stage of decay. The use of a master stream appliance as an offensive tool is an extreme procedure, but high-rise and large-area firefighting are extreme situations. Prepare yourself for the extreme, worst case scenarios.

Overall Strategic Plan

A fire department's strategic plan for fire attack operations at a typical high-rise or standpipe-equipped building fire should include multiple engine companies, at least six that will likely be from the first and second alarm assignments in most jurisdictions. Once again, in any city or jurisdiction, including the largest cities with the most resources, upon confirmation of a working fire in any high-rise building, an immediate request for additional resources (second alarm) is the most important action that can be taken by the first arriving fire officer. The high-rise event is extremely labor intensive, and the resources needed will be massive. Call for help early and often. The DFD includes four engine companies on our first alarm assignment to a high-rise fire. The second alarm includes an additional three engine companies, for a total of seven.

As for the engine companies, first, plan on using at least two engine companies, generally the first two engine companies due to arrive for the purpose of fire attack, stretching and advancing the first, or primary, attack line on the fire floor. Second, plan on using at least two more engine companies, generally the third and fourth due engine companies, for the purpose of backup, stretching and advancing the second, or backup, line onto the fire floor, or, in some situations, directly up to the floor above. Third, plan on using at least two more engine companies, generally the fifth and sixth engine companies to arrive, for the purpose of stretching and advancing the third line onto the floor above the fire floor, or, in some situations, depending on how the second handline was used, this line can be

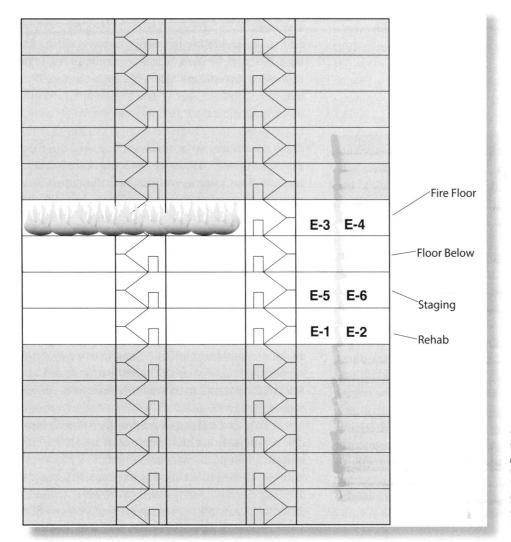

Fig. 11–21 This diagram shows the positions of the six engine companies that will be rotated on and off the fire floor, and used to operate one 2 ½-inch attack line

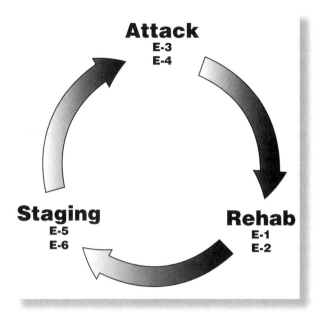

Fig. 11–22 The resources of six engine companies is necessary to maintain positive, forward progress with one 2 ½-inch attack line at extended high-rise/standpipe operations

directed to backup the primary attack team on the fire floor. In many situations, these later-arriving engine companies may be needed to provide relief for engine companies on the primary attack line.

In situations where the fire is extremely large and has control over a significant portion of the fire floor, six engine companies will be needed to maintain forward progress with one handline, at this, a so-called *"campaign"* fireground operation. That is, two engine companies for the initial attack, two engine companies to relieve them, and an additional two engine companies located in staging for relief (fig. 11–21). This is the three-pronged approach to fire attack. As the first two engine companies run low on air, they are immediately relieved by two more engine companies, with the first two going to Rehab. Two more engine companies need to be in staging, ready to relieve the companies attacking the fire. The two engine companies in Rehab will have to re-hydrate and evaluate members for injury or medical problems, and, ideally, get back to staging as soon as possible to relieve those engine companies currently attacking the fire (fig. 11–22).

The chief officer in command of a serious high-rise fire must be mentally prepared to request and utilize a minimum of six, four-man engine companies within the first 15 minutes of the fire fight. In addition, if there is not immediate, positive results toward controlling the fire, the incident commander must be mentally prepared to call for additional resources in the form of a third, fourth, and additional alarms (or whatever terminology your fire department uses to request additional resources) immediately, in order to provide for an additional six engine companies to relieve those committed to critical operating positions.

This resource guideline may seem excessive to some, but a brief review of any serious high-rise fire event, in any city or jurisdiction, will quickly support this guideline. Also, keep in mind, we are talking about firefighters who will likely have an operating time of 10 minutes or less. The typical fire department has members equipped with SCBA air cylinders rated for one half hour. That rating is completely incorrect. The half hour air cylinder, if it is completely full, will not even come close to providing a half hour of work time.

I have conducted studies that concluded that firefighters in top physical condition, when working in a fire environment, could work for about 10 to 12 minutes before their low-air warning alarm sounded. An additional three to five minutes of air was available for egress. Ultimately, if forward progress is to be maintained at a serious high-rise fire, relief companies will need to be in position, just below and outside of the fire area before low-air warning alarms require the primary attack engine companies to retreat.

The high-rise firefight can be an extremely punishing event. High-rise building construction, modern fire loading, ventilation difficulties, and possible wind currents can all create extremely high heat conditions that are immediately debilitating to operating firefighters. Even those firefighters in top physical condition will not be physically able to complete multiple rotations out onto a typical fire floor. Anticipate losing at least 20% of the firefighters to fatigue, injury, or, in many cases, a serious medical problem before, during, or after their first assignment. And, as a rule of thumb, make every effort to limit firefighters to a maximum of two air cylinders before they are relieved and sent to long-term rehabilitation, preferably outside the building.

When all of the factors associated with high-rise operations are put together, it becomes very clear how an extremely large number of resources will be needed to safely and successfully control the fire event. Also, these previous resource guidelines are based on manpower levels of at least four members per fire company, including the officer. Those jurisdictions that operate with three or less firefighters per company will need to increase the total number of fire companies, as appropriate, to compensate for lower manpower levels per company.

Summary

The basic fireground function performed by the engine company is of utmost importance. Engine companies should plan on using powerful weapons to capitalize on the short window of opportunity to stop a potentially serious fire in a high-rise or standpipe-equipped building. The teamwork of at least two engine companies working together to advance one handline will make the application of more powerful weapons, such as the 2½-inch handline and portable master streams, a reality. Finally, the incident commander should plan on using several engine companies to complete the fireground functions and achieve the critical goals at a high-rise fire and/or standpipe operation.

CHAPTER 12

Water Supply

Water supply is the first of several engine company fireground responsibilities. For fireground operations in high-rise and standpipe-equipped buildings, the water supply component becomes much more complicated than at our more frequent operations, such as a fire in a single-family dwelling. It isn't just a simple matter of stretching a preconnected attack line, charging it with tank water, and having the second due engine company establish a sustained water supply.

Properly supplying a typical standpipe system with water involves several various, interrelated elements. In this chapter, I discuss the procedures that should be followed to ensure that attack teams have the appropriate volume of water, at the necessary pressure, to mount an effective attack on a fire deep inside a standpipe-equipped building or on the upper floors of a high-rise.

Types of Standpipe Systems

The various different types of standpipe systems were identified and defined in chapter 4. In order to properly supply a standpipe system, the engine company pump operator must have at least a basic understanding of the specific system involved. Identifying whether the system is wet or dry, automatic or nonautomatic, guides the pump operator to the proper procedures he must complete.

Automatic wet systems

Most modern high-rise and standpipe-equipped buildings have very comprehensive fire protection and suppression systems that typically include a fully-automatic wet standpipe system. These modern systems are designed to be self-sufficient, and water supply is typically built into the system. This sometimes includes an on-site water supply (fig. 12–1) and/or water from a

Fig. 12–1 An on-site water supply of two 11,200 gallon tanks located on an upper floor of a Denver high-rise

municipal water system. In most modern high-rise buildings, particularly the taller ones, there are one or several built-in fire pumps that are designed to provide the necessary water pressure to properly supply the standpipe system. In fact, if the fire suppression system in these modern buildings is working properly, fire department water supply operations, using fire department pumpers, are generally set up and prepared for implementation only as a backup or redundancy, in the event that the buildings system fails to operate properly.

Non-automatic dry systems

On the other hand, there are numerous older buildings in many cities and jurisdictions, and some new buildings, that have much less sophisticated fire protection and suppression systems. For example, buildings with nonautomatic dry standpipe systems will require immediate support from fire department pumpers in order to establish and maintain a water supply for the standpipe system. These systems present numerous potential problems for operating units, many of which were addressed in chapter 4.

Regardless of the type of system and how it is supplied with water, it's ultimately the responsibility of the engine company to ensure that a sustained water supply has been established by the fire department (be it the primary and only water supply or a backup to a built-in system), and that the standpipe system is properly supplied. This should be done for all fires at high-rise and standpipe-equipped buildings.

Once again, some systems will require immediate supply and pressurization of the standpipe, whereas others are designed to operate without support. However, regardless of what type system is in place, fireground operations at high-rise and standpipe-equipped buildings must include the establishment of a sustained, external water supply, as well as the completion of all necessary components to supply the standpipe system, even if water is never actually pumped into the system from fire department pumpers.

Fireground Operations

Upon arrival at a high-rise or standpipe-equipped building, the first arriving engine company should begin the process of establishing a sustained water supply for the standpipe system. Typically, the officer and firefighters will enter the building with the appropriate equipment and become part of an investigation/fire attack group. The engineer/pump operator will be left outside to begin the process of establishing a water supply.

The engineer/pump operator

It is important to emphasize that the engineer/pump operator, along with all other members, must consistently operate in a proper mind-set and prepare for a serious fire every time. Much of our time spent at high-rise and standpipe-equipped buildings, in big cities and small communities, involves something other than combating an actual working fire. In my career, I have responded to thousands of incidents at high-rise and standpipe-equipped buildings, but only a fraction of those were working fires. The engineer/pump operator, along with all members, must be prepared for a fire and operate as if there is one, until the investigation proves otherwise. That doesn't mean that the engineer/pump operator should blindly pump water into every standpipe system every time one is encountered, but it does mean that he should always be preparing to do so.

It's easy to spot a professional engineer/pump operator from one who is just pointing the pumper and shuttling the crew to and from the firehouse. When I arrive at my tenth automatic fire alarm of the day, even if it's at a building that has had previous malfunctions, it is always gratifying when I see the engineer/pump operator outside his pumper, sizing up, conducting reconnaissance, and preparing to operate.

The good engineer/pump operator gets out of the rig, even on a cold winter day, and identifies the location of the fire department connection (FDC), preparing to supply the system, even on automatic alarms, most of which prove to be false. He inspects it for damage and determines whether or not it is in proper working order. In addition, he identifies the location of the closest hydrant, and the second closest hydrant in the event that the first one is out of service or not operating properly.

The engineer/pump operator can and should also take a quick look at the exterior of the building, including a complete 360-degree recon if practical and possible, specifically looking for any signs of smoke or fire. Last, he can make any adjustments to the position of his pumper apparatus, spotting it as close to the FDC as possible.

I recall a particular fire that occurred several years ago in a high-rise multiple dwelling located in the Capital Hill neighborhood of Denver. The fire occurred at 1211 Vine Street and was initially dispatched as an automatic fire alarm investigation.

The DFD sends one engine company, one truck company, and a district chief to investigate automatic alarms in high-rise multiple dwellings. This is a reduced assignment based on the high number of automatic alarms that occur throughout the city of Denver on a daily basis. If any second source calls are received reporting smoke or fire, the assignment is immediately upgraded to a full high-rise fire assignment that includes four engine companies, two truck companies, one heavy rescue company, one hazardous materials company, and two district chiefs, for a total of 34 personnel.

Responding to this fire alarm, the first due companies were Engine Co. 8 and Truck Co. 8, along with District Chief 4; there were no second source calls. Needless to say, all of the responding units and personnel had been to this building numerous times before, for everything from burnt food to system malfunctions. It can be very easy for any of us to become complacent about automatic alarms in these frequent flyer buildings, but as I said in chapter 1, we must continually fight the deadly disease of complacency and all its associated baggage.

Upon arrival at 1211 Vine Street, having arrived from the north, Engine and Truck 8 observed that nothing was showing from the north (D-Delta side) and the east (A-Alpha side) of this high-rise multiple dwelling. Crew members from both companies went inside the building to investigate the alarm, with the exception of the engineer/pump operator from Engine Co. 8.

Unfortunately, too many engineers/pump operators stay in the rig and wait, expecting and anticipating the words, "burnt food" or "system malfunction," to come over the rig radio. That night the engineer/pump operator on Engine Co. 8 set an example that all engineers/pump operators should follow. He got out of the rig, and went to locate the FDC. He wanted to confirm its location, identify any obstructions, and check the exterior of the building. When he got to the south side of the building (B-Bravo side), he immediately saw a heavy volume of fire pushing from several windows at the third floor. He immediately communicated this to the crews inside, and proceeded to stretch supply lines in order to establish a water supply. That's professionalism, based on good habits, that should be followed at every incident.

The engineer's/pump operator's standpipe kit

In order for the engineer/pump operator to complete many of the essential tasks associated with establishing a sustained water supply at a high-rise or standpipe-equipped building, he must be equipped with the proper tools. A specific tool kit should be preassembled and carried on the apparatus in a convenient location. Many of the most frequently needed tools can be assembled and carried in a single tool bag. This is referred to as the *engineer's/pump operator's standpipe kit*.

The following tools and equipment should be considered the essential items necessary for an engineer's/pump operator's standpipe kit:

1 – Lightweight canvas equipment bag

2 – Lightweight spanner wrenches

1 – Lightweight, 18-inch aluminum pipe wrench

1 – Long needle-nose pliers

1 – Long flathead screwdriver

1 - Knox key for unlocking Knox FDC caps

1 – Lightweight 2½-inch double male adapter

1 – Lightweight 2½-inch double female adapter

1 – 2½-inch Siamese, with clapper valves

1 – Lightweight 2½-inch cap

1 – Pair of leather work gloves

1 – Flashlight

4 – Sets of 12–foot tubular webbing with locking carabiners attached

4 – 2½-inch rubber gaskets

The Fire Department Connection (FDC)

For the engineer/pump operator, even the very best and most dedicated will likely not be able to remember the location of every high-rise and other standpipe-equipped building's fire department connection (FDC). The exceptional engine companies will compile a list of FDC locations in their first due response area, and keep this in the cab of the rig for reference. The engineer/

pump operator may or may not have an idea as to the location or locations of the FDC or FDCs at a particular building. Identifying the location of the FDC is step one. When pre-fire plan information is not available, and there is no past experience to fall back on, the engineer/pump operator must search for the location of the FDC. Generally, it is a good idea to start on the address side of the building near the entrance and work around the outside perimeter of the building in a clockwise rotation.

Firefighter Zach Bousman of DFD Engine Co. 23 developed an excellent reference book that includes FDC locations, and basic pre-fire plan information for all high-rise and standpipe-equipped buildings in Engine 23's first due response area. This book also includes an aerial photograph of the building or in many cases, the complex of several buildings (fig. 12–2).

Fortunately, at many high-rise and other standpipe-equipped buildings, there will be a visual and audible device (typically a strobe light and horn) located above the FDC to indicate its location (fig. 12–3). When the building is in alarm, these devices should be operating. In the absence of these indicating devices, the engineer/pump operator will have simply to walk around the perimeter and attempt to locate the FDC.

Fig. 12–2 DFD Engine Co. 23 uses a reference book that includes pre-fire information and an aerial photograph of all high-rise and standpipe equipped buildings in their district

Address
Denver Housing Authority (DHA)
20-STORY MD — FIRE-RESISTIVE — 200 x 75

Annunciator Panel	Fire Command Center (1st Fl.)
Basement	None
Best Hydrant	NW Corner
Second Best Hydrant	SW Corner
Fire Command Center	1st Floor around Corner from Elevators
DHA Key Box	Command Center
Elevator Service/Shutoff	Roof
FDC/Standpipe Connection	NW Corner
Hose Stretch	6 Lengths from Standpipe Outlet to Furthest Apartment on Floor Above
Standpipe Hose Valve Outlets	East and West End of Public Hallway
PRV's	Powhatan Factory Pre-Set
Knox Box	Outside Main Entrance
Roof Access	West Stairwell
Utilities	Main Floor East End

The engineer/pump operator should remember not to overlook areas remote from the building, such as entrances to the complex, or parking area, especially in multibuilding complexes where the same FDC might supply the standpipe for more than one building (fig. 12–4). Also, it is a good idea to look for other identifiers, such as fire hydrants, post indicator valves, wall indicator valves, or outside stem and yoke valves, any of which might be located close to the FDC.

Once the FDC is located, and prior to connecting supply lines to the FDC, the engineer/pump operator must first ensure that all trash, debris, and any other loose items are removed from inside the FDC (fig. 12–5). Some items will be difficult to reach and remove; therefore, a long tool may be needed, such as long needle-nose pliers, which is recommended equipment for the engineer's/pump operator's standpipe kit (fig. 12–6).

Engineers/pump operators must be very careful and protect themselves accordingly when probing into an FDC. Dirty pieces of broken glass can cut a member and lead to infection, but that's minor compared to the potential harm a needle stick from a drug user's disposed hypodermic needle could create. Use a flashlight (especially at

Fig. 12–3 Audible and visual warning devices (horn and strobe) located above the FDC to mark its location

Fig. 12–5 FDC should be inspected for trash and debris, which must be removed

Fig. 12–4 FDC located remote from the building

Fig. 12–6 Long needle-nose pliers can be used to safely remove many pieces of debris

night), gloved hands, and tools to probe into the FDC. Be careful not to push debris further into the FDC and past a clapper valve, as that could create a clogging problem within the system, inside the attack hoseline, or at the nozzle.

The clapper valves should also be inspected—check to see if they are operating properly and if they will seat properly upon pressurizing the system. If one of the two clapper valves is broken or not operating properly, but the other is operating properly, hook up to the side with the broken clapper valve first, allowing for the good clapper to do its job on the opposite side. This allows for one supply line to be attached and charged, supplying the standpipe system as quickly as possible, and then the engineer/pump operator can stretch and connect a second supply line (fig. 12–7).

In the event that both clapper valves are broken on a FDC, the engineer/pump operator can use a 2½-inch cap to plug one side and then hook up to the other. However, it is recommended that you attach supply lines to both sides from the outset, in order to maximize water supply to the system. If one side is capped and the other side is charged with one supply line, the capped side becomes useless, and a supply line cannot be attached unless the line supplying the opposite side is first shut down.

A better solution to the problem of a broken clapper valve is for the engineer/pump operator to attach a 2½-inch siamese, equipped with clapper valves to the side of the FDC that has the broken clapper valve. (A 2½-inch siamese equipped with clapper valves should be carried as standard equipment on the pumper and inside the engineer's/pump operator's standpipe kit). First, the clappered siamese will prevent water from discharging out the FDC. Second, and most important, it provides a second and third inlet to the FDC, which can greatly increase the overall water supply potential to a building's standpipe system.

Next, the female swivels should be checked to see if they are operating properly, and if the rubber gaskets are in place. I have found that more often than not, one or both of the female swivels is not operating properly, and is in fact frozen in place. In this situation, there are a couple of options. First, the pump operator can twist the supply line several times (approximately four to five times for couplings with national standard thread) before threading the male connection onto the FDC. Once the connection has been threaded on a few turns, the hose can be twisted back the other direction threading the hose onto the outlet completely.

A much more efficient method, which I prefer, would be to use fire department adapters that should be carried inside the engineer's/pump operator's standpipe kit. Specifically, a double male adapter can be threaded onto the frozen female swivel of the FDC. After that, a double female adapter should be threaded onto the double male, and once that's completed you have a properly operating female swivel to which the supply line can be attached (fig. 12–8).

Most buildings are equipped with at least two inlets on the FDC. However, in some situations, one of the inlets might be completely inoperable, clogged with trash and debris, have damaged threads, or perhaps be equipped with a cap that cannot be removed. Furthermore, some buildings might only have one inlet to begin with. In these situations, the sharp engineer/pump operator is proactive, and as previously mentioned, will utilize a siamese appliance from his pumper, one that is equipped with clapper valves, to create at least two inlets (fig. 12–9). That way, at least two supply

Fig. 12–7 A clappered siamese fire department connection (FDC) allows for one side to be charged with water quickly, before hooking up to the other side.

lines can be hooked up to the building's FDC, thus creating a water supply redundancy. Once again, it is a very good idea to routinely place a fire department clappered siamese on one inlet of the FDC in order to create a third inlet, allowing for an additional water supply redundancy from another pumper.

Apparatus placement

At most of our fireground operations, apparatus placement should typically favor the truck company, as the use of their aerial device generally rests heavily on their ability to get as close to the building as possible. Therefore, engine companies must exercise good discipline and *spot* their apparatus short, past, or across from the fire building, or anywhere out of the way so the aerial apparatus can utilize as much area as necessary to gain the best position for aerial use.

However, this rule should be slightly altered at fires in high-rise and standpipe-equipped buildings. The truck company aerial apparatus placement should still take priority, if appropriate, specifically if there is an obvious life safety issue, such as occupants in need of rescue from windows within reach of the aerial device. However, if there is not an obvious and immediate need for an aerial device, the engine company should then have a green light to place their pumper apparatus in the best position possible to supply the building's standpipe system. If a truck company needs to spot its apparatus in front of an FDC to accomplish a rescue, so be it. In this situation, the engine company will have to stretch supply lines around the truck company apparatus in order to supply the FDC, and the supply lines will unfortunately be a bit longer than what would be ideal.

There are several good rules to keep in mind when spotting a pumper at an FDC to supply a standpipe. First, make every attempt to spot the pumper as close to the FDC as possible,

Fig. 12–8 Double female adapter threaded onto the double male, thus creating a working female swivel for the FDC.

Fig. 12–9 Clappered siamese from the pumper used to create three inlets

preferably within 50 feet, or one length of hose. This ensures that the stretch of supply line to the FDC is as short as possible, thus minimizing the overall friction loss in supply lines and exposure to falling debris.

In addition, the discharge outlets that will be used to supply the standpipe system should be as remote as possible from the pump panel of the pumper apparatus, as well as the engineer/pump operator and any other members operating in this area. The typical pumper has a pump panel on the driver's side of the apparatus at street level, midway between the front and back of the rig. Therefore, when the apparatus is configured in this manner, using the discharge outlets on the opposite side of the rig to supply the FDC would be best.

The reason for this configuration is that the hoselines used to supply the standpipe system might be pressurized to extremely high pressures, especially when supplying the upper floors of a tall high-rise building, and therefore should be hooked up to the pumper in a location away from the engineer/pump operator. Should a section burst, or if a hose jacket separates from a coupling, or if any other connection fails for whatever reason, the engineer/pump operator and other operating members will be afforded the protection of distance, and the pumper apparatus serves as a barrier.

The location where the hose is attached to discharge outlets should be based on the location of the pump panel. Most pump panels are in the same location, with a few pumpers designed with pump panels located midship on the upper part of the rig, accessed by a walkway. Pump panels can also be found at the rear of the apparatus on some pumpers. Regardless of the location, engineers/pump operators should make every effort to attach hoselines supplying the standpipe system to discharge outlets as remote from the pump panel as possible. Apparatus placement from the outset of an event should be based on spotting the pumper so that the FDC is on the opposite side of the pumper from the pump panel.

Securing the supply lines

There is one other component regarding the protection of operating members from high pressure supply lines. In addition to attaching these hoselines to discharge outlets remote from the pump panel, they should also be secured in multiple locations prior to charging them with water. Specifically, the

Fig. 12–10 Hoselines supplying an FDC secured to a substantial object at the pumper, at the building, and in the middle.

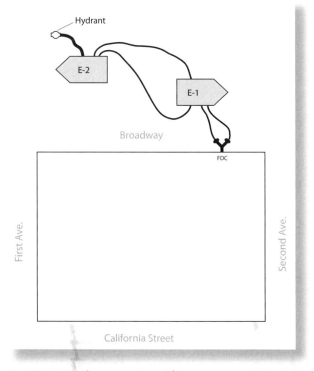

Fig. 12–11 Tandem pumping with two pumpers. E-1 at the FDC supplying the standpipe system with two supply lines and E-2 at the hydrant, supplying water to E-1.

supply lines should be secured with some sort of tether near the pumper near the building, and at the midpoint of the supply lines between the pumper and the building. The tether should be secured to the supply lines at these locations, with the other end secured to a substantial object, which could be the pumper, the building, or any number of other substantial objects, such as a parking meter, telephone pole, and so on (fig. 12–10).

When this is properly and proactively completed, if there is a burst section, hose separation, or connection failure, the hoseline will not whip around out of control, possibly striking an operating member, which could cause a serious or fatal injury. A supply of high test strength webbing and locking carabiners works well to secure the supply lines. This equipment should be kept in the engineer's/pump operator's standpipe kit.

General Pumping Considerations

With the hose layout to the FDC completed and secured, the engineer/pump operator from the first due engine company can prepare to receive water from the second due engine company engineer/pump operator located at the hydrant. I recommend that the overall primary water supply for a high-rise and or standpipe-equipped building be completed by using at least two separate fire department pumper apparatus, completing a short relay, called *tandem pumping*, from the source, through at least two pumps, and then into the building's standpipe system. This is the procedure adopted and used by many fire departments, including the DFD (fig. 12–11).

The reason for this is twofold: First, for operations at tall high-rise buildings, the pressure needed to properly supply the upper portions of a standpipe system will likely be very substantial. Sharing the workload of achieving these high pressures between at least two pumps increases operational safety and success. Second, because at least two pumps from two separate pumpers are used, there is a built in redundancy, providing a backup in case one of the pumpers has a mechanical failure.

Ideally, a good rule of thumb for supplying a standpipe system is to utilize one standard two-stage pumper rated to deliver 1,000, 1,250, or 1,500 gpm for every 10 flights of vertical elevation. Some fire departments, especially those in very large cities with mega-high-rise structures, have specified and purchased several high-pressure pumpers for water supply operations at these extremely large and tall high-rise buildings (fig. 12–12, and fig. 12–13).

Second due engine company

The second due engine company engineer/pump operator should spot his apparatus at an appropriate hydrant, preferably as close to the first due pumper apparatus as possible, and supply the first due pumper at the FDC with the appropriate pressure. Fire departments that operate at high-rise and standpipe-equipped buildings should specify pumper apparatus with at least two-stage pumps.

Fig. 12–12 High-pressure pumper assigned to FDNY Engine Co. 65 in midtown Manhattan

The pumper at the FDC should be in the pressure (impellers in series) position, and the pumper at the hydrant should be prepared to operate in volume (impellers parallel) or pressure (series). Generally, the pumper at the hydrant should begin in the volume position. If water has to be supplied to a significant vertical height (more than 10 flights), both pumpers should be in the pressure or series position. Therefore, water travels through at least four impellers before entering the standpipe system.

Third and fourth due engine companies

At confirmed working fires in high-rise and standpipe-equipped buildings, I recommend that two of the later-arriving engine companies be used to establish and maintain a separate, secondary fire department water supply. In essence, this is a separate pumping system to serve as a backup and redundancy to the overall primary fire department water supply operation. My preference is to utilize the third and fourth due engine companies for this operation; however, individual fire departments must determine what resources to use for this operation, based on the priorities of a given event.

Keep in mind, the officers and firefighters from these engine companies should be used for operations inside the building, as outlined in the chapter 11, with the engineers/pump operators and pumpers used for the secondary water supply. The primary goal here is to establish a secondary water supply as soon as possible at working fires that have not been brought under control quickly, so as to not have all the water supply eggs in one basket.

If there is a second FDC for the building, then the third and fourth due pumpers should supply the building via tandem pumping at that FDC in a similar manner as previously described (fig. 12–14). However, if there is not a second FDC, as a water supply redundancy and safety measure, the third and fourth due engine companies (pumper apparatus) should set up and prepare to supply the sole FDC with one additional (third) supply line. Also, an additional supply line (fourth) should

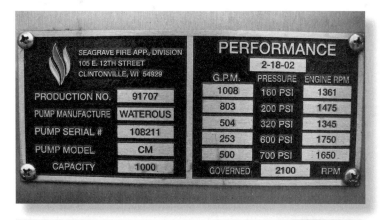

Fig. 12–13 The pump panel of FDNY Engine Co. 65's high-pressure pumper. Note the inset with gpm and pressure specifications

be stretched to an interior standpipe hose valve outlet, preferably located on the first floor (fig. 12–15).

In either situation, the third and fourth due pumpers should take water from a different hydrant than the one used by the second due pumper, preferably a hydrant that is being supplied by a separate water main. If there is a second, separate FDC, it should be supplied in the same manner as the first FDC. If there is only the one FDC, the third and fourth due pumpers should complete all hookups and charge all lines with water up to the pumper closest to the FDC. The third supply line going to the FDC, and the fourth supply line going to the interior standpipe hose valve outlet, should remain dry, and the engineer or pump operators of the third and fourth engine companies should stand fast and wait to supply the standpipe, but only if there is a failure in the first water supply system (the one that includes the first and second due pumpers). Also, the third and fourth due engineers/pump operators should leave their respective pumper apparatus out of pump gear until they are actually called upon to supply the standpipe system.

Bypassing the FDC

There is an additional method that may be used to supply some standpipe systems. This method can be used to create an additional redundancy or backup water supply, but it can and must be used in situations where the FDC is damaged and/or inoperable. This supply method involves using an interior standpipe hose valve outlet to *backfill* the standpipe system riser.

The procedure for this operation involves stretching a supply line into the building and to a convenient standpipe hose valve outlet, preferably on the first floor of the building. A minimum of one 2½-inch line can be used, but it is preferable to use at least one 3-inch supply line. A 3½-inch supply line,

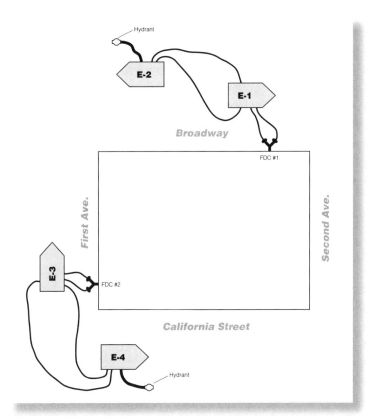

Fig. 12–14 For buildings with two FDCs, E-1 is at one FDC, supplied by E-2. E-3 is at a separate, second FDC, and is being supplied by E-4.

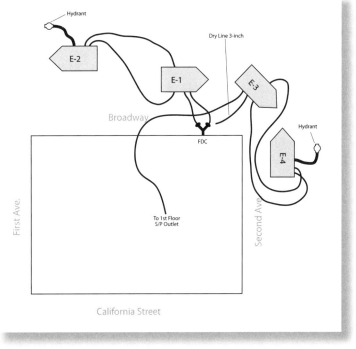

Fig. 12–15 For buildings with only one FDC, E-1 is at one FDC, supplied by E-2. E-3 is located near E-1, with a supply line stretched to the FDC, and a supply line stretched to the interior of the building, first floor standpipe outlet. E-3 is being supplied by E-4.

if available, would be even better. In fact, for all standpipe water supply operations, the preferred minimum size supply line is 3-inch.

The supply line is stretched up to the hose valve outlet and connected. Adapters will typically be needed, specifically a double female, and any necessary reducers (fig. 12–16). Ideally, if a 2½-inch siamese with clapper valves is available, and if there is enough room to connect it to the outlet, it should be used so that a second supply line can be attached (fig. 12–17). A 2½-inch, 60-degree elbow should be used as necessary to prevent kinking of the supply line feeding the interior standpipe hose valve outlet.

Once the supply line is connected, the line can be charged, and the hose valve outlet opened. With appropriate pressure, the standpipe riser can be supplied with water or backfilled, using this method. One word of caution, as I discussed in chapter 4, if the hose valve outlet is equipped with a pressure-reducing valve (PRV), the standpipe cannot be backfilled using this method. This is because the PRV is also a check valve, and, once the hose layout is pressurized, the valve will close, which will not allow water to pass by the valve and enter the standpipe riser.

Pumping into a standpipe system

Once the supply lines have been laid out and all connections made, the engineers/pump operators are ready to supply the system with water. However, this is not simply a matter of opening a few valves on the pump panel and letting it rip. Pumping water into a standpipe system, especially at high pressures, is an operation that should be completed carefully and with significant finesse.

Most standpipe systems are not designed to withstand extremely high pressures. In addition, we often times find ourselves pumping water into a standpipe that is very old, perhaps in marginal condition, and one that has not received the appropriate maintenance and testing over the years.

My good friend and mentor, Regis Donovan is an engineer (pump operator) assigned to DFD Engine Co. 12. Regis is a third-generation firefighter, with close to 30 years experience. He has operated at countless fires, and has been on the supply side of many a standpipe operation. Regis taught me years ago that a standpipe should not be automatically supplied at every fire or potential fire. Supplying the system is a conscious decision based on specific information and direct orders from the incident commander and/or division supervisors in charge of the fire attack.

Furthermore, Regis explains his operation to include a quick mathematical calculation of needed water pressure, followed by a slow introduction of water into the standpipe, and finally a very cautious increase in the pump discharge pressure. As Regis

Fig. 12–16 Some standpipe systems can be supplied by attaching to an interior hose valve outlet on a lower floor using adapters, including a double female.

Fig. 12–17 If room permits, ideally, a clappered siamese should be placed on the interior outlet to allow for two supply lines.

Donovan puts it, the pump operator should slowly "creep" the discharge pressure up to the required and appropriate level, constantly maintaining communication with units operating inside the building. Ultimately, if the standpipe system is overpressurized and something within the system breaks, the water supply problem becomes exponentially more serious.

Hydraulic calculations

Generally speaking, the hydraulic calculations necessary to properly supply a standpipe system involve the following:

1. The nozzle pressure and friction loss for hose and any appliances used for the attack line or attack lines inside the fire building
2. The head pressure due to elevation, approximately 5 psi per floor, or 0.5 psi per foot of elevation
3. The friction loss in the standpipe system itself; 25 psi is a generally accepted figure for this
4. The friction loss in the supply lines from the pumper to the FDC, which is generally very small and negligible (when using short supply lines, preferably one length or 50 feet, and supply lines at least 3-inch in diameter).

Upon arrival at most high-rise fires, it is not always clear at the outset what the exact location of the fire is, specifically the precise fire floor location. Ongoing size-up will help clarify this; however, initially the engineer/pump operator might not know the exact location that he should be pumping to. Therefore, when the exact location or fire floor is not immediately known, engineers/pump operators should be taught to pump to an approximate location. Specifically, engineers/pump operators should pump water to an approximate range of floors, estimating the fire floor location, give or take two floors.

A good rule of thumb, which is fast and relatively accurate, involves using the figure 100, plus 5 per floor, for every floor up to the fire floor. For example, a reported fire on the twenty-fifth floor would require 100 psi, plus an additional 120 psi for the 24 floors of elevation, for a total pump discharge pressure of 220 psi.

I developed a standpipe pump chart used by the DFD for quick reference that pump operators can utilize when supplying standpipe systems (fig. 12–18). It is based on the utilization of 2½-inch hose for attack line, and low-pressure, 50-psi smooth-bore nozzles. This chart goes up to 65 floors, which coincides with the tallest buildings in the city of Denver.

Protecting supply lines and engineers/pump operators

Key to the water supply success at any significant high-rise fire event is to properly and proactively protect the supply lines going into the building's FDC, and, most importantly, to protect the engineers/pump operators responsible for the delivery of that water. This protection is from a wide range of falling debris, particularly broken glass, sometimes very large pieces, from building windows. When this debris rains down from above, it

DENVER FIRE DEPARTMENT
FRICTION LOSS PUMP CHART
HIGH-RISE STANDPIPE OPERATIONS

150' 2-1/2" Hose 1-1/8" Smooth Bore Tip

FLOORS	PUMP PRESSURE
1 to 5	125 PSI
6 to 10	150 PSI
11 to 15	175 PSI
16 to 20	200 PSI
21 to 25	225 PSI
26 to 30	250 PSI
31 to 35	275 PSI
36 to 40	300 PSI
41 to 45	325 PSI
46 to 50	350 PSI
51 to 55	375 PSI
56 to 60	400 PSI
61 to 65	425 PSI

Fig. 12–18 Quick reference pump chart used by the DFD to supply standpipe systems in high-rise buildings

is very powerful and represents an extreme hazard to operating members below. It doesn't take much glass to injure or kill an operating member or to sever a supply hoseline.

Keep in mind that falling debris is not just limited to small, medium, and large pieces of window glass. Any number of different items can create a hazard for firefighters and equipment in the street below. That, unfortunately, may include human bodies, which represent a very massive and dangerous falling object. Of the 343 firefighters, that we lost on 9/11 some were struck and killed by falling debris, including human bodies. In fact, some firefighters were struck and killed when they were walking up to the building entrance on their way to operate inside, before the towers collapsed.

The events of 9/11 were, of course, very extreme, and although we must remain vigilant and prepare for a similar event, we hope nothing like it will ever occur again. However, falling debris at several serious high-rise fires and some relatively minor high-rise fires has caused significant injury to firefighters and severed countless hoselines in the past. Operationally, we must include methods to provide as much protection as possible for engineers/pump operators and supply hose.

Success and safety here revolves around implementation of protective measures. To begin with, spotting the pumper apparatus as close to the FDC as possible means that there will be shorter supply lines, and, thus, less overall exposed surface area of the hose. Once the supply is underway, members should be assigned to cover the supply lines with something substantial in an effort to protect the hoselines. Substantial objects could be many things, but specifically EMS backboards and or ground ladders with salvage covers laid on top of them are often times immediately available and can provide a fair amount of protection.

Members assigned to cover the supply lines might find the assignment unnecessary at the outset of an event before debris begins to fall. Once again, protection should be initiated as early as possible, because completing such an operation after the debris starts to fall is extremely dangerous, and members could easily be injured or killed. With limited resources at the outset of an event, and perhaps throughout the duration of a serious high-rise fire, it might be difficult to provide additional resources to complete this operation. Therefore, the engineers/pump operators may have to complete this initial "first aid" hose protection themselves, after a sustained water supply into the buildings standpipe system has been initiated.

The protection of operating members, specifically engineers/pump operators, will require a much more substantial barrier than that which might work for the supply lines. As problem solvers, operating members should utilize whatever means is necessary to create a barrier to protect engineers/pump operators from falling debris. This can include placing engineers/pump operators inside the enclosed cab of the pumper apparatus as necessary and appropriate. However, this will be difficult and impractical, as the engineers or pump operators will need to be close to the pump panel to monitor to pump and ensure that water is being properly supplied to the standpipe system. Nevertheless, after charging the system, and setting the pressure relief valve, the engineer/pump operator may have to leave the pump panel for emergency shelter. The DFD has included in our high-rise procedures the utilization of our collapse rescue unit as a resource at serious high-rise fire events. This unit comes with a significant supply of lumber and large sheets of plywood, which can be used to create a protective barrier. The 4-foot-by-8-foot plywood sheets can be used immediately to cover significant portions of the of the supply line. Six sheets would work to almost completely cover a short stretch of supply line from the pumper to the building's FDC.

A much more elaborate and substantial barrier could be assembled at a base location, away from the building and the hazard of falling debris. Once assembled, it can be brought up to the point of operation, in smaller pieces if necessary, and quickly placed in service. There are plenty of talented carpenters on any fire department who could quickly frame up a small protective structure to be used as a barrier for an engineer/pump operator in harm's way.

Summary

It is the responsibility of the engine company to ensure that there is a sustained water supply at all structural fire events. That includes events that occur in high-rise and standpipe-equipped buildings. Water supply at these type buildings is much more complicated to establish and maintain, but it is absolutely essential for overall fireground operational success and safety. The procedures and recommendations made in this chapter will help achieve that operational success and safety.

Chapter 13

Truck Company Operations

Up to this point, I have focused a significant amount of attention on engine company operations. Most of my fire service friends and colleagues would probably consider me to be more of an engine company guy than a truck company guy. That is likely due to the fact that most of the teaching, mentoring, and training that I provide revolves significantly around the topic of engine company operations.

However, I am also a tireless advocate of good, proactive *truck company* fireground operations. In fact, I spent the first part of my career working almost exclusively on truck companies, and on the Denver Fire Department's (DFD) one rescue company. In fact, my first permanent assignment to an engine company occurred only after I was promoted to the rank of captain. Because of that truck and rescue company experience as a young firefighter, I recognized early in my career the value and importance of the fireground support functions that are the responsibility of the truck company.

A good-natured competition between engine company and truck company personnel has existed for many years. Firefighters assigned to double company firehouses can attest to the daily banter and practical jokes executed between the engine and truck guys. The central theme revolves around who really has the more important job, with members taking any and every opportunity to down play the significance of the other company's job.

In reality, all kidding aside, members of both companies truly respect and appreciate the job one another does to achieve the overall fireground mission. I believe that the very best firefighters have spent significant time assigned to *both* engine and truck companies, gaining experience performing both of these critical fireground functions. The engine company members clearly recognize the value of a good truck company when they are making a tough push inside a dangerous building, and, remarkably, conditions quickly improve due to aggressive topside ventilation. The truck company members recognize the value of their engine company counterparts, as they are protected by the engine company during their support operations, allowing them to complete aggressive searches above and close to the fire area.

A good fire department emphasizes the importance of both engine and truck company fireground operations. Value should be placed on both of these critical fireground functions, with the coordinated teamwork between the two being the most important factor for fireground safety and success. Ultimately, neither is more important than the other, but rather, there is a synergy that occurs when the truck company supports the engine company, thus leading to a safe and successful operation.

In essence, the truck company performs and provides support to the engine company. Without that support, engine company functions become much more difficult, if not impossible to accomplish. When we say, "the fire goes as the first handline goes," we must include the fact that the first handline, in many situations, isn't going anywhere without critical truck company support.

In a nutshell, a good truck company provides the following support:

1. Gets the engine company into the fire building
2. Locates the seat of the fire
3. Finds out where the fire is going and takes action to stop it
4. Provides access for the engine company to the seat of the fire
5. Directs the engine company to the seat to the fire
6. Initiates an aggressive primary search
7. Establishes secondary egress when and where possible
8. Conducts an aggressive check for extension of fire and smoke
9. Provides appropriate ventilation when and where possible
10. Controls the utilities
11. Conducts secondary and final searches

In this chapter, I will be using the term *truck company*. Some fire departments use the term *ladder company*, but generally, the functions performed are the same, regardless of what the company is called. It is also very important to acknowledge that many fire departments, due primarily to budgetary limitations, have a limited number of truck companies, and many more fire departments have no dedicated truck companies at all. I point this out so that those readers who work in organizations without truck companies will simply consider the information in this chapter as *fireground support functions*.

Fig. 13–1 Fire departments can effectively utilize a wide range of ladders at low-rise building fires.

For most fireground operations, it isn't a vehicle with an aerial ladder mounted on top that is most important, but rather, the personnel (firefighters) who perform the various critically important fireground support functions. So, focus on the completion of these various support functions, regardless of where the manpower to accomplish it comes from, and, if appropriate, simply substitute the term *fireground support function* when I use the term *truck company*.

Basic Truck Company Functions

The previous list of truck company fireground support functions is a very brief summary of the support functions that are necessary to help an engine company achieve fireground success. The truck company has, of course, a general list of broad functions that must be completed at most fireground operations. Unlike the engine company, the truck company functions do not consistently occur in the same order. In other words, where the engine company establishes a water supply, selects a weapon, stretches and advances a line to locate, confine, and extinguish the fire, almost always in that order, the many critical fireground support functions are not always completed in the same order, as the priority of each is based on the circumstances of a given fire.

For example, arriving at a serious fire in a low-rise multiple dwelling, with several occupants showing at windows, and heavy smoke pushing from the rooms behind them, exterior rescue via ground, aerial, or tower ladder becomes the obvious and immediate priority. Whereas, a fire in a one-story commercial building late at night more than likely will require immediate forcible entry support from the truck company in order to get the engine company into the building to initiate attack.

In his book, *The Fire Officer's Handbook of Tactics*, Chief John Norman gives us a convenient acronym that lists seven of the most important truck company fireground support functions. That acronym is L O V E R S U.[1] Those support functions are as follows:

L – Laddering
O – Overhaul
V – Ventilation
E – Entry (Forcible Entry)
R – Rescue (Search and Rescue)
S – Salvage
U – Utilities

In this chapter, I will discuss each of these critical truck company fireground support functions and how each relates to our operations in high-rise and multistory low-rise, standpipe-equipped buildings. I also expound on some of these topics in greater detail in subsequent chapters.

Laddering

Our ability to utilize ladders on the fireground is contingent on the vertical height of the building, the length of the ladder, and the location of our objective. At low-rise buildings, we can potentially be quite effective with ladders (fig. 13–1). Good discipline, with proper apparatus placement from the outset, is critical to achieve successful and effective use of ladders.

Numerous barriers can prevent our ability to spot aerial apparatus in the best position: parked cars, including in some cases, police cars; overhead electrical wires, trees, and building setbacks, just to name a few. Ultimately, we have a finite amount of ladder available to us, usually in the 100-foot range. A steady diet of training, coupled with good apparatus placement, can yield positive results when using ladders at low-rise buildings (fig. 13–2).

As for high-rises, if the fire is located on the lower floors of the building, aerial ladders can be a valuable tool (fig. 13–3). We must

Fig. 13–2 Truck companies should regularly train on laddering operations at buildings in their first due response areas.

Fig. 13–3 Truck companies performing horizontal ventilation for a lower floor fire at this high-rise multiple dwelling.

always anticipate and prepare for potential use of aerial ladders, even at a true high-rise. However, once it has been determined that the fire is well above the reach of our aerial devices, most of these vehicles should be left at a base location a safe distance away from the fire building. That is, unless there is specific equipment located on the rig that is too heavy and/or too large to transport up to the fire building from base. In this case, the aerial apparatus (rig) will need to be located as near to the fire building as is safely possible.

Overhaul

When I first came into the fire service 25 years ago, I went through the fire academy and was taught something called *salvage and overhaul*. I didn't realize it at the time, but this class really didn't provide a comprehensive and thorough understanding of the topic of overhaul. In his book, John Norman was the first one that I know of who really gave an appropriate and precise explanation of overhaul. Rather that putting it on the back burner and considering it just an extension of salvage, Norman broke the overhaul operation down into two, very important and distinct categories.

Precontrol overhaul. The most important of the two overhauls, *precontrol overhaul*, is truly a critical fireground support function that, if done properly and aggressively, will ensure fireground success. Conversely, when it's not done, fireground failure is almost certain. Precontrol overhaul is a fancy way to describe our actions related to checking for and stopping the uncontrolled extension of fire. We must remember that every fire is a six-sided, multidimensional object that will travel in all six directions if left unchecked. Because of the significant exposure concern at a typical high-rise fire, the potential for extension of fire and smoke is tremendous, especially when the fire occurs on a lower floor.

Commercial occupancies. For operations at commercial high-rise buildings, extension of heat, fire, and smoke can occur via several different means. One of the first procedures that should be undertaken by the truck company personnel is to expose the plenum space on the fire floor, by pushing up or pulling down one or more drop ceiling tiles (fig. 13–4). When possible, this should take place at the entrance to the fire floor, just inside the floor from the stairwell, which will often yield information to help us determine whether there is any horizontal extension on the fire floor.

In addition, truck companies must be very diligent and proactive in checking for vertical fire extension in the commercial high-rise building. Focus on utility areas, such as vertical pipe chases, especially areas of poke-through construction, and specifically when these poke-throughs have not been appropriately sealed with a fire stopping material (fig. 13–5).

Fig. 13–4 At the entrance to the fire floor, truck companies should proactively open the plenum space to check for horizontal fire extension.

Keep in mind that any of the numerous vertical shafts, such as elevator hoist-ways, stairwells, especially open stairs in older buildings, along with electrical and plumbing utility shafts, and HVAC ducts, can all be convenient avenues of vertical fire extension and smoke spread. At the 1988 First Interstate Bank Building fire in Los Angeles, several hours after the main body of fire had been brought under control, fire companies found that fire had extended up and into a storeroom on the 27th floor, more than 10 floors above the last floor of heavy fire involvement.[2] This fire was caused by vertical extension, and serves as an excellent example and reminder of just how severe the fire extension problem, especially vertical fire extension, can be in a large high-rise building.

Like our search operations, checking for extension should begin on the fire floor, looking for horizontal spread, followed by the floor above looking for vertical spread, and then the top floor. Don't overlook the perimeter areas where auto exposure can take place via an unprotected curtain wall gap. Everything between the floor above and the top floor will eventually have to be thoroughly searched for any victims and any extension of fire. Any areas of concern will have to be opened up and checked; and that is truly the essence of precontrol overhaul.

Residential occupancies. For operations in residential high-rise buildings, like commercial buildings, there are several avenues by which heat, fire, and smoke can extend. Residential high-rises don't typically have a drop ceiling and plenum space in the public hallway, but, more often, a concrete, compartmentalized construction, generally lessening the severity of horizontal extension. However, in order to limit construction costs, areas such as kitchens and bathrooms are typically stacked vertically. Therefore, the utility shafts for plumbing and other utilities run the entire vertical distance of the building, creating numerous convenient areas for vertical extension (fig. 13–6).

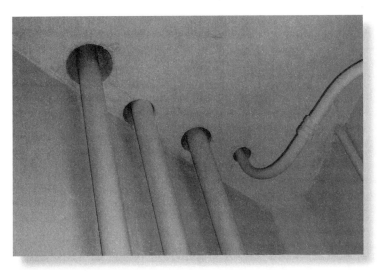

Fig. 13–5 Unprotected areas of poke-through construction can be avenues of vertical fire extension in high-rise buildings.

Precontrol overhaul operating procedures at residential high-rise fires, should begin with the fire apartment, and then any exposure apartments located next to the fire apartment. If fire extended into the public hallway, extension into other apartments is possible and must be checked. Next, the apartment on the floor above that is located directly above the fire apartment should be checked. From there, the top floor should be checked.

In addition, because windows in a residential high-rise are not typically as strong as windows in a commercial high-rise, and because they can be

Fig. 13–6 The numerous vertical arteries conducive to fire extension are clearly indicated by the vast number of vents on the roof of this low-rise multiple dwelling.

opened—in fact some will be open—unobstructed extension via auto exposure is a significant concern. One or more floors above the fire floor may have to be checked for extension of fire via auto exposure. Remember, we're not just concerned about fire extension here, but also the extension of deadly smoke into one or more apartments above the fire floor that could easily occur through open windows. This vertical migration of deadly smoke has caused many injuries and deaths in high-rise buildings, oftentimes occurring quite remotely from the fire floor. Even if there are little or no signs of fire extension to the floor(s) above, aggressive searches should be completed, due to possible smoke extension.

Post-control overhaul. This is the overhaul that is typically associated with salvage operations. Just as the name states, this is a post-control operation, as are salvage procedures, in most circumstances. Once again, we are most concerned with the precontrol procedures, proactively identifying areas of fire extension and stopping it. This post-control phase is designed to remove anything that could cause a rekindle, while working to salvage any items of value.

Ventilation

Ventilation is another one of the critical fireground support functions performed by the truck company. Basic ventilation involves the removal of heat, smoke, and other nasty products of combustion from the fire building. However, this is much easier said than done in large, low-rise, standpipe-equipped buildings, and especially difficult in tall high-rises. There are many aspects associated with air movement and ventilation inside these large, tall, and complicated buildings. The topic of ventilation, along with air movement is addressed in chapter 15.

Forcible entry/exit

The truck company fireground function of forcible entry is another critical support component. Providing access to the fire area for the engine company, opening up and accessing floors, rooms, and other areas in order to search for human life and check for fire extension, often begins with forcible entry. It is also critically important to keep in mind that getting in is sometimes easier than getting out. Forcible entry should and must include forcible exit. This can include forcible exit from a stalled elevator car, out onto the roof from a stairwell, or out of a large mechanical room.

Commercial occupancies. Generally, the forcible entry and exit problems will be more complicated at commercial occupancies than residential occupancies. Truck companies should develop and implement tool assignments that are occupancy specific. Operations in commercial occupancies should include basic hand tools, including a set of irons (Halligan and striking tool), but also more powerful tools, such as gasoline-powered rotary saws with metal-cutting blades.

Many commercial buildings, especially high-rises, have on-site security 24 hours a day. Security personnel will often be available to open entrance doors, provide additional keys, and support our operations. However, in the absence of on-site security personnel, preplanning these buildings

Fig. 13–7 Several sets of interior master keys are often contained inside a large lock box located inside the fire command center of a high-rise building.

and identifying the location or locations of key boxes, such as an exterior *Knox box*, is essential to operational success. Many of these large commercial buildings have multiple sets of master keys, usually contained in a large lock box located inside the fire command center (fig. 13–7).

Master keys are a tremendous asset, and will greatly increase the efficiency and speed of our operations. Unfortunately, most buildings do not provide enough sets of master keys, at least not enough for major emergencies such as fires, where multiple companies will be performing various different but interrelated operations simultaneously and remote from one another. When there are a limited number of master keys, they should be distributed carefully, focusing on getting at least one set of master keys to each major operational division. For example, a set of keys to the fire attack group, a set of keys to the upper search and evacuation group, and a set of keys to the lobby control unit/logistics section.

Even with the benefit of master keys, there will likely still be many secure and inaccessible areas inside a typical large commercial high-rise building. Truck companies will need to be well-equipped with a wide range of forcible entry tools and techniques in order to access all areas of the building.

Keep in mind, many high-rise buildings, particularly modern commercial high-rises, have electronic locking mechanisms on many doors, such as stairwell doors. These devices allow for the doors to be unlocked via a switch, or some unlock automatically due to activation of the fire alarm or specific zones of the fire alarm system (fig. 13–8).

Residential occupancies.
In residential occupancies, the forcible entry problem is generally less complicated, due primarily to unsecured and open public areas, such as the public hallway. Access to the building at the main entrance can sometimes be accomplished using master keys from an exterior Knox box. Generally, if there is a working fire in the building, some occupants

Fig. 13–8 This switch, located in the fire command center of a large commercial high-rise, unlocks all of the stairwell doors when activated.

will likely be at the lobby level and will open any locked entrance doors for arriving units. Many of the upscale buildings have on-site security, while others may have an onsite manager. In addition, many residential occupancies have an intercom system where individual apartments can be called from the foyer at the lobby level. After identifying ourselves as the fire department, most occupants will "buzz you in" to the building.

Once fire companies have made access to the building, the next barrier typically encountered will be the stairwell door leading to the fire floor. These doors may or may not have locks on them. Generally, locked stairwell doors are more common in commercial high-rises than residential. Good pre-fire planning will help identify these barriers, allowing us to obtain master keys and identify forcible entry needs ahead of time.

After gaining access through the stairwell door onto a fire floor, the next forcible entry problem will likely be the door to individual apartment units. Once again, many buildings will have master keys that provide access to all floors and apartment units. Depending on the building, the door to the apartment may or may not be heavily fortified. However, with the exception of public housing and buildings located in high crime areas, most apartment doors are not that heavily fortified and forcible entry is general not that complicated.

Tools should include a set of irons, and a hydraulic forcible entry tool, such as the rabbit tool or Hydra-Ram (fig. 13–9). These powerful hydraulic tools have significantly increased our operational efficiency, effectiveness, and speed on the fireground, especially in multiple dwelling occupancies. Keep in mind, it is critical that when forcible entry takes place, specifically on apartment doors, the integrity of the door must be maintained. This will allow operating units to close the door and cut off any extension of heat, fire, and/or smoke. If we are going to accomplish a successful *protect in place* strategy with any occupants, we must be able to close the door to their respective apartment in order to protect them. We will not be able to do this if we compromise the integrity of the door because of sloppy forcible entry procedures.

Fig. 13–9 Forcible entry tools for residential high-rise operations should include a set of irons, and a portable hydraulic forcible entry tool, such as a rabbit tool or Hydra-Ram.

Search and rescue

Life safety and the rescue of any savable human life is our number one priority on the fireground. In order to accomplish this, an aggressive search must be conducted. Like all of our basic fireground functions, search operations include both a primary search and a secondary (redundant) search. This is another topic that requires specific and comprehensive attention, so I devote chapter 14 to search and rescue.

Salvage

For most of our operations, specifically the more frequent ones that occur in smaller buildings such as single-family dwellings, performing salvage is usually completed in the post-control phase. As a chief officer, I continually emphasize that concerns of water damage and salvage operations are all post-control issues and should not be on the minds of firefighters while they're working to control a serious fire. However, for fireground operations in large standpipe-equipped buildings, especially high-rises, there are specific aspects of salvage that must be proactively completed in the precontrol phase. Specifically, there may be a need to aggressively protect critical building utilities, particularly the electrical service, from collateral damage caused by the runoff of water being used to fight the fire (fig. 13–10).

Fig. 13–10 Collateral damage caused by thousands of gallons of runoff water must be addressed by proactive salvage to protect critical areas, such as electrical vaults and rooms.

The loss of building utilities, specifically electrical service, could seriously cripple our operation. It's imperative that the location of electrical rooms and vaults, specifically the ones below the fire floor, be identified and protected from water damage. There will already be a massive commitment of resources to numerous critical fireground functions. Most fire departments will be stretched thin, with little or no resources to spare. However, every effort must be made to assemble a team, preferably a *Salvage Group*, to identify and protect critical building utilities locations, with an emphasis on the areas nearest to and immediately below the fire floor first.

Members will have to be creative, using any and all fire department equipment and building resources available to protect these areas from water damage. Certainly a large supply of salvage covers will be beneficial, but much of the work will have to be completed using various items found in the building, and by assembling makeshift barriers to stop water and/or divert it away from areas susceptible to water damage.

This is not going to be a popular assignment, but keep in mind that most of the operations at a serious high-rise fire will fall into the category of *support*. Certainly the fireground functions related to life safety will take precedence. However, assembling a team to form a salvage group, in order to protect and maintain the utilities powering critical building systems is of the utmost importance. Maintaining electrical power to most areas of the building, including power to elevators as well as public address and smoke control systems, gives operating forces a significant, long-term advantage during the battle.

Utilities

The utilities in a typical high-rise building include electric and water. Also, some buildings, particularly those that are located in climates subject to cold weather, may also have natural gas and/or steam for heating. Controlling the utilities at a large high-rise building is a complicated operation that must be completed with considerable finesse. This is not just a simple matter of having one member flip the main shutoff on the electrical box and rotate a valve on a gas meter a quarter turn during a 360-degree recon of a fire building.

The main issue with utilities control during fireground operations at high-rise buildings and other large low-rise and standpipe-equipped buildings is that we will need the utilities in order to operate. Specifically, we need to at least partially maintain the electrical service, especially in those areas that are not affected by the fire. This is of critical importance so that we can operate and use elevators (if it's safe to do so), have lighting to see, and keep critical building components like fire and life safety systems operating.

Utilities control should focus on the areas that are involved in fire, immediately threatened by the fire, and/or threatened by potential collateral water damage. In most large buildings, certainly most high-rises, there will typically be electrical rooms on each floor that will allow for isolation of all or portions of that floor's utilities (13–11). Ultimately, we will generally have the ability to at least selectively control the electrical power to specific areas of the building, as appropriate. In

Fig. 13–11 Many large high-rise buildings have electrical rooms on each floor that allow for selective control of the electrical utility.

most situations, the electrical utility will be the first priority with regard to control.

For those buildings that have natural gas, this too is a utility that must be controlled. Particularly if the natural gas supply travels through an area that is involved in fire, or one that has the potential to become involved with fire (fig. 13–12). However, once again, control must be done with finesse, and every attempt must be made to specifically isolate the natural gas supply and maintain it as necessary, for such things as emergency utilities such as an emergency electric generator powered by natural gas.

Fig. 13–12 This low-rise multiple dwelling has a natural gas supply line running along the ceiling in the public hallway. Control of this utility is critical, especially if fire conditions extend into this hallway.

Summary

Numerous support functions must be completed at fires in large, standpipe-equipped and high-rise buildings. These support functions are often completed by firefighters from truck companies. Everything from forcible entry to search and rescue fall into the category of support functions. Most of the work that must be completed in order to bring a high-rise fire under control will not be directly related to fire suppression, but, rather, to the many functions that support the overall fire control and rescue effort. Regardless of where these resources come from, they must be requested early, in a proactive manner, in order to establish and maintain the support necessary to achieve operational safety and success.

References

1. Norman, John, 2005, *Fire Officer's Handbook of Tactics*, *Third Edition*, PennWell Corporation, p. 135.

2. United States Fire Administration (USFA), Technical Report Series, First Interstate Bank Building Fire, Los Angeles, California, 1988, p. 6.

CHAPTER 14

Search and Rescue Operations

In chapter 13, I discussed how the truck company is responsible for accomplishing a long list of fireground support functions. Each and every one of those is extremely important in order to achieve the overall mission in a successful manner. However, clearly, the most important of all fireground functions is that of *rescue*.

Once again, the size, coupled with the significant time and distance factors associated with large standpipe-equipped, low and high-rise buildings, makes every fireground operation in these buildings much more difficult to accomplish. This difficultly is especially true with search and rescue operations.

Compared to our daily operations in smaller, less complicated buildings, searching the massive areas of a large high-rise building and rescuing one or several occupants is a monumental task. Various factors, especially complications created by logistics and exposures, contribute to our search and rescue problem. A large low-rise building might have several hundred occupants, and a large high-rise building might have several thousand. A serious fire in any of these buildings, especially on a lower floor, with deadly smoke rapidly extending laterally and upward and contaminating large portions of the building, is a critical problem. Where do we begin to address and prioritize the overwhelming search, rescue, removal, and/or evacuation of numerous occupants?

Identifying Who Needs to be Rescued

Upon initiating search and rescue operations in these large, complicated buildings, we must first realize that the rescue, removal, and/or evacuation of hundreds, perhaps thousands of people, is quite simply, in most situations going to be impossible. Clearly, there will be those who are obviously in critically dangerous positions, and must truly be rescued immediately. However, many occupants might not have to be rescued. In fact, moving them might actually place them in greater danger than if they were left alone. Therefore, as part of a good, comprehensive search and rescue plan, we must include several different strategic options when dealing with building occupants.

First, when developing a rescue plan, we must identify exactly who needs to be rescued and who doesn't. At most fireground operations in large low-rise and high-rise buildings, there will be a significant number of occupants. Of those occupants, many will be unaffected by the fire or smoke and are perfectly safe if they simply remain in place.

Protect in place

As part of our operations, we must first attempt to identify who can be *protected in place*, and then make every attempt to communicate with them. Most modern high-rise buildings have public address (PA) systems that allow for communications throughout the building. Many of these systems are located inside the fire command center, and a systems officer or his designee should be placed in charge of making frequent announcements in order to communicate with occupants.

Another strategic communications option that is likely available in most communities is *Reverse 9-1-1*. Modern technology has given us many tools, and one of those allows for the 9-1-1 emergency telephone communication system to actually be used in reverse. That is, occupants of a high-rise building, or any building for that matter, could be contacted by 9-1-1 operators who could actually give instructions on evacuation or advise occupants to remain in place. Reverse 9-1-1 has worked very effectively in many communities faced with a wide range of emergency events. It could certainly be applied to our fireground operations, especially those that occur in large high-rise buildings. Ultimately, using reverse 9-1-1 could significantly enhance our ability to communicate with occupants of a fire building, thus making for a much smoother overall operation.

Unfortunately, the events of September 11, 2001, have changed how occupants of a tall high-rise building might actually respond to directions asking them to remain in place. After the first tower of the World Trade Center (WTC Tower 2) was hit on 9/11, occupants of the other tower (WTC Tower 1), at that time unaffected, were told to stay in place. These well-intentioned instructions were given in order to protect these occupants and keep them out of harm's way in an attempt to protect them from the falling debris caused by the damage and fire in Tower 2, and to keep from adding thousands more people to an already congested evacuation that was culminating at street level. At the time of these instructions, it was unknown to building security (or any of us) that the incident at WTC Tower 2 was actually a terrorist attack, and that WTC Tower 1 would be attacked shortly.

Ultimately, many occupants might not follow instructions to remain in place. With that known, we must prepare for it; nevertheless, when a protect in place strategy can be used, we should attempt to use it. For example, when a fire occurs on the upper floors of a tall high-rise building, at least initially, many of the occupants on the floors below the fire floor can and should be protected in place. A good rule of thumb is to protect in place those occupants that are more than three floors below the fire floor. This rule is most applicable for operations in commercial occupancies.

Once again, this is an *initial* procedure, and is designed to minimize the congestion in stairwells, making it easier to get operating forces aloft, and providing room to establish staging and rehab areas. Keep in mind, for serious fires that are not brought under control in a reasonable period of time, and specifically where the structural integrity of the building might become compromised, we must then begin the evacuation of those who we were initially protecting in place.

When occupants are located above the fire floor, protecting in place is obviously going to be much more difficult to accomplish. Generally, it will be easier in residential occupancies than commercial occupancies to protect occupants in place who are located above the fire floor because of the compartmentalization typical of residential occupancies. As for commercial occupancies, because of a lack of compartmentalization along with central heating, ventilation, and air conditioning (HVAC), smoke and toxic products of combustion may more easily migrate to locations above the fire floor. However, many modern buildings do have built-in air management/smoke control systems that might work to isolate the smoke and protect occupants, thus, potentially allowing us to protect a greater number of occupants in place.

The three floor rule

Generally, in commercial high-rise buildings, we can attempt to protect in place those occupants who are more than three floors above the fire floor. I prefer to use the *three floor rule*, removing those who are three above and three below, and protecting the rest in place, as a good general rule. However, in addition to the floors immediately above and below, the three floor rule also includes the occupants located on the top three floors. So, removal should include three below, three above, and the top three, for a total of nine floors, initially.

For residential occupancies, we can attempt to protect in place many more occupants, including, in many situations, those on the floor above the fire floor, but only after a thorough evaluation by members assigned to the floor above. If fire has extended to any apartments on the floor above, or if there is a heavy smoke condition, we will have to reevaluate the protect in place strategy.

Evacuation

For those occupants that cannot be safely protected in place, and those that must be removed for a variety of reasons, evacuation is the next strategic option. The evacuation of numerous occupants from a large high-rise building is once again going to be a very difficult logistical operation. We must attempt to control any evacuation, including limiting the number of occupants to be evacuated, when and if possible, and giving them specific directions. As previously stated, many occupants might simply disregard our directions, but we must make every attempt to coordinate a safe and successful evacuation.

Once again, when available, a public address (PA) system and/or a reverse 9-1-1 can and should be used to direct the evacuation of occupants. Of utmost importance is the need to direct occupants to the evacuation stairwell that must be established by initial attack companies immediately after establishing an attack stairwell. The evacuation of occupants should only require direction from fire department personnel, with little or no hands-on assistance. For those occupants who do need hands-on assistance, that falls under the category of a removal or rescue.

Removal

Some occupants, such as the elderly, physically disabled, and those who are injured or otherwise nonambulatory, will have to be removed, assisted by firefighters. For example, a person confined to a wheelchair will hopefully be able to reach a refuge area, usually with the assistance of a designated buddy (usually only in commercial buildings). From this location, such as inside an evacuation stairwell on a floor landing, firefighters will have to remove this person to a safe location at least four floors below the fire floor.

Rescue

The most extreme strategic option is that of actually rescuing occupants. Situations that require a true rescue include those where occupants are located within an *immediately dangerous to health and life* (IDLH) atmosphere, including heavy smoke or high heat. Occupants may be disoriented, close to being overcome by smoke, or completely unconscious. Depending on the size and weight of the occupant in need of rescue, especially those who are unconscious, at least two firefighters, and in most situations, one or more companies of firefighters, will be needed to effect the rescue. This represents an extreme situation, where we must have the very best firefighters who are in top physical condition and mentally prepared in order to successfully complete the rescue.

Prioritizing Search Areas

Just like all of our other fireground functions, search operations must be prioritized in order to make the problem more manageable. That priority, at least for multistory low-rise and high-rise buildings, is fairly clear cut. Certainly, any specific reports of persons in distress or obvious visual size-up information regarding occupants in immediate need of rescue should be addressed first.

Almost all serious fires in low- and high-rise buildings will include many of what I refer to as "small brush fires" that must be addressed as quickly as possible, and in some cases immediately. Reports of victims trapped above a fire floor in a stairwell or elevator are an extremely high priority.

The incident commander will likely receive radio messages from the dispatcher passing on information from 911 calls. Perhaps it's an occupant of an upper floor apartment, stating that her apartment is rapidly filling up with smoke, she's elderly, and her husband is in a wheelchair. These occupants may in fact be in immediate danger or not. Nevertheless, their perception of immediate danger is real, and we must help them. A search and rescue team, ideally an entire truck company, or at the very least a team of two firefighters, will have to be assigned and sent to investigate and confirm whether or not these occupants need to rescued, removed, evacuated, or protected in place.

As a chief officer and incident commander, I have experienced at several fires that overwhelming feeling of having to address numerous problems simultaneously, without enough initial (first alarm) resources to do it. At one particular fire in a high-rise multiple dwelling, I was quickly inundated with radio reports from my dispatcher, as well as verbal reports at the command post from police officers, paramedics, and civilians reporting people in distress throughout the fire building. This was within the first 10 minutes of the incident, when all of my first alarm companies were committed to fire attack and search on the fire floor and floor above. My second alarm companies were on the way, but none were on the scene yet.

When this happens, you must prioritize the reports and handle each one as quickly as possible. A determination must be made as to what areas of the building create the greatest danger to occupants. In addition, if you receive multiple reports of the same problem, such as several reports of victims trapped in a stairwell, that should be given a very high priority.

At the previously mentioned fire, I was able to break the numerous problems down into a much more manageable size, by using the incident command system (ICS). I assigned my second due chief to the Fire Attack Group, and my third due (second alarm) chief to establish an Upper Search and Evacuation Group. Finally, I assigned the fourth due (third alarm) chief to operate in a separate wing of the building altogether. Each of these chiefs, with the companies assigned to them, took a huge chunk of the problem away from me and made my overall command responsibility and duties much easier to accomplish. Eventually, all areas of the building were successfully searched, and all occupants accounted for.

The search and rescue operation at a large multistory fire building is based on establishing which areas create the most severe hazard to occupants. In any multistory building, whether it's a low-rise, or a very tall high-rise, the areas to be searched are generally prioritized as follows:

1) Fire floor
2) Floor above
3) Top floor

In addition, you must be very aware of all the vertical channels that allow smoke to communicate upward, specifically, areas that might contain fleeing and trapped occupants, especially stairwells and elevators. Early in the operation, teams must be assigned to thoroughly search all stairwells above the fire floor, with the first priority being the fire attack stairwell. Any and all elevators located above the fire floor must also be located, opened, and searched. The first priority must be those elevators that are unaccounted for and stalled in hoist-ways.

Fire Floor

Just like any fireground operation, the immediate fire area creates significant danger to any occupants in or near this location. Barring any specific information to the contrary, clearly the fire floor is the highest priority for search and rescue. The resources needed to complete this task are very much dependent on the volume of fire, area or areas of involvement, and whether the building is a commercial or residential occupancy.

Residential occupancy

The residential occupancy is generally going to present a less severe overall fire problem than the commercial building. However, do not underestimate the significant life hazard associated with residential fires, especially those that occur during late night and early morning hours when most, if not all, of the occupants are home and asleep.

If we're lucky, the fire in a low or high-rise residential building will be confined to the original fire apartment. In this situation, the strategy should focus on an immediate search of the fire apartment, while protecting in place the occupants of the other apartments, if possible. The fire companies and/or Fire Attack Group operating on the fire floor must search the fire apartment, followed by a search of any apartments on either side of the fire apartment. In addition, depending on smoke conditions in the public hallway, any apartments directly across from the fire apartment should be searched as soon as possible. In many situations, especially when smoke and fire have extended out into the public hallway, every apartment on the fire floor will need to be searched.

When the fire has already extended out into the public hallway, usually due to the fire apartment door being left open, the operation becomes much more complicated. Occupants of adjoining apartments on the fire floor are in real and immediate danger, especially if any have opened the doors to their apartments. Until the fire in the public hallway has been controlled, we have no choice but to protect these occupants in place until we can access these apartments and complete a search. The only other option is an exterior rescue using aerial devices, but this is only applicable in low-rise buildings or if the fire is on a lower floor of a high-rise and within reach of an aerial device.

Successfully protecting occupants in place on the fire floor is dependent on whether they have kept the doors to their respective apartments closed or not. Unfortunately, many will open their doors upon hearing audible horns from the fire alarm, only to be met by high heat and heavy smoke conditions. Many occupants, upon encountering the heat and smoke, will panic, leaving their door open, while attempting to either retreat back into their apartment, or attempting to flee via the public hallway. Of those who attempt to flee via the public hallway, most will proceed toward the elevators, purely out of habit. In fact, many won't even know where the stairs are located. Many of these occupants will be quickly overcome by the rapidly developing smoke condition, and firefighters can expect to find them anywhere, from just inside their apartment to the elevator vestibule or anywhere in between.

Some buildings, especially new, modern high-rise residential buildings and hotel occupancies, have automatic closing devices on the apartment/hotel room doors, required by most codes. This can be a benefit because the door is designed to automatically close when occupants retreat back into their apartment, protecting them from the fire and smoke. However, this is a double-edged sword—the automatic closing device can also trap occupants in the smoke-filled atmosphere of the public hallway when they attempt to retreat that direction, as their door closes and locks them out.

Ultimately, all areas of the fire floor will need to be thoroughly and systematically searched, as quickly as possible. During this search, forcible entry may be required to access many apartments. Basic fundamentals and forcible entry rules should be applied. Specifically, it is absolutely critical that, if doors to individual apartments have to be forced open, members remember to *maintain the integrity of the door*. Strictly adhering to this forcible entry rule is of utmost importance, because the apartment door is the only barrier between any occupants and the smoke. If we want to maintain the ability to protect occupants in place, we must be able to close the door to their apartment.

A great deal of finesse must be used when forcing apartment doors in order to maintain their integrity. Barbaric methods, using large striking tools such as a sledge hammer to force an apartment door, although sometimes necessary, should generally be avoided. We are not a police SWAT team serving a search warrant. Furthermore, conventional methods using striking and prying tools are very difficult to apply, specifically in the low visibility typical of heavy smoke conditions. One of the best tools to use for forcing apartment doors are the new variety of portable hydraulic forcible entry tools, such as the *rabbit tool* and *Hydra-Ram*. Generally, these work well on most apartment doors, except those that are heavily fortified. These tools require minimal striking to create a purchase for the jaws, and, once the jaws are in place, most residential apartment doors can usually be forced open with little effort.

The search operation for a single apartment, when the fire is confined to that apartment, will be relatively easy to complete. With a hoseline in place, and an engine company protecting the search team, they have an umbilical cord which will lead to safety. If the apartment is relatively small—less than 1,000 square feet, typical of most apartments—then a simple right or left hand search, using the apartment wall as a compass will usually work well, and provide for firefighter safety. However, if the search has to begin at the entrance to the fire floor from the stairwell, especially if a hoseline is not yet in place, then a search rope should be used. Search ropes are essential when searching floors above the fire floor that are charged with heavy smoke conditions.

Commercial occupancy

Most commercial occupancies present a much more difficult and complicated search problem than the residential buildings. This is due primarily to the large size and the floor layout that is typically found at most low- and high-rise commercial buildings. Most commercial floors, although designed around an open floor plan, are actually set up with short partitions creating numerous, perhaps hundreds of, small individual work areas (fig. 14–1). This creates the absolute worst-case scenario for operating members, because the lack of full floor-to-ceiling partitions allows fire and smoke to spread throughout the floor. However, at floor level, these partitions create a complicated maze of barriers, making search operations very difficult, dangerous, extremely time consuming, and very labor intensive. This maze of partitions is often found on a floor that is 10,000, 15,000 or even 20,000 or more square feet. That's a huge area, and under heavy smoke conditions, it will take several fire companies and numerous search teams a long period of time to complete a primary search on just this one floor.

Searching the fire floor presents many added complications, especially if there was heavy fire involvement. Powerful fire streams from engine companies, and truck companies pulling ceiling tiles to expose the plenum space, will often result in a debris field, including the aluminum framework of the drop ceiling. This, along with wires, cables, and other debris creates a very dangerous and difficult terrain that can quickly trap a firefighter in zero visibility.

Many commercial buildings have significant compartmentalization on various floors. Typically floors with multiple tenants will be somewhat compartmentalized, with several doors leading to various rooms, closets, and offices. These floor plans are usually easier to search than the floors with maze-like partitions, but the extremely large area, coupled with numerous unusual and odd-sized rooms, make searching these floors very difficult and time consuming.

Standard equipment for search operations in low- and high-rise commercial buildings must include search ropes and thermal imaging cameras. In addition, the resources needed to fully search all of the floors above in a tall high-rise building with a serious fire on a lower floor will be significant. Prepare for and anticipate using numerous companies and search teams.

Keep in mind, search operations using ropes are very complicated and can be extremely frustrating in heavy smoke conditions with numerous barriers. Truck companies must conduct regular training drills on search and rescue operations of large, complicated areas, using thermal imaging cameras and search ropes. Attempting to complete a search operation with ropes, having spent little or no time training and preparing for the operation, is a recipe for disaster. In fact, search ropes in the hands of untrained members, can actually be detriment, and can easily lead to firefighter entanglement and entrapment. More on the use of search ropes later in this chapter.

The Three-Phase Search Operation

Most of our fireground operations are set up with built-in redundancies. For example, our engine company operations include a minimum of two handlines, typically referred to as the *primary attack line* and the *backup line*. Such procedures are fairly standard across the American fire service, and are proactive measures designed to maximize firefighter safety, especially in the event of a burst length of hose or other unforeseen system failure in the primary attack line.

Everything we do on the fireground should be followed by a secondary or backup procedure. Search operations are no different. Most fire departments, as part of their standard operating procedures, complete at least two separate searches during the operation. These are typically referred to as the *primary* search and the *secondary* search.

Primary search

The primary search is exactly what its name implies, a primary or first search. The primary search has many specific characteristics, including the goal of completing this search in a systematic manner and as fast as possible. During the primary search, firefighters are looking for a variety of things, most importantly, any victims. However, often overlooked is the fact that a good search team is also looking for and identifying any areas of fire, fire extension, and smoke conditions, especially in areas remote from the main body of fire. Search teams literally serve as a reconnaissance team for the incident commander and/or division supervisor, gathering critical information and communicating that as quickly as possible.

Basic fundamentals of the primary search include looking for any victims in the most likely locations, specifically in paths of egress for adult occupants. In low- and high-rise buildings, the fire attack stairwell is a critical location that has been a deadly trap for countless victims. This location is one of the main egress paths within a large multistory building, and all stairwells must be searched, top to bottom, as soon as possible, with the areas above the fire floor given priority. In addition, the countless other egress paths, including elevators, must also be searched.

Other egress paths include hallways and various areas that may be designated as refuge areas, such as elevator vestibules. It very much bothers me that areas such as an elevator vestibule have been designated as refuge areas in some buildings. In one particular building in downtown Denver—a very large, commercial high-rise building—disabled occupants are assigned a buddy, who has been instructed to assist the disabled person to the elevator vestibule, a so-called refuge area. Yes, this area is behind fire doors that are supposed to activate upon a fire alarm, but, nevertheless, this procedure leaves me very uneasy, as the disabled occupant and their buddy may become completely cut-off from help and any egress given a serious fire. If a serious fire and smoke condition exists on the same floor as their refuge area, coupled with an elevator failure, these occupants have no way out. In addition, it will be extremely difficult for us to get to them to effect their rescue.

I am still actively trying to change this procedure, which has been taught and accepted in many buildings as appropriate, and, unfortunately, endorsed by many fire prevention authorities. I would suggest educating occupants to move disabled persons to the appropriate stairwell (a designated evacuation stairwell), or a refuge area with direct access to an evacuation stairwell, where they are less likely to be cut off and not totally reliant on an elevator for rescue. Yes, the nonambulatory disabled person confined to a wheelchair will present a significant and labor intensive challenge to remove and carry down several flights of stairs, but it can be done. Most importantly, when they are located inside the stairwell or refuge area with stairwell access, we can get to them and remove them in the safety of an evacuation stairwell. Furthermore, we may only have to move them a short distance in the stairwell, generally, four floors below the fire floor. At that location, we might be able to protect them in place, or access an elevator in a safe location to complete their evacuation from the building.

For search operations in a residential high-rise, specifically within an apartment, our search operations will be very similar to those we conduct on a daily basis at our more frequent operations. With the exception of limited secondary egress, specifically when we are above the reach of aerial devices, the search area is someone's home, and typical procedures can be followed. Once again, adult occupants will likely be found in more obvious locations, but if any of the occupants are children, especially small children, they may be in any number of hiding places.

In both residential and commercial high-rise buildings, if the occupants of any area became cut off by fire and smoke conditions, they may retreat to a number of locations, but the most likely will

Fig. 14–1 A typical floor area of a commercial building, with numerous small individual work areas, making search operations extremely difficult.

be near windows. Most civilians are very unclear as to our operational procedures and equipment capabilities. Their frame of reference includes numerous Hollywood movies where incredible rescues are made using ropes, helicopters, and large air bags. Furthermore, many will attempt to retreat to a window, in hope of breaking it and reaching clean air. An early search of the outside parameter of a floor area, especially if windows are seen broken from the outside, may result in successfully finding a victim, and hopefully accomplishing a rescue. Also, one of the more obvious locations would be exterior balconies that are very common in many residential high-rise buildings. These areas can be a very effective refuge area for victims, but become very dangerous if a rapidly moving fire is directly below.

Secondary search

The secondary search is designed as a redundancy to look for and identify any victims that may have been missed during the primary search. This search will be much more time consuming than the primary, as members are literally "leaving no stone unturned." This search must include all areas, and is completed in a much slower, more methodical manner.

On some occasions at our more frequent fires, I have had well-intentioned companies report to me "secondary search negative," but this report is received only a short time after the completion of a primary search, and often by the same company that completed the primary search. My point is this: far too many firefighters and officers don't completely understand the concept of a secondary search. When the secondary search in any building is completed and comes back negative only moments after the primary search was completed, then it wasn't a secondary search at all, but, rather, another primary search.

This problem can be addressed effectively through good training, but also by adjusting operating procedures to ensure that a company or team, other than the one that performed the primary search of a specific area, is assigned to complete the secondary search. The goal is to have a fresh search team with an unbiased perspective, having not yet seen any of the areas to be searched, complete the secondary search. This enhances the probability of a very thorough, comprehensive secondary search of all areas.

For large low- and high-rise buildings, all of the searches will be very time consuming. Where the primary search will take perhaps several companies and search teams a long time to complete, the secondary will most likely take longer, perhaps much longer, depending on conditions. Adding to the overall fireground problem, using different companies and search teams to complete the various phases of the search operation, along with the time factors involved, will place additional strain on most fire departments because they are likely already taxed to provide sufficient resources.

Final search

Many proactive fire departments have added to their overall search operation a third and final phase, called a *final* search. Most notably, the Chicago Fire Department (CFD) has implemented the final search, after experiencing a very deadly fire in the Cook County Administration Building. At this fire, several fatalities occurred in the attack stairwell, above the fire floor. The CFD has taken very proactive steps to enhance their overall high-rise firefighting operations, including a strong focus on early and redundant searches of all stairwells. The CFD's new procedures include a final search. The CFD also assigns *Rapid Assent Teams* to proactively and quickly search stairwells, specifically the attack stairwell, at the outset of a serious high-rise fire.

In addition to the primary and secondary searches, fire departments should include a final search at all major fireground operations, specifically for fires that occur in large low- and high-rise buildings. This final search, which should be conducted in a similar manner as the secondary search, provides an additional redundancy to ensure that absolutely no occupants have been left unaccounted for anywhere in the building. It also serves as a final confirmation of no hidden fire or smoke extension anywhere in the building. A final search should be included as part of a three-phase search operation, specifically for large low- and high-rise buildings.

Equipment for Search Operations

In order to successfully complete a search operation in these large, complicated buildings in an efficient and timely manner, fire departments must utilize every resource possible, including the very best equipment. There are many equipment items that will help firefighters achieve operational success during a search. Some of the most essential equipment items are search ropes, thermal imaging cameras, and methods to label areas that have been searched.

Search markers (door straps)

A primary search in a small single family dwelling may include a kitchen, living room, and a couple of bedrooms. This search will usually be completed quickly and require little effort to account for all of the areas that have been searched. On the other hand, large low- and high-rise buildings will have vast areas, and numerous rooms behind countless doors. There will be an absolute necessity for a simple system to account for what areas have been searched and what areas have not.

For years, many fire departments have used some sort of device to aid in search operations. The DFD uses a simple device called a *door strap/search marker* which serves multiple purposes. Some departments refer to this as a *latch strap*. The basic concept involves using the search marker to accomplish the following:

1. To keep a door from locking behind the search team
2. To identify areas that are currently being searched
3. To identify and account for those areas that have already been searched

Specifically, when a search team is using the marker, they place it over both the inside and outside door knobs of the entrance door to area being searched. This serves to keep the door from locking behind the search team, and also shows that the area is currently being searched (fig. 14–2).

After completion of a primary search, the search team removes the search marker from the inside door knob, but leaves it in place on the outside door knob (fig. 14–3). In this position, it is easily seen or can also be felt in low visibility conditions, and indicates that a primary search has been completed in the room or area behind the door.

A different company or search team will utilize a separate search marker on the door, once again, first to keep the door from locking behind the search team, and second to indicate a search in progress. After completing this secondary search, the second search marker is doubled up on the outside door knob to indicate completion of the secondary search (fig. 14–4).

The door strap/search marker used by the DFD is an inexpensive homemade device that is very easy to make. Using scrap rubber from an old inner tube, the search marker is cut out using a template. Two holes are cut into the small rubber piece, creating something that resembles the mask used by the Lone Ranger. I know I'm getting old when several of the firefighters reading this book will ask, "Who's the Lone Ranger?" Firefighters in Boise, Idaho, will recognize the term, as these devices are actually referred to as "Lone Rangers" on the Boise Fire Department.

Taking it a step further, the DFD has companies actually stencil their respective company designation in four places, on both sides of the search marker, which helps with the overall accountability of the search operation. Also, because it doesn't cost much to make several of these search markers, most fire departments embrace the idea. Many firefighters assigned to truck companies will carry several of the search markers either in a turnout coat pocket, or on the outside of the coat

Fig. 14–2 A DFD search marker placed over both door knobs keeps the door from locking behind operating members, and indicates that a search is underway.

Fig. 14–3 A DFD search marker attached only to the outside doorknob indicates completion of the primary search in that area.

Fig. 14–4 A DFD search marker doubled up, and attached to the outside doorknob only, indicates completion of the secondary search in that area.

attached to a small carabiner (fig. 14–5). This is ideal, keeping several search markers immediately available for use during an extensive search operation.

Alternative search marking system

The door strap/search marker system works very well in most situations. However, an alternative marking system may be applicable in some situations, and it is always good to have a plan B. Prior to using the search markers, I used a marking system that uses various different types of markers, such as chalk, grease pencils, and permanent markers. In fact, many firefighters routinely carry a marker or two and a stick of chalk in the pocket of their turnout coat. Most common are large pieces of chalk and grease pencils. These markers can be used for a variety of purposes on the fireground, including marking doors to indicate a completed search.

The system used by the DFD as a secondary or alternative system is similar in concept to the system used by USAR teams at major emergency events, where numerous buildings must be marked after being searched, as part of an overall search accountability system. For fireground operations, an appropriate marker can be used in the following manner:

A company places their unit designation on the left-hand side, lower third of a door, upon entering to complete a search, for example, DFD Tower Ladder Co. 23, is abbreviated as Tr-23. DFD Rescue Co. 1 is R-1, and so on (fig. 14–6a). Upon completion of a primary search, as the company exits the room or search area, a member places a single line through the company designation at an angle (fig. 14–6b). Ideally, as previously stated, it is best to have a different company or search team from the one who completed the primary search complete the secondary search. In doing so, there will be another company designation placed on the right-hand side, lower third of the door. Upon completion of the secondary search, as the company exits the room or search area, a member places a large X through the company designation (fig. 14–6c). Once again, the company designation by itself indicates that company or team is behind the door in that area conducting a search. A one-line slash through the company designation indicates completion of the primary search, whereas an X through the company designation indicates completion of a secondary search.

It is a good idea to have a variety of different markers, in order to have something that will work on different surfaces. For example, a grease pencil works well on a metal door, but not very well on a wood door, whereas, a permanent marker works well on the wood, but not metal surfaces. Chalk is ideal for writing messages on the concrete walls typically found in many stairwells.

Although using the markers is a good alternative and plan B, I very much prefer and recommend the door strap/search marker method over writing markers. The door strap is much easier to use and less cumbersome, especially in a low or zero visibility environment. Nevertheless, it is a good idea to have a small supply of writing markers inside the standpipe kit, so they are immediately available when needed. In addition,

Fig. 14–5 Truck company firefighters can carry several search markers on a carabiner in an easily accessible location.

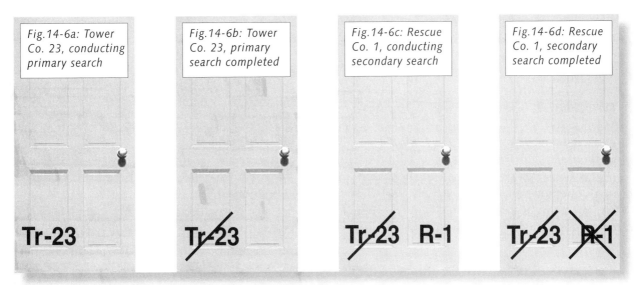

these markers should be placed in a rigid container, such as a small Tupperware container, to protect them from damage, especially the chalk. It is also a very good idea for individual firefighters to get in the habit of carrying a couple of different types of markers with them in a turnout coat pocket, so they are immediately available, if needed.

Improvised methods

In a perfect world, we would not need to have a plan B, C, and D. However, the fireround is far from perfect, and loaded with problems and complications. A system of maintaining an inventory or accounting of what areas have been searched is critical, especially in these very large buildings. The door strap/search markers will work well if there are doors at the entrance to each area to be searched. Writing markers are certainly a good alternative for some applications, but there are numerous other situations where neither will work.

Most notably are the large, open floors, typical of a commercial high-rise, with numerous individual cubicles and workstations. These areas don't have doors or doorknobs to hang search markers on, or to write on with markers. Therefore, firefighters must be innovative, and use various improvised methods to indicate what areas have been searched.

A very good method that is particularly suitable in this circumstance is to use a common office desk chair to mark areas that have been searched. Even in low or zero visibility conditions, firefighters can utilize a chair, placed on its side across the entrance area to a cubical or work station, to indicate a completed search (fig. 14–7). Office chairs work very well because they are portable, relatively lightweight, and one can usually be found at each individual workstation.

Fig. 14–7 A typical office chair laid on its side, can be used to indicate a completed search in areas where the door strap/search marker cannot be used, such as at the entrance to small work areas and cubicles.

There is certainly a wide variety of different methods firefighters might use to indicate a completed search in these small work areas. Regardless of what is used, remember to keep it simple. Furthermore, come up with a plan, and make sure it is standardized throughout your organization so that all members have a clear understanding of what it means. Even if a specific method is developed on the spot and utilized in the heat of battle, the procedure should be shared with other operating units, particularly those who will be continuing a search where another company or search team left off, and for those conducting secondary searches.

Thermal imaging cameras (TIC)

In the fire service of the twenty-first century, modern technology has provided us with a wide range of valuable tools and equipment. One of those is certainly the thermal imaging camera (TIC). Although there are definite limitations, and an over-reliance on such tools can lead to serious problems, the TIC is an invaluable tool for large area search operations. The first generation of TICs were large, heavy, and cumbersome devices, especially when compared to the new, compact, lightweight versions in use today (fig. 14–8, and fig. 14–9).

The TIC has become standard equipment in most fire departments, and is routinely carried by truck company firefighters into most fire buildings. The TIC can tremendously increase a search team's overall efficiency and effectiveness. At a single-family dwelling fire, the TIC is a very valuable tool. At large low-rise and high-rise building fires, especially large commercial buildings, it is essential.

However, and, once again, as good as this tool is, firefighters must never overlook its limitations, or underestimate how devastating equipment failure could be if members become overly reliant on it. I was taught years ago that one should never, ever fall into the so-called *superman syndrome* when using a TIC.

Every search operation must be based on fundamentals, including orientation via a wall, and/or search rope, which will lead members to safety. The TIC should be used *in addition* to basic methods, not in lieu of them. Proceeding deep into the dark and dangerous environment of a fire floor, relying solely on the guidance of a TIC, can give a firefighter a tremendous sense of confidence. However, that confidence and secure feeling will be quickly shattered if there is a sudden equipment failure or if the battery goes dead.

For large area searches, in addition to using a search rope or ropes, I would recommend using at least two TICs, one at the entry point, and one at the lead search position. The member operating the TIC at the entry point can sometimes, depending on the floor layout, communicate with and direct search teams over a wide area, having an overall perspective. The member operating the TIC at the lead search position can use the TIC to provide

Fig. 14–8 Early versions of thermal imaging cameras were large and cumbersome.

Fig. 14–9 New, modern models of thermal imaging cameras are small, lightweight, and easy to use.

more precise guidance as the search team moves forward, deeper onto a floor.

Like everything else we do, the amount of time dedicated to quality training will be the greatest determining factor in operational success or failure. All members, but especially those assigned to truck companies and responsible for search operations, should be involved in regular drills on the use of TICs. In addition, a solid maintenance program, which includes checking the TIC daily and after each use, along with a proactive schedule to rotate batteries, will ensure optimal performance in the heat of battle. Members must verify that batteries are fully charged and have a clear understanding of battery duration. Also, spare batteries should be charged and immediately available, preferably with the TIC operator, perhaps carried in a turnout coat pocket. Members should regularly practice removal of the (dead) battery from the TIC, and replacing it with a different (fully charged) battery in zero visibility conditions.

Fig. 14–10 For search rope operations in a commercial high-rise building, the rope should be anchored outside of the search area, inside the stairwell, and attached firmly to the handrail.

Search Rope Procedures

Search operations in large, standpipe-equipped buildings, especially warehouses, and the large open floor areas of commercial high-rise buildings, can only be completed by using search ropes. There are a wide variety of suggested methods and procedures that can be used when searching with ropes. Regardless of what method is used, the most important component is training. Simply put, if members are not well trained and have not conducted regular training drills performing searches with ropes, they should not use them. To attempt a complicated search with ropes, having little or no training, is extremely dangerous.

Here are some recommended procedures for search rope operations. First, a search operation of a large area using ropes should include a large team of preferably two or more (four-man) companies. However, with limited resources, the search team must be comprised of a minimum of one fire company, and at least four members. I base these recommended procedures on a realistic search team size of four members, which is a resource that most fire departments should be able to provide.

The main search rope, or *tether*, should be anchored to a point well outside of the search area, either completely outside of the building for operations at a large one-story warehouse, or anchored to a stairwell handrail inside a commercial high-rise building (fig. 14–10). Just like a hoseline hooked up to a standpipe outlet on the floor below, we want our main search rope set up to immediately guide members out to safety.

Next, one member should be positioned at the main entry point to the large area to be searched. He should be equipped with a TIC to provide overall guidance for the searchers, and a portable radio to communicate with team members and division supervisors (fig. 14–11). In addition, the search team should be equipped with several high intensity lights, with at least one positioned at the entry point to serve as a beacon, guiding members back to safety. There are many different types of quality lights available on the market today. Some of the best include an automatic, flashing on/off mode for the primary light, and very bright, flashing LED lights, in various colors, located on the back of the unit.

With one member set up at the entry point, three members can then proceed out into the search area to begin the search. The lead member, preferably the company officer, carries the primary rope bag serving as a guide rope or tether. In addition, he should be equipped with a TIC, to provide additional guidance as the team searches (fig. 14–12). As the rope pays out, it should be kept as tight as possible. The other two members can be periodically sent out laterally, away from the main search rope, on tag lines, to search specific areas. When the team members are operating on tag lines off

Fig. 14–11 One member of the search team should be positioned at the entry point, equipped with a thermal imaging camera.

Fig. 14–12 The member leading the search team, usually the company officer, should also be equipped with a thermal imaging camera to guide the searchers.

the main rope, the search leader can use the TIC to observe and guide the searchers.

Summary

Saving a human life is by far one of the most satisfying experiences a firefighter will ever have. Life safety and rescue is our number one priority. The opportunity to accomplish such a tremendous feat doesn't happen every shift. In fact, it's likely a very rare occurrence for most firefighters. However, when that rare opportunity presents itself, you want to be prepared. Talk is cheap, and preparation is only achieved through a consistent diet of quality training. Rapidly and efficiently searching an extremely large warehouse or upper floor of a commercial high-rise is an extremely formidable task. It can be completed, but only by the most dedicated, well-trained, physically and mentally prepared members.

CHAPTER 15
Air Movement and Ventilation

Of the numerous complex components associated with fireground operations in large low-rise, and especially in tall high-rise buildings, air movement and ventilation are the ones I consider to be most complicated. Simply put, there are many gray areas, lots of unknown factors, and ongoing unusual and unpredictable air movement. Some tactics and procedures that work well for us at our more frequent operations in smaller buildings can greatly exacerbate our problems if applied at large high-rise buildings. Ventilation procedures used at operations in these large, complicated buildings require a great deal of conscious thought and deliberation, with continued evaluation as to their effectiveness. It's not just a simple matter of opening the roof or firing up a fan (smoke ejector or positive pressure blower).

Air movement in these large buildings can been broken down into three broad categories:

1. Natural air movement within and outside the building;
2. Mechanical air movement produced by the buildings heating, ventilation, air conditioning (HVAC) system(s);
3. Operational air movement created by active ventilation procedures, both natural and mechanical, introduced as part of the fire department operations, using fire department equipment, and or the building's built-in smoke control/air management system(s).

Natural Air Movement

The success or failure of fire department operations at large, low-rise, standpipe-equipped buildings, but especially at tall high-rise buildings, is greatly influenced by the fire department's knowledge of air movement and ventilation within these large structures. Every minute of every day, there are unique air movement characteristics occurring inside all large buildings, especially tall high-rises. Most of these naturally occurring air movement phenomena are completely out of our control. However, by working to study and understand these various air movement concepts, we will be able to more effectively coordinate our operations around them, and, in some cases, allow the natural air movement to work in our favor.

Stack effect

One of the most misunderstood and misapplied terms associated with high-rise air movement is *stack effect*. Stack effect is defined as "the vertical, natural air movement throughout a high-rise building caused by the difference in temperatures between the inside air and the outside air. Positive stack effect is characterized by a strong draft from the ground floor to the roof. Positive stack effect is more significant in cold climates because of the greater difference in temperature between the inside and outside of the building. The colder the weather and the higher the building, the greater will be the stack effect. Negative stack effect can also occur in the reverse direction in hot climates, but is not as dramatic because the difference in temperature is not as great. Stack effect is not caused by the fire, but it causes the smoke to spread throughout the building during a fire."[1]

"Stack effect is responsible for a significant amount of smoke movement in a high-rise building fire. The magnitude of stack effect is a function of:

1. Building height.
2. Air tightness of exterior walls.
3. Air leakage between floors.
4. Difference in temperature between inside air and outside air."[2]

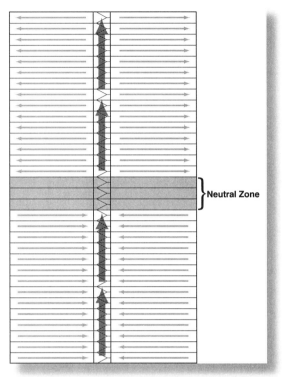

Fig. 15–1 Movement of air, above and below the neutral pressure plane, when there is a positive stack effect

The air movement in a building, including the phenomena of stack effect, is not something that one can clearly identify and visualize. However, there are characteristics and some indicators that can help firefighters determine whether there is a positive or negative stack effect occurring inside a building, and evaluate the strength of the draft. For example, on a cold winter day in Denver, with an outside air temperature of 35° Fahrenheit and an interior building temperature of 75° Fahrenheit, there will be a very obvious and strong positive stack effect. On the other hand, a hot summer day in Los Angeles, with an outside air temperature of 95° Fahrenheit and an interior building temperature of 75° Fahrenheit, there will likely be a relatively strong negative stack effect (also referred to as a "reverse" stack effect).

Another basic method that can be used by firefighters to evaluate the potential stack effect inside a building can be accomplished when members enter the building. As firefighters open the doors at the ground or lobby level, a powerful current of air drawn into the building indicates a strong, positive stack effect. In fact, often times, these entrance doors can be very difficult

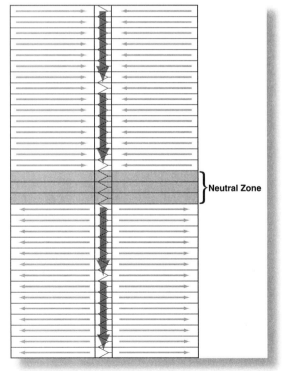

Fig. 15–2 Movement of air, above and below the neutral pressure plane, when there is a negative or reverse stack effect

to open, due to the force of a powerful stack effect occurring inside and creating a negative pressure within the building at the lower levels.

Most commercial high-rise buildings have double sets of doors or revolving doors at the main entrance. As people enter and exit the building via these double doors or revolving doors, the doors are much easier to open and there is little indication of a draft. Firefighters must keep this in mind, and remember that when an entrance has double doors, both doors must be opened in order to evaluate the stack effect. One firefighter will have to hold one of the doors open, while another firefighter opens the second door, thus allowing for a continuous opening in order to evaluate the draft and stack effect. For buildings with revolving doors, there will always be normal swinging doors beside the revolving door. These side doors can be used to complete an initial evaluation of the building's stack effect.

Fig. 15–3 Some commercial high-rise buildings have some windows that can be opened, usually with a special key.

Neutral pressure plane

In high-rise buildings, there will also be a *neutral pressure plane* that also affects the movement of air on a given floor or floors. The neutral pressure plane (NPP) is defined as "the level within a building at which the barometric pressures both inside the building and outside of a building are equal. The NPP may extend over several floors."[3]

When there is a *positive stack effect* occurring in the building, air will move out away from the core of the building toward the exterior on the floors above the NPP, and air will move in toward the core from the exterior on the floors below the NPP (fig. 15–1). Whereas, when there is a *negative* or *reverse stack effect* occurring in the building, air will move in toward the core from the exterior on the floors above the NPP, and air will move out away from the core toward the exterior on the floors below the NPP (fig. 15–2).

Within the NPP, the stack effect is actually neutral, meaning that there is neither an active movement of air upward or downward. Obviously, the NPP is invisible, so the exact location and area (number of floors) involved will be unknown without a comprehensive evaluation. Furthermore, the location and floors involved will not remain static, but will change depending on numerous environmental factors outside of our control. Also, the location of the NPP will be different from one building to the next.

Ultimately, we are not specifically concerned about that exact location of the NPP, but rather whether air will move inward or outward on a given floor, especially the fire floor. To determine this, members can evaluate air movement by making a small inspection opening on the floor below the fire floor. This can be done by opening a window if possible, such as will likely be the case in a residential high-rise, or by making a small hole in a window by carefully breaking the glass (if possible) in a commercial high-rise building.

Keep in mind, most commercial high-rise buildings have very thick and strong window glass that will be very difficult to break. However, some commercial high-rise buildings also have windows that can be opened, but it usually requires a special key (fig. 15–3). There are also many other buildings that have predesignated windows, usually located a specific distance apart—for example every 50 linear feet—that are made of tempered glass, and are thus much easier and safer to break. Many buildings have labeled these tempered glass windows to indicate their presence to operating firefighters. I will address the specifics and precautions of breaking windows later in the chapter.

The effects of wind

The negative effects that a wind condition can have at a serious high-rise fire cannot be overstated, and should not be underestimated. There have been numerous high-rise fires throughout the country where a wind condition seriously hampered firefighting operations. Many of these resulted in injuries and deaths, including several firefighters.

Although there may be no wind, or just a slight wind condition at street level, this is not a reliable indication of what the wind condition is, or will be when you arrive at the 25th floor. Operating firefighters should always anticipate a wind condition that will likely be significant on the upper floors of a high-rise building. These wind conditions can greatly complicate firefighting operations, and, in some cases, can make a direct attack upon the fire almost impossible.

Most notable are situations where windows have failed on the windward side of a fire building with strong wind conditions. This creates a positive pressure on one side of the building, or perhaps partially on two sides, which will forcefully drive heat, fire, and smoke back into the building, and directly into the face of an approaching attack team. A strong stack effect will also exacerbate this wind condition, by forcefully drawing it toward the numerous vertical shaft-ways, specifically the attack stairwell. Attack teams may have to retreat, and, in many cases, have had to resort to a flanking attack on the fire.

At the Polo Club High-Rise fire in Denver, I remember operating on an attack line for an extremely long period of time. We were making absolutely no headway in the hallway on the seventh floor in an attempt to push our way into apartment #712. That was with two handlines, operating side by side, for over one half hour. I didn't realize it at the time, but an extremely strong wind condition, coupled with a powerful stack effect, was making control of this fire impossible.

District Chief Mike Miller was the incident commander, and he was making plans to attempt a defensive attack via tower ladders from the exterior. However, there was no way to get apparatus positioned on what was the fire side of the building, due to other structures. Our only option was to fight the fire from inside. My company officer that day was Firefighter Joe Cipri, who was acting lieutenant. He was a highly experienced firefighter, and he showed his knowledge when he suggested a *flanking attack* from an adjoining apartment. With companies struggling to hold the fire with handlines in the hallway, a wall between the fire apartment and an adjoining apartment was breached, allowing companies to direct water into the fire apartment from a flanking position. Because of this, the fire was finally brought under control. The power of wind, coupled with a strong stack effect, was clearly evident at this fire.

Stratification of smoke

Another natural condition that can occur at a high-rise fire is the *stratification* of smoke. The taller a building is, the greater the potential for stratification of smoke, especially when the smoke is coming from a fire on a lower floor. In addition, if the fire is controlled quickly by sprinklers or a

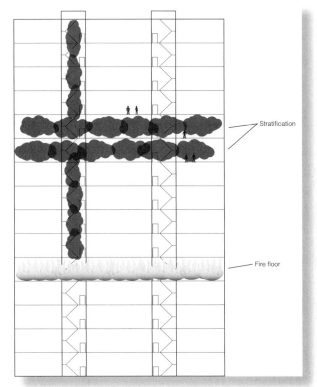

Fig. 15–4 The stratification of smoke on the upper floors of a high-rise building can be deadly to both occupants and operating firefighters.

rapid fire department attack, this too can increase the probability of smoke stratification, as there is no longer any thermal energy being produced by the fire to heat the smoke.

✶As the smoke rises inside a tall high-rise building, it cools down and can eventually cool enough to where it is equal to or less than the ambient air temperature inside the building. At this point, having lost its buoyancy, the smoke stops traveling upward, stratifies, and begins to spread out laterally (fig. 15–4).

✶The danger of stratified smoke is significant and can sometimes be underestimated. Operating members can get tunnel vision, focusing primarily on the fire floor. When the fire is brought under control quickly and is no longer producing deadly smoke, it can give us a false sense of security. We can easily overlook the fact that there may still be plenty of deadly smoke in the building.

Cold, stratified smoke still contains high levels of carbon monoxide, and, although it is not under pressure and may look fairly benign, it is still very deadly. Many lives have been lost, including those of firefighters operating high above the fire floor, when they were overcome by cold, but deadly, smoke.

There are several keys to operational success when dealing with the stratification of smoke. First, fire companies assigned to an Upper Search and Evacuation Group must quickly identify areas of smoke stratification, and report this back to command officers. Once areas contaminated by smoke are identified, a proactive and aggressive search of those areas must be initiated and completed as quickly as possible. Furthermore, especially if the fire is under control, plenty of resources should be dedicated to form ventilation groups, and implement the necessary procedures to clear the building of this deadly, stratified smoke.

Mushrooming of smoke

The *mushrooming* of smoke at the upper level of a fire building is something that firefighters encounter every day at our most frequent fires. This will rapidly occur in a one- or two-story house, as the smoke has only a short distance to travel before it runs into a barrier. Smoke can also mushroom in large low-rise and high-rise buildings, as well. This is especially true when a fire occurs on an upper floor, as the smoke has less distance to travel before reaching the top floor. A serious fire, producing significant smoke and high heat, can quickly lead to severe mushrooming of smoke on the top floor, and can quickly bank down and build up dangerous smoke conditions on several upper floors (fig. 15–5).

Once again, fire companies operating on the upper floors must quickly identify this problem, report the condition to command officers, and initiate procedures to correct it. The search priority in a multistory building of fire floor, floor above, and top floor is directly related to the issue of vertical smoke travel, especially the mushrooming of smoke on the top floor, or several upper floors. Clearly, one of the most obvious tactical options is vertical ventilation by opening up the building—literally taking the lid off—and giving the pressurized, deadly, mushroomed smoke a place to go.

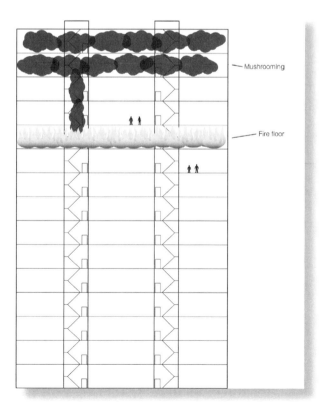

Fig. 15–5 A fire on the upper floor of a high-rise building can quickly lead to the mushrooming of deadly smoke on the top floor or several upper floors.

Mechanical Air Movement (HVAC)

In most high-rise buildings, mechanical air movement is accomplished through a central heating, ventilation, and air conditioning (HVAC) system. HVAC systems are designed to process and treat the air inside a building, in order to control the temperature, humidity, and cleanliness of the air. The air is distributed to areas throughout the building, and then returned for reprocessing.[4]

The overall air processing systems in large low-rise and especially tall high-rise buildings fall into two categories: 1) central HVAC systems, where air is supplied to more than one floor (most common in large, uncompartmentalized commercial high-rise buildings), and 2) non-central HVAC systems where air is supplied to only the floor where the HVAC equipment is located, (most common in compartmentalized residential high-rise buildings).[5]

A central HVAC system has several components, including the processing system, the supply system, and the return system.

The processing system

The HVAC processing system(s) is typically located on the mechanical floor(s) of a high-rise building. Depending on the size of a high-rise building, there will be one or perhaps multiple mechanical floors. The components of a processing system include:

1. Outside air supply dampers
2. Equipment for heating, cooling, filtering, and humidifying the air
3. Supply air fans
4. Smoke and heat detectors
5. Duct work to the supply air shaft
6. Duct work from the return air shaft
7. Return air fans
8. Exhaust air dampers
9. Mixing dampers

The supply system

The components of the supply system include:

1. A supply air shaft
2. Fire dampers
3. Supply air ducts
4. Air diffusers

The return system

The components of the return system include:

1. Return air inlets
2. Return air plenums
3. Fire dampers
4. Smoke detectors
5. Return air shaft

In large buildings, especially tall commercial high-rise buildings, there will typically be multiple HVAC systems, with each individual system supplying specific zones of a building. One zone will consist of multiple floors.[6]

A central HVAC system can significantly contribute to our operational problems. Left unchecked, it can actually move smoke from one location of the building to another. The older systems are most susceptible to creating problems for us, and, historically, the rule of thumb was to actually shut the HVAC system down immediately upon arrival at a fire in a high-rise building. This procedure may, in fact, be appropriate for some buildings, especially those with old, antiquated central HVAC systems.

Operationally, we must remember that when a building has a central HVAC system, any reports of smoke may or may not be close to the actual fire area. A central HVAC system can inadvertently collect smoke-filled air from one part of a building and transport it to another area, perhaps quite remote from the actual fire area. Certainly, the newer, modern systems are more sophisticated, with built-in features designed to prevent such smoke contamination. However, even new systems can fail, and, operationally, we must be prepared for it.

In conjunction with most new, modern central HVAC systems are built-in smoke control/air management systems that work to control the movement of smoke. Many of these systems are very sophisticated and have tremendous capabilities. I consider these to be part of the last broad category, operational air movement.

Operational Air Movement

The final category of air movement within large low-rise and tall high-rise buildings is what I refer to as *operational air movement*. This category includes mechanical air movement via built-in smoke control and air management systems under the direction of fire department command officers. In addition, operational air movement includes natural and mechanical air movement via fire department operations.

Built-in smoke management systems

Most modern high-rise buildings are equipped with sophisticated smoke control/air management and smoke removal systems. This is actually a fire and building code requirement in most jurisdictions for new high-rise buildings. These modern systems allow for the control of air movement from a remote location, which is typically the fire command center in the lobby. By simply flipping a switch on a panel, we can control the air movement on a specific floor of a high-rise building (fig. 15–6).

In addition, most of these systems are automatic, designed to work in conjunction with the building's fire alarm system. Upon activation of a fire alarm, many are designed to stop air intake on the fire floor or floor of alarm, while pressurizing one or two floors above and below. For example, a smoke detector is activated on the 22nd floor of a 43-story, commercial high-rise building. Upon activation of

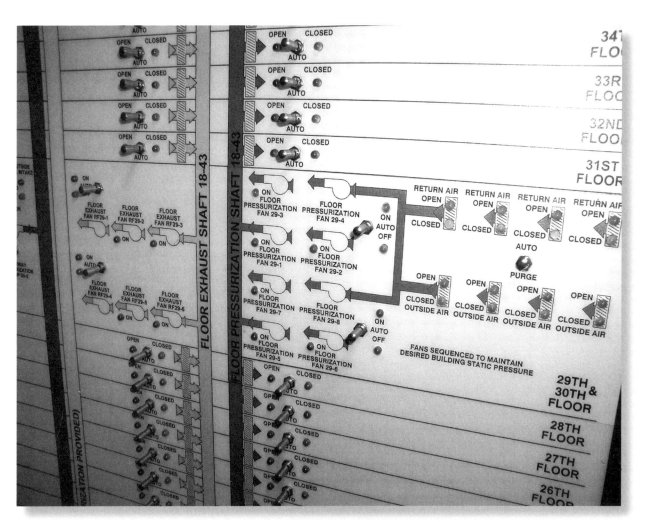

Fig. 15–6 A typical smoke control/air management panel located in the fire command center of a large high-rise building

this alarm, the building's air management system automatically shuts down the air supply to the 22nd floor, but keeps the air exhaust open. This is obviously designed to eliminate the possibility of feeding the fire with air (oxygen), while allowing for the release of any smoke or other products of combustion. As the air supply to the suspected fire floor or floor of alarm is shut down, the system actually pressurizes the floor above (23rd floor) and the floor below (21st floor) by opening the air intake and closing the exhaust (15–7). This is designed to actually sandwich the fire floor in an attempt to confine the fire to the floor of origin. It is not a bad idea in theory, if it works as designed, and only if there are not any unknown factors that might cause the system to negatively affect our operations.

As I mentioned in chapter 7, smoke management systems must recognize any access stair locations. Specifically, floors with access stairs, such as an access stair between floors 36 and 37, must be recognized by the smoke control/air management system as one floor. Therefore, whatever smoke control measures are automatically or manually activated, the same thing will occur on both the 36th and 37th floors.

If in doubt, shut it down

These modern smoke control/air management systems are designed to control air movement and provide for a safe and precise removal of any smoke or other products of combustion. However, they don't always operate as designed and or intended. Ultimately, during fireground operations, we must monitor these systems very closely and with a pessimistic view of their reliability. If the officer in command, based on information from operating units and or division and group supervisors, believes that the system is not helping our operations, and especially if it appears to be hurting our operations, then it should be ordered shutdown immediately.

This has historically been the recommended procedure at high-rise fire events, that is, to immediately shutdown the HVAC systems upon arrival. However, these modern systems can, in some cases, make our operations much easier, but this must be determined by a thorough and ongoing size-up.

Manual control

Most modern smoke control/air management systems are also designed to be controlled manually. Key to our operational success here is to assign a *systems officer* as early in the event as possible, in order to monitor and control the air management system and other building systems. In addition, we must make every effort to locate someone from the building engineer's staff, preferably the building engineer himself, and place that individual with the systems officer as a resource throughout the event. This includes familiarization with the buildings smoke control/air management system and the ability to operate it, manually controlling air movement throughout the building, as necessary.

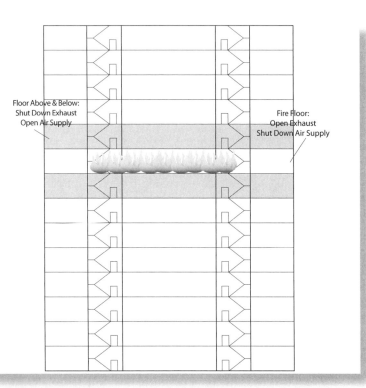

Fig. 15–7 A typical smoke control/air management system has the capability to shutdown the air supply to the fire floor and pressurize the floors above and below.

Fire blanket

There are a number of procedures that have been developed and are being evaluated to help combat severe, wind-driven fires. One of those is the *fire blanket* (high-rise building fire curtain), which theoretically is designed to be dropped from the floor above in order to cover windows, creating a barrier and thus preventing the wind from entering the building and forcefully pushing heat, fire and smoke further into the building. The FDNY helped develop this tool and is equipped with it (fig. 15–8).

In order for the fire blanket to work, firefighters must be thoroughly trained in its use. Remember, we are attempting to apply human solutions to problems created by Mother Nature. The concept of deploying a large fire-resistive cover on the exterior of a building, under severe wind conditions, will certainly be a formidable task.

Fig. 15–8 The high-rise fire blanket deployed and covering windows

Breaking windows

At our most frequent operations in low-rise buildings, firefighters should be given the authority and responsibility to break window glass as necessary. I have always told the fire officers and firefighters that work under my command that they are in a better position (inside the building) than an incident commander (outside the building) to determine whether or not they need to break a window. I also reiterate that breaking a window at our most frequent fireground operations is for the primary purpose of establishing a secondary egress. The horizontal ventilation is actually a secondary purpose for breaking a window.

However, for fires in large low-rise multistory buildings, and especially tall high-rise buildings, the breaking of window glass must be coordinated and approved by the incident commander. First, the breaking of windows is generally not going to be for the primary purpose of secondary egress, especially at the upper level of a tall high-rise, beyond the reach of aerial ladders.

There are, in fact, many extreme dangers associated with breaking window glass at these multistory low-rise and tall high-rise buildings. First, when a window is broken, gravity takes over and the glass falls with powerful force to the street below. This glass will powerfully strike whatever it hits, and can be extremely destructive. Of primary concern are any operating members or civilians located below the falling glass. In addition, fire apparatus and equipment in the street below, especially hoselines supplying the building's standpipe system, are particularly susceptible to damage.

During operations in large low-rise and tall high-rise buildings, if windows must be broken, it should only be done after careful deliberation and approval from the incident commander or his designee. All such actions must be completely communicated, and should include the removal of any civilians located below, and the protection of any operating members and equipment.

Members breaking the glass should do it with finesse. Make every attempt to bring the window glass back into the building (if possible), as opposed to letting it fall to the ground. The use of tape, even simple duct tape, can be used to hold the window together, allowing members to bring it back into the building after it has been broken. Many buildings, especially commercial high-rises, have a thin layer of material on the window designed to shade the interior of the building from the sun. This material sometimes helps to hold the window together, making it easier to pull back into the building.

In addition, operating firefighters should make every effort to determine if there are any windows that can be opened, or any tempered glass windows, before breaking any windows. Also, members must be cognizant of strong wind currents that could easily pull members and, in some cases, literally suck them out a window to their death.

Elevators

It is important to note that specific features of the building can greatly contribute to the smoke movement. Specifically, elevators can easily push smoke from one location to another. Although modern high-rises require elevator hoist-ways to be pressurized, that is only going to occur after the building goes into alarm, or when the smoke control/air management system is activated. Prior to that activation, smoke can migrate into an elevator hoist-way, and when the elevator moves, such as during a Phase I fire service recall, it can literally push the smoke to what might be a location remote from the fire floor. Firefighters must be cognizant of this, and understand that there may be smoke at one or several locations that are remote from the actual fire location.

Fire/smoke towers

Although I have already addressed this previously in chapter 7, it is important to reiterate. I am referring to the smoke movement that will naturally occur via a stairwell that is a fire/smoke tower. The stack effect draft that naturally occurs in all high-rise buildings will certainly exist in a fire/smoke tower. This will be a benefit in most situations, as the deadly heat, smoke, and fire gases will be drawn up and out of the building, protecting occupants evacuating via this stairwell (fig. 15–9).

The concern revolves around fire department use of these fire/smoke towers as an attack stair. This should be avoided, due to the fact that the natural stack effect will, in many cases, forcefully draw the heat, fire, and smoke toward the stairwell, and can make fire attack from this position extremely difficult, dangerous, and, in some cases, impossible. This is because both of the doors, typical of a fire/

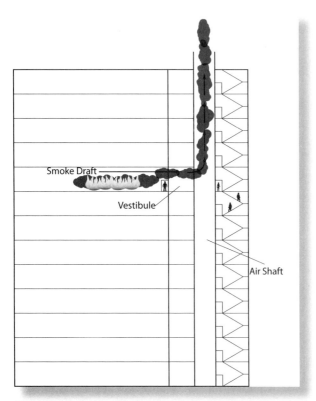

Fig. 15–9 A fire/smoke tower is designed to direct smoke away from evacuating occupants.

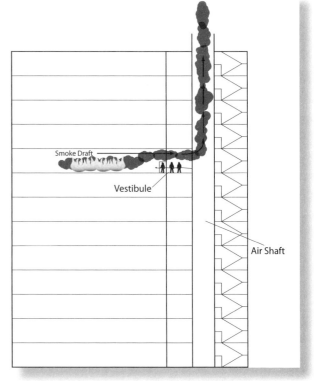

Fig. 15–10 Fire/smoke towers should not be used as an attack stair, as heat, fire and smoke can be drawn toward an attack team.

Fig. 15–11 Old style, electric smoke ejectors. Small unit approx. 5,000 cfm and large unit approx. 10,000 cfm

smoke tower, will have to be chocked open in order to advance a hoseline out onto the fire floor from the stairwell. With the doors left open, there is no longer a barrier to prevent the draft into the stairwell. However, when the fire/smoke tower is used for its intended purpose as an evacuation stairwell, the doors are not left open, but usually automatically close as fleeing occupants pass through them. Thus, the fire/smoke tower works as designed, and keeps the stairwell from becoming contaminated with heat, smoke, and fire gases. Bottom line make every effort to avoid using a fire/smoke tower as an attack stair (fig. 15–10).

Natural vertical ventilation

Earlier I addressed the area of smoke mushrooming on the top floor(s) of a multistory building or high-rise. This can be a deadly problem, especially with an uncontrolled fire, as the smoke banks down, affecting several of the upper floors. Clearly, the obvious need to open the building up at the top, above this deadly smoke, is of utmost importance. However, from a ventilation standpoint, operating members must complete this task with a comprehensive understanding of the potential negative impact it might have on operating forces on and near the fire floor.

Specifically, we are talking about the powerful draft of stack effect that might be greatly exacerbated by opening the building up at the top level. In many situations, especially if deadly smoke is mushrooming at the top floors, and if occupants are in danger, there will be no choice but to open up. However, anytime this is done, it must be communicated to all command locations, especially the officer or chief in charge of the fire attack group. Communication must be maintained, so if the vertical ventilation is having a negative or dangerous impact on operating forces, it can be closed up.

Operating members assigned to complete such vertical ventilation must remember to apply the basics, including maintaining the integrity of the door or other opening, if possible, so that it can be closed, if necessary. Firefighters have also been taught to remove a roof bulkhead door completely from its hinges, so as to maintain an effective ventilation opening. In some case, such as at low-rise building fires, this may be acceptable and, in fact, important to do. However, I would suggest not doing this at larger buildings, especially tall high-rises, as we might have to quickly close the door if a powerful and dangerous draft is created that compromises the position of operating members on the fire floor.

Positive pressure fans

Many high-rise buildings will not be equipped with a smoke control/air management system. Even in buildings that are equipped with these systems, we must be prepared for a system failure or the need to augment the building's ventilation system with our equipment. Most fire departments are equipped with powerful fans designed to move a significant volume of air. The oldest models that have been around for years are referred to as *smoke ejectors* (fig. 15–11). These square units come in two common sizes, and are capable of moving approximately 5,000 and 10,000 cubic feet of air per minute (cfm) respectively.

The newest models of fans are the very powerful electric and gasoline-powered, positive pressure ventilation (PPV) blowers (fig 15–12). These units come in a wide range of sizes, and the larger units can move close to 20,000 cfm. Although these PPV blowers come in electric and gasoline models, the advantage of the gasoline models is that they can be as much as 40% more powerful than the electric models. However, the gasoline units are noisier than the electric units, and they do produce exhaust, including carbon monoxide. The electric units are much quieter and do not emit exhaust.

Currently, there is excellent research and development in the American fire service regarding the use of positive pressure ventilation for high-rise firefighting operations. The National Institute of Standards and Technology (NIST) is leading the research, with the assistance of numerous fire service leaders who are highly experienced in high-rise operations. Fire officers from across the country, including high-rise experts from the Chicago Fire Department and FDNY, have attended two recent ventilation tests in high-rise buildings.

Central to this research is how fire departments can control air movement and smoke management during high-rise building fires using positive pressure blowers. The first of two tests was conducted in a vacant, high-rise building in Toledo, Ohio. This was an excellent opportunity, thanks to Chief Skip Coleman and the Toledo Fire Department, and served as a good starting point for data collection. Unfortunately, the tests in Toledo did not involve live fire, therefore data collection was limited.

Subsequent testing was conducted in a vacant Chicago Housing Authority high-rise. These tests included live fire, and provided invaluable data regarding realistic smoke movement driven by thermal energy in a high-rise building. During this weeklong testing, numerous fires were set in different locations of the building to evaluate smoke movement.

A friend of mine, Bob Hoff, who is the Assistant Deputy Commissioner of Operations for the Chicago Fire Department (CFD), shared some preliminary information from the tests with me. In one scenario, the CFD used their mobile ventilation unit (MVU) (fig. 15–13), (which is a large positive pressure blower mounted on a light duty truck chassis) to simulate a 15 to 20 miles per hour (mph) wind driven fire in a fourth floor

Fig. 15–12 Large, positive pressure blowers can be very effective for high-rise ventilation.

apartment. The CFD's MVU can be elevated, which allowed for it to be placed up high, just outside the fourth floor apartment windows. Based on past experience throughout the American fire service, a 15 to 20 mph wind is actually fairly tame compared to what has been encountered at some very powerful wind driven fires. However, even this relatively light wind condition created an extremely powerful wind driven fire that clearly illustrated how destructive a force this can be. The test fire in question pushed through the apartment and out into the public hallway with such blowtorch like power that it burned through two sheets of 5/8-inch sheetrock board within a very short period of time (approximately 4 to 6 minutes). Unfortunately, this wind driven test fire also destroyed several pieces of NIST's expensive data collection equipment.

Ultimately, fire service leaders and NIST hope to demonstrate the value of and establish procedures that might help fire departments combat the effects of wind by using portable positive pressure blowers to support the fire attack. In addition, they hope to refine procedures used to pressurize stairwells and help remove smoke on various levels of high-rise buildings that are not equipped with sophisticated smoke control/air management systems.

This a good start in the right direction to achieve a more comprehensive understanding of air movement in high-rise buildings and effective ventilation procedures. However, before any conclusive evidence can be applied to our high-rise operations, there remains a significant amount of testing to be done. The use of PPV on the fireground can be a benefit, but must be used with discretion and within the scope of a well coordinated plan. Furthermore, attempting to overcome a powerful wind condition is an extremely difficult proposition: we are attempting to fight Mother Nature. On going testing and evaluation will hopefully continue to yield positive results that can be applied at future high-rise fires.

Operational guidelines

When areas or entire floors of a high-rise building are contaminated by smoke, we can implement positive pressure ventilation procedures to help remove the smoke. The utilization of vertical shafts, specifically one of the stairwells, will be necessary, in most cases, to move the smoke. Generally, one or two blowers located at ground level can move enough air to start venting a contaminated stairwell. Additional blowers placed on stairwell landings every 10 floors will help augment the movement of air.

A combination of vertical and horizontal ventilation can be used to clear smoke-filled floors. One stairwell can be pressurized with air using PPV blowers for vertical ventilation, and then additional blowers can be set up to positively pressurize a fire floor via horizontal ventilation. If windows can be opened, they can be used to selectively remove smoke from floors contaminated by smoke. For high-rises with nonoperable windows, both stairwells could be used to move the smoke, but only after the stairwells have been thoroughly searched and any occupants have been removed from harm's way.

Summary

Air movement and the ventilation of smoke within a high-rise building can be extremely complicated. Firefighters must begin with an understanding of natural air movement phenomena,

Fig. 15-13 This is the mobile ventilation unit (MVU) used by the Chicago Fire Department

especially stack effect, along with a building's air movement (HVAC) and smoke control/air management systems. Smoke contamination on the upper floors of a high-rise building has been deadly at numerous high-rise fires to both civilians and firefighters. We must continually work to develop the best methods to aggressively ventilate the smoke before it can harm occupants of the building and operating firefighters.

References

1. Chapman, Elmer F. 1984, Ventilation Principles for High-Rise Buildings, with New York Firefighters (WNYF), 4th Issue, 1984, p. 11.

2. Ibid. p. 11.

3. Ibid. p. 10.

4. Chapman, Elmer F. 1983, HVAC and the Fire Chief, with New York Firefighters (WNYF), 2nd Issue, 1984, p. 11.

5. Ibid. p. 11.

6. Ibid. p. 11.

CHAPTER 16

Rapid Intervention Team (RIT) Operations

There are various names used throughout the American fire service to describe the fireground operation of firefighter rescue. Probably the most commonly used term is rapid intervention team, or RIT. Some fire departments prefer to use rapid intervention crew, or RIC. The term FAST, or firefighter assist search team, is also used by some fire departments. The term that I am most familiar with is *rapid intervention team* or RIT, which is used by the Denver Fire Department (DFD).

Certainly what's most important is not the name, title, or phrase used to describe the operation, but rather the operation itself. Specifically, the fact that additional resources have been proactively called for before they are needed, are on scene at an incident, and are dedicated to the fireground operation of firefighter rescue, should the need arise is key. For consistency, I will use the term RIT, as that is what I am most familiar with, and it is certainly used widely across the American fire service.

RIT History

During the decade of the '90s, and into the beginning of the 21st century, the whole concept of *rapid intervention* was developed and has evolved into what it is today. From a time when even the term *RIT* was foreign to most fire departments, to today when RIT is not only a common term to most, it is now the topic of countless articles, books, and even Web sites dedicated to RIT operations. Terms like "two in, two out" are used by many in the fire service, and most fire departments operate with some form of RIT in a standby position at all fireground operations.

The decade of the '90s certainly included a strong commitment to the subject of rapid intervention, as well as overall firefighter safety and survival. However, with all of our advancements, we still have a tendency to get stuck in one place, specifically, the single-family dwelling mind-set. A great example of this would be our tendency to apply single-family dwelling RIT procedures to large commercial and high-rise buildings. Our own history has proven time and again how this misapplication can result in tragedy.

Using RIT at Larger Buildings

At a small- to medium-sized, single-family dwelling, a RIT of one fire company, with a minimum of four personnel, is considered sufficient by most fire service authorities. However, once again, as history has proven, even a RIT of that size might not be enough to effect the rescue of a downed firefighter. That one-company RIT with three or four personnel doesn't even begin to address the potential firefighter rescue problems at a commercial building, especially large commercial buildings. How well prepared is the average fire department for RIT operations at a high-rise building, especially a large commercial high-rise?

Some of the more proactive fire departments in the American fire service are moving toward larger teams for RIT operations, especially at large commercial buildings. What was once referred to as a rapid intervention team has now evolved into a *RIT group* with several personnel assigned and, in many cases, multiple companies. Suffice to say, at a high-rise operation, there may be a need to establish multiple RITs within a RIT group, with multiple companies assigned to it and preferably led by a chief officer. This is very doable for large fire departments with significant resources, but it becomes a considerable challenge for the small- and medium-sized departments.

Yes, this is much easier said than done. To begin with, most fire departments, even some of the big ones will be overwhelmed at the beginning of any significant high-rise fire event. At the outset, the ability to establish and maintain even a small RIT will be a luxury that most can't afford, that is, if they're going to accomplish some of the basic goals such as putting the fire out and searching the building.

Keep in mind, we have learned a lot and evolved significantly over the past decade. However, that evolution took place over a long period of time and occurred after accomplishing numerous small steps in the right direction. Several years ago, I remember responding as a RIT company to a serious working fire with first and second due companies operating inside the fire building. The chief in command/ordered my company (the RIT) to stage a block from the fire and stand by for further orders. Yes, very ridiculous, but true. Fortunately we have evolved, and most organizations are today applying the concept of RIT in the proactive manner for which it was designed. There is, however, a significant learning curve, and we still have much work to do. We must address the absurd mentality that a RIT of 2, 3, or 4 firefighters is sufficient for a fire in a large high-rise building; it is *not*!

Operations at Low-Rise Buildings

Placing the RIT in a proactive position where they can be most effective in the event they need to be deployed was an important step in the evolutionary process, specifically for our most frequent operations at low-rise buildings. Several years ago, I developed a concept called RIT RECON, and have used it with great success since then. It is primarily designed for our operations in small- and medium-sized buildings, such as fires at single-family dwellings, and small commercial buildings. The RIT RETCON has been adopted by the DFD as part of our RIT procedures.

RIT RECON

The idea of RIT RECON is to have the RIT be proactive, with the goal of identifying and, if possible, eliminating some potential hazards that could lead to a need for RIT deployment if not addressed proactively. The central theme is information gathering or reconnaissance and proactive measures to increase firefighter safety and survival. At operations in low-rise buildings, I have my RIT complete a reconnaissance for me.

The R E C O N is an acronym which I have developed, and is defined as follows:

R – Rescue of firefighters
E – Egress for firefighters
C – Construction of the fire building
O – Outside survey (360 degrees)
N – Nasty hazards (hazardous materials)

R – Rescue of Firefighters. The RIT should quickly identify the primary location or locations of operating fire companies and firefighters. This can be done in many ways, including listening to the radio transmissions and assignments given on the tactical channel prior to arrival. The incident commander or group supervisor should communicate as much information as possible to the RIT regarding the location of fire companies, firefighters, assignments given, and so on. This should be done via face-to-face communications at the command post or the division location, if possible.

Furthermore, the RIT should ask questions to gain as much information as possible. In conjunction with the information regarding the location of firefighters, the RIT should attempt to identify any potential hazards, including items that might create an entrapment for firefighters, or any other problems that could affect firefighter operational safety and survival. Finally, the necessary steps should be taken to eliminate, or at the very least reduce, these hazards.

E – Egress for Firefighters. The RIT is assigned to proactively remove security grates or burglar bars covering windows or other means of egress at our low-rise and, specifically, ground-based operations, thus establishing a secondary means of egress for firefighters operating in the interior of a building. Also, the RIT can be used to proactively throw ground ladders to locations above grade, such as second and third floor windows, whenever firefighters are operating in those above grade locations. For the upper floors of a low-rise building, an aerial ladder could and should be used to proactively establish secondary egress within the RIT RECON. These specific examples might not apply to a high-rise building fire, but the concept remains the same: identifying, establishing, and maintaining egress for firefighters.

C – Construction Type. As part of their information gathering, the RIT during their RECON should attempt to identify what the principal type of construction (structural component) of the fire building is, whether or not a truss or trusses serve as part of the construction, and whether or not lightweight components are utilized as part of the construction. For high-rise operations, this means opening up the plenum space on a floor somewhere below the fire, and visually inspecting and identifying the construction features of the building.

O – Outside Survey. The RIT should attempt to complete a 360-degree survey of the outside (if possible), with a focus on identifying the approximate dimensions of the fire building. Also, identify if the fire building is an irregular shape. Both of these will be very beneficial at high-rise operations and operations in standpipe-equipped buildings—this information will help determine a proper course of action, the amount of area that could potentially be involved in fire, and the necessary resources required to complete various operations including search. During this 360 survey, the RIT also identifies the four primary exposures to the fire building, their locations in relation to the fire building, what they are, and if they are threatened or could potentially become threatened. This is, of course, most applicable at low-rise operations.

N – Nasty Hazards. The RIT should attempt to identify any nasty hazards that exist, such as any hazardous materials placards (specifically an NFPA 704 Hazardous Materials placard). At high-rise operations, this information may be obtained at the fire command center via review of any material safety data sheets (MSDS) for the building (fig. 16–1). Identify if any electrical hazards exist, such as electrical service threatened by fire, wires already burned through and down on the ground, and so forth. Also, identify if there is any natural gas, propane, or any other gas threatened. Ultimately,

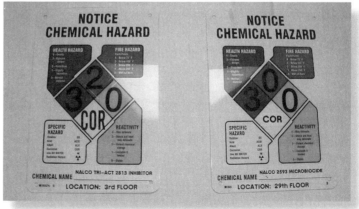

Fig. 16–1 A RIT can identify the type and location of hazardous materials within a high-rise building by reviewing materials safety data sheets (MSDS) located in the fire command center.

attempt to identify if any other nasty hazards or potentially nasty hazards exist that could negatively affect the overall operation, or compromise firefighter operational safety and survival.

Once again, although designed and used primarily for operations at low-rise buildings, specifically our most frequent fires such as single-family dwelling and small commercial building fires, many of the components in the RIT RECON can and should be applied to operations at larger buildings, and during our high risk, low frequency events, such as for fires in high-rise and standpipe-equipped buildings.

High-Rise RIT

When it's a not so typical fire like a high-rise, we must evolve even further. Having a single RIT at the command post either outside the building or perhaps in the lobby, with a serious fire on the 20th floor, is a great example of a serious shortcoming. Remember, the first word in the RIT acronym is *rapid*! If we are going to have any chance at all in rapidly helping another firefighter or firefighters in need, we have to be reasonably close to them to begin with.

At a serious high-rise fire, there will obviously be several different operations going on simultaneously in order to accomplish our basic operational goals and objectives. Two significant operations include fire attack and search; therefore the overall operation should include a RIT to support each of these specific operations.

Fire attack RIT

Let's start with the fire attack. This will likely be the first operation to get underway, hopefully. It will certainly be a critical area, placing operating firefighters in a very hazardous environment. Therefore, it only makes sense that we place a RIT as close to this operation as possible for potential deployment (fig. 16–2).

Keep in mind, this is most likely not going to be the only, or sole RIT for the entire operation. This is merely one of several RITs that will potentially be placed in several critical locations and operating under the command of a RIT group supervisor, preferably a chief officer. Furthermore, the fire attack RIT might include several companies at this one position.

There is a big difference between having several fire companies committed to and operating on a large, 10,000, 15,000, or 20,000 square-foot floor of a commercial high-rise versus two engines and a truck operating on the upper floors of a residential high-rise, with eight 750-square-foot apartments. We must remain flexible and dynamic with regard to resource deployment for RIT operations at a high-rise. Always be prepared to expand the size of a RIT, depending on the potential needs of a given incident.

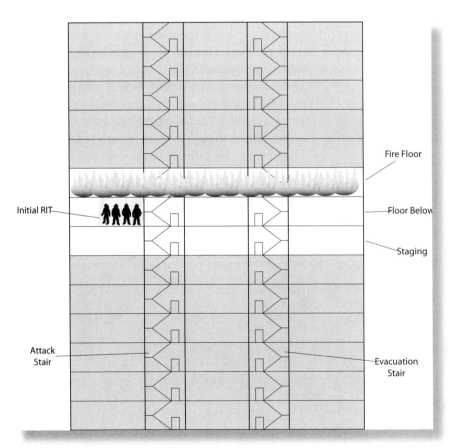

Fig. 16–2 An initial RIT should be located as close as possible to the fire floor.

The first RIT to be established at a high-rise fire or a fire in a large, standpipe-equipped building should be placed as close as possible to the primary fire attack location, keeping in mind that there will potentially be extension of the fire, and, in some cases, the possibility of other fire locations within the building. However, in our most common situations, there will likely be a clearly defined fire attack location (fire floor) early in the event. Getting a RIT to an unaffected area below that fire location, but as close as possible to it, and as quickly as possible should be a high priority. With a probable interior staging location two floors below the fire floor at high-rise events, the floor below the fire floor becomes the obvious choice to locate the fire attack RIT (fig. 16–3).

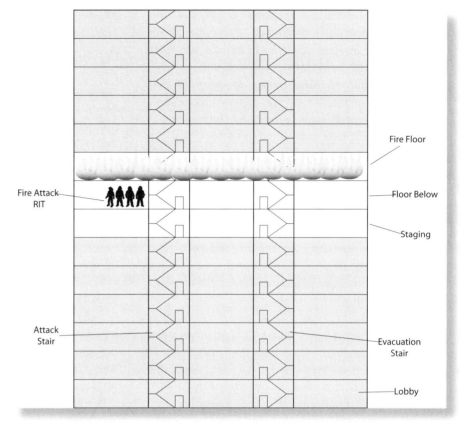

Fig. 16–3 The fire attack RIT can be located on the floor below the fire floor.

Upper search and evacuation (USE) RIT

As the event unfolds, numerous other operations will take place. There will be a need to search not only the fire floor, but everything above, including stairs and elevators. The myriad of support operations that will have to be completed will be significant. For each of those operations, any that have to be performed in an immediately dangerous to life and health (IDLH) atmosphere that is contaminated by smoke and/or heat will require the additional support of a RIT. Where to locate the RIT in these situations is not always a clear and easy answer.

The fire that occurred at the First Interstate Bank Building in Los Angeles in 1988 is a good example. The fire started on the twelfth floor of a 62-story commercial high-rise building. The fire rapidly extended upward, and, at its peak, multiple floors were involved. In this real-life high-rise fire event, LAFD had to deal with 50 floors above the initial fire floor. This was a very large commercial high-rise building, with thousands of square feet per floor. The 50 floors above had to be searched, and all areas had to be checked for extension. The firefighters involved in these type operations can and have become victims themselves at numerous high-rise fires. There must be a RIT ready and in position to deploy in the event that one or several of our own becomes lost, trapped, or somehow incapacitated.

Where should a RIT be placed when we are talking about extremely large and spread out areas, such as numerous floors above the fire floor? Obviously, the safest area is below the fire, but if that's the 11th floor and a firefighter becomes trapped on 42, you can take the *rapid* out of RIT.

The RIT cannot be placed in harm's way itself, that is, if we expect them to be able assist and possibly rescue another firefighter. But we need to think and identify the possible locations where a RIT could be located, preferably as close to operating firefighters as possible, but still in a relatively safe location, at least free of smoke and heat.

Some areas to consider would be the evacuation stair and any number of potential floors above, if these areas are free of smoke and heat conditions (fig. 16–4). Also, one must not overlook the roof as a potential area to place an additional RIT, but, long before the event, the logistics associated with this operation must be completely resolved.

Roof RIT (air support)

The fastest way to place firefighters on the roof of a high-rise building is via a helicopter. For those few fire departments that have their own air support division, they are well ahead when it comes to this operation. A regular training program and the establishment of airborne fire companies is a critical component associated with fireground success and firefighter safety.

For the majority of fire departments that don't have their own helicopter, the first step would be to identify who does, what their capabilities are, and if they are willing to assist you on a moment's notice. Last but not least, can you trust them with your people? A comprehensive discussion of air support operations, including the insertion of a RIT on the roof, is addressed in chapter 18.

Mutual and automatic aid agreements

Placement of the RIT(s) at a large scale high-rise operation is going to be a complex and demanding task. At most high-rise operations there will be a need for more that just one RIT. There may even be a need to dedicate a full alarm assignment to RIT at a major high-rise fire. This is easier said than done for most fire departments that are already operating with limited resources due fiscal constraints and/or to years of budget cuts. Fire service leaders must be creative and think of outside resources, establishing mutual aid and automatic aid agreements prior to the major events. Being prepared for this eventuality will pay dividends on the fireground.

The concept of *task forces* and *strike teams* used in the state of California is a good example for all of us to follow. This was originally designed and is primarily utilized for large scale wildland firefighting operations, but it can also be utilized for serious campaign fires in structures, specifically high-rise fires.

In many metropolitan areas with large numbers of high-rise buildings, there are often times many separate cities and jurisdictions, and, thus, several different fire departments. The Denver metropolitan area is a good example. The

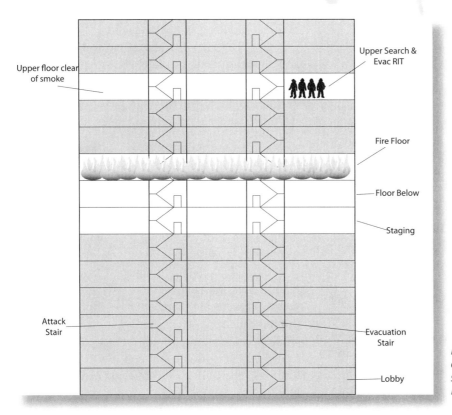

Fig. 16–4 An upper floor or the evacuation stairwell, if clear of smoke, are potential areas to locate a RIT.

DFD is a large fire department, but serious high-rise fires in the past have significantly taxed the resources of this organization.

Within the Denver metropolitan area are numerous small and medium-sized fire departments that collectively have substantial resources. The bottom line is we are much more powerful working together as one team, than we are separated by jurisdictional lines. I look forward to the day, hopefully in my career, when I can ask my dispatcher to send me several strike teams for assistance at a serious high-rise event. This concept could help make multiple RITs and the long list of numerous other high-rise fireground functions a reality for the countless fire departments that are restricted by limited resources.

Fig. 16–5 Firefighters assigned to a RIT must be equipped with long-duration SCBA.

RIT Basics

The basics of RIT used at low-rise operations, can also apply to RIT operations at a high-rise fire. The same tools and techniques that are used for most RIT operations at our most frequent fires can be applied to the operations at a high-rise or standpipe-equipped building. However, it never hurts to reinforce the critical items and equipment.

Long-duration SCBA

The need for long-duration breathing apparatus at high-rise operations is critical to RIT operations. Specifically, fire departments that are expected to or may potentially operate at high-rise events must be equipped with long-duration SCBA. For most fire departments, if they are so equipped, that equipment is going to consist of 1-hour air cylinders (fig. 16–5). This should be considered the minimum for all members assigned to RIT operations.

There is also a wide range of long duration SCBAs with operational times in the four-hour range. These are the rebreather-type SCBAs, and should be considered for high-rise operations, especially for RITs (fig. 16–6).

RIT tools

Other essential RIT tools include forcible entry hand tools, as well as hydraulic forcible entry tools such as the rabbit tool or Hydra-Ram, and power tools such as circular saws with metal cutting blades (fig. 16–7). In addition, powerful

Fig. 16–6 Some long-duration SCBA, such as the rebreather-type units, are rated to supply air for up to four hours.

lights (hand lanterns), thermal imaging cameras, and tag line ropes should be considered essential RIT equipment for high-rise operations. Keys to open elevator hoist-way doors, as well as general building master keys, can also help facilitate RIT operations.

Resources needed for RIT operations

During high-rise operations, the deployment of a RIT might end up being for something as simple as finding a lost firefighter and guiding him to safety. On the other hand, a RIT might and has been deployed to rescue a completely incapacitated 225-pound firefighter in full gear located a long distance from the evacuation stair. The latter of these scenarios will obviously be a much more complex, and extremely demanding operation, both physically and mentally, for all those involved.

Furthermore, contemporary lessons in the American fire service have demonstrated that a RIT operation might well be a multiphase operation, requiring numerous resources. As the Phoenix Fire Department (PFD) learned from the tragic death of Firefighter Brett Tarver, rapid intervention is not always rapid. Firefighter Tarver became disoriented while operating in a large commercial building fire. Multiple firefighters had to be deployed to affect Tarver's rescue, and some of those firefighters became victims themselves. The PFD conducted a very proactive and comprehensive investigation of this tragic incident, which included extensive RIT training and data collection. The PFD found out, and shared with the American fire service, that the resources needed to rescue a single, downed firefighter are much greater than we had previously thought. In addition, they found that we must prepare for and anticipate that one or several members of the rescue teams might get into trouble, become lost, trapped, run out of air, and thus become victims themselves.

First, especially for a RIT operation in a large high-rise or standpipe-equipped building, it will require several firefighters just to locate the downed firefighter. Once the firefighter is located, several other firefighters may be needed to stabilize him and prepare him for removal. And finally, several more firefighters will have to complete the rescue and removal. Most disheartening is the fact that during this multiphase process, we should prepare for and expect that some of the rescuers may also become victims themselves, thus further compounding the overall problem.

If elevators have failed, or have been deemed unsafe to use, you must also consider the complicated logistics associated with removing an unconscious firefighter from several floors above ground. In some cases, it might be easier and faster to take the victim firefighter up rather than down. But once again, this would ultimately require the use of air support and the ability to evacuate the victim from the roof as safely as possible—another easier-said-than-done proposition for most fire departments.

Door strap/search marker

In chapter 14, the concept of using door strap/search markers is discussed. These devices, used by the DFD, and different variations used by other fire departments, are excellent tools for search operations in large buildings.

Over time, the evolution of this simple device included the fact that it can be a tremendous benefit during RIT operations. Specifically, when a search team is using the device, they place it over both the inside and outside door knobs of the entrance door to the area being searched. This serves to keep

Fig. 16–7 A variety of forcible entry tools is essential for high-rise RIT operations.

the door from locking behind the search team, and also identifies that the area is currently being searched. In a situation where there is a firefighter or group of firefighters who have given a *mayday* or otherwise called for help, the search marker can be an excellent way to assist in locating those in need of help.

For a RIT to locate a missing firefighter or group of firefighters in a large high-rise building can be like looking for a needle in a haystack. However, where search markers are used by the missing firefighter or group of firefighters, the RIT can quickly narrow its search. A search marker left hanging on an outside door knob indicates that a primary search has been completed (fig. 16–8). The RIT could quickly bypass that area. However, when the RIT locates a search marker that is not hanging, but is wrapped around the door and hooked onto both door knobs, this identifies an area where firefighters are located (fig. 16–9). Thus, the search marker can be a benefit to facilitate rapidly locating a firefighter or firefighters who are missing, trapped, or are in distress.

Keep in mind, the previous examples using search markers to narrow the search for firefighters are based on operations in areas with some compartmentalization, specifically entrance doors to rooms. It is on these doors that the search marker is used. However, for many situations, there may be only a few entrance doors to much larger, uncompartmentalized open areas, such as a trading floor in a large commercial high-rise. In these situations, RIT and search operations will be extremely difficult and complicated, and the search markers will provide less assistance.

Extreme operations

A serious fire in a large, commercial high-rise office building will present numerous problems. With the potential for building systems failing throughout the operation, specifically elevators, a RIT deployment to locate and rescue an injured, trapped, or otherwise incapacitated firefighter represents an *extreme* operation.

In these situations, we will need to call upon our A-team, and they must bring their *A* game. Firefighters in top physical condition, who are also prepared mentally and physically for this extreme operation, will give us the greatest potential for success. These firefighters need to be very innovative, prepared to implement and utilize a wide range of solutions and makeshift equipment to solve these extreme problems.

Tools, equipment, and methods that might be applicable at ground-based RIT operations may not be realistic or practical in the high-rise arena. To begin with, equipment such as a Stokes litter could be too heavy and cumbersome to carry aloft. Once an unconscious, 225-pound firefighter is loaded onto the litter, carrying it with the victim onboard down several flights of stairs can be extremely difficult. Negotiating corners and turning at each floor landing would be very cumbersome, if not entirely impossible to accomplish.

Fig. 16–8 A search marker left in this position indicates that the primary search is completed and members are no longer located here. The RIT could quickly bypass these areas when looking for a search team that is lost or in distress.

Fig. 16–9 A search marker in this position will indicate to the RIT that this is a location with operating firefighters, perhaps the ones who are in trouble.

When I was the Captain of DFD Rescue Co. 1, my crew and I practiced several methods to remove firefighters from the upper floors of high-rise buildings via the stairs. One tool that had been used with great success at numerous ground-based operations was the SKED rescue sled (fig. 16–10). It showed tremendous promise as a potentially viable method to evacuate a firefighter by lowering the loaded firefighter down each flight using a simple rope system, in conjunction with the hand rail. At the stair landings, turning a large member loaded in the SKED is difficult and cumbersome, but it can be done.

Improvised carrying devices

Ultimately, the rescue and removal of a firefighter or civilian from the upper floors of a high-rise building will likely require that firefighters quickly improvise a portable carrying device from whatever may be immediately available. RIT firefighters should keep in mind the wide range and variety of potential carrying devices that exist in a typical high-rise building. Items such as lightweight office chairs can be used to carry victims down several flights of stairs. Also, small throw rugs, covers, or fire department salvage covers might be useful to load and carry a victim.

Firefighters, being the true problem solvers that they are, will identify and implement whatever is necessary to get the job done. There are countless successful methods out there, and many more will be developed in the heat of battle at future events. All firefighters must continually share their new and innovative ideas and methods that are discovered under extreme conditions and utilized with great success.

Superman carry

Years of training and experience has shown that probably one of the best and fastest methods to remove a firefighter or other victim from the upper floors of a high-rise building via the stairs

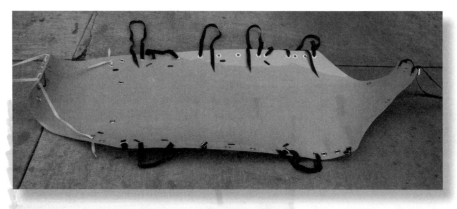

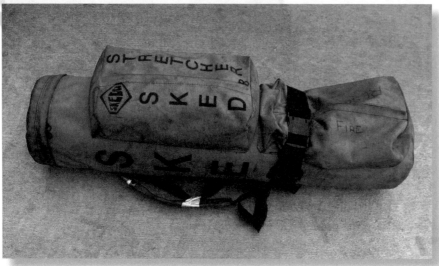

Fig. 16–10 The SKED rescue sled is a lightweight device that can be very applicable for high-rise RIT and rescue operations.

is to simply carry them. I said simply, but there will be nothing simple about this extremely physical evolution.

DFD firefighters have developed and introduced the *Superman carry* that physically strong firefighters can utilize to remove an incapacitated firefighter or civilian. The Superman carry requires the rescue firefighter to loosen his SCBA harness and move it slightly to the right side of his body. The other members lift the victim up and load the victim onto the rescue firefighters back. The rescue firefighter grasps the victim's hands in front to securely hold him. The other members will support this operation, stabilizing the victim and supporting legs, as well as protecting the rescue firefighter from slipping and falling.

The Superman carry can typically be used for one or two flights per rescuer, and then the victim must be laid down, and another firefighter will have to take over. Rotating in such a manner, a team of four physically strong and mentally prepared members can evacuate an unconscious victim down several flights in a relatively short period of time.

When I say a team of four firefighters could do this operation, I mean real *firefighters*. Not those who pretend, wear the uniform, and collect a check every other Friday. The firefighter who developed this carry is a very large and extremely strong man. Those who would comment that this is simply impractical are the exact kind of people who you should expect can't and won't be able to do it. Remember, we're talking *extreme* operations. We need extremely strong, dedicated, and prepared firefighters to complete the extreme assignments.

General RIT Strategies

There are some general RIT strategies that can be applied to fireground operations in high-rise and standpipe-equipped buildings. These general strategies can serve as a starting point for RIT implementation, keeping in mind that this is an ever-evolving process. New ideas and better methods will continually surface, with the best of those proven during extreme operations. For now, I would suggest the following procedures.

Commercial high-rise operations

As soon as possible after the major operational components have been put in place, namely fire attack and search operations, at least one fire company, preferably a truck company, should be assigned to the position of fire attack RIT. This company should be equipped with the necessary tools and equipment, and proceed to a position just outside of the fire attack area. This will typically be on the floor below the fire, or inside the fire attack stairwell just outside the entrance to the fire floor.

Keep in mind, common sense and operational efficiency will not allow for a RIT at a serious high-rise fire event to be placed into a stationary position, and not move or perform any tasks at all. Rather, the RIT should maintain its primary responsibility of firefighter rescue should the need arise, while assisting with numerous support tasks. For example, from the floor below or inside the attack stairwell, the RIT should assist with the movement of the attack hoseline or hoselines. The company officer in charge of the RIT must maintain discipline, having all members of the RIT staying together as a unit, with no freelancing or committing to other specific assignments without direct orders from the incident commander or a group supervisor.

For all high-rise fires, all of the areas above the fire floor are considered serious exposure concerns. All areas of each and every floor, along with stairwells and elevators above the fire floor must be searched for victims and fire extension. An upper search and evacuation (USE) group will be implemented, preferably with a chief officer in charge as the USE group supervisor.

With numerous companies likely assigned to this USE group, and potentially spread out over a wide area of several floors, a strategically-placed USE RIT will be essential. The easiest way to initially provide this USE RIT is to give a company a dual role assignment, with RIT as the primary responsibility.

I would suggest assigning the USE RIT to proceed to the upper search and evacuation (USE) area via the evacuation stairwell. This stairwell will need to be completely searched, thus, at least one company, preferably a truck company, can be assigned to the USE RIT, with a collateral assignment of searching the evacuation stairwell. This will place the USE RIT in the area of potential deployment, while conducting and completing another essential assignment.

If the evacuation stairwell was kept clear of smoke and heat, this USE RIT will be able to complete the stairwell search without placing themselves in an IDLH atmosphere. In the event that conditions change, and smoke and/or heat is encountered, or if this USE RIT locates a victim or victims during their search that must be rescued or removed, they should communicate this to their group supervisor, at which point they will no longer be considered a RIT. Another company will need to be assigned to the USE RIT, as the previous company assigned to this had to be reassigned to another critical task, or because the search became much more complicated and dangerous due to smoke and heat conditions.

Some readers may be critical of this philosophy, thinking that this RIT will become physically exhausted just climbing stairs and searching the stairwell. This is and will be extreme. We need physically and mentally prepared firefighters for this and every assignment at the serious high-rise fire event. The full-time real-estate agent who goes to the firehouse on his day off will not be an asset at this, or any other emergency event for that matter. This one's for the professional firemen, no doubt about it!

As the operation evolves, at least one later arriving company, such as a unit responding on an extra alarm (preferably a truck company), should be assigned to lobby RIT in the building lobby to be deployed or relocated as necessary (fig. 16–11). Operational efficiency and common sense dictate that this lobby RIT can and should assist with other support functions, while standing by for possible RIT deployment. For instance, this lobby RIT, as a unit, can assist the lobby control unit with various duties, but they must remain intact and prepared for deployment, if necessary.

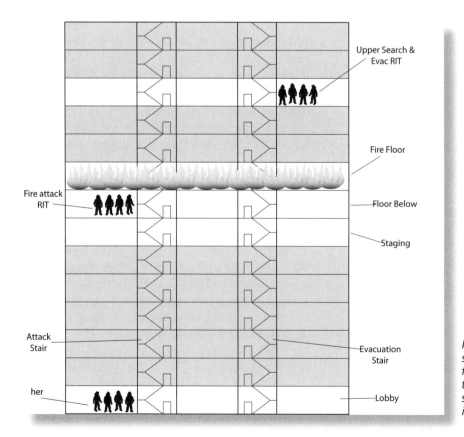

Fig. 16–11 Ideally, multiple RITs should be assigned at high-rise fires and located just below the fire floor, within the upper search and evacuation area, and in the lobby.

Finally, taking this extreme RIT operation to an entirely new level, one or more companies may have to be assigned to a position of roof RIT. This of course requires the insertion of members onto the roof, if possible. This will involve specific units, usually from the department's special operations units who have had previous training conducted on a regular basis in the area of air support operations and high-rise insertions. Once again, dual roles, or in many cases, multiple roles, may have to be assigned, but the primary function would be that of RIT, from the advantageous position of the roof (fig. 16–12).

Ultimately, we are ideally attempting to place a RIT in position at the main fire area, above that area if possible, and below. Most fire departments will not have the luxury of assigning a RIT or multiple RITs to various critical areas with the sole responsibility of rapid intervention. Assisting and supporting other operations, while remaining intact and prepared for the primary assignment of RIT, is the objective.

Residential high-rise operations

Because most residential, high-rise fires are typically less demanding and labor intensive than the commercial jobs, the RIT component becomes less of a challenge as well. Like the commercial operation, the first RIT needs to be a fire attack RIT in position near the main fire area. Because of fewer companies operating on the fire floor, and the advantage of compartmentalization typical of a residential high-rise, a RIT of only one company per area should be sufficient, at least initially.

Depending on smoke contamination, possible extension upward, and whether fire companies assigned to areas above are in an IDLH atmosphere, a RIT may or may not be necessary above. However, once a USE group supervisor reports smoke or heat conditions above, a USE RIT should be assigned.

At all residential operations, at least one additional RIT should be assigned and placed at the lobby level as a lobby RIT. Once again, like the commercial operation, the lobby RIT can be relocated and/or assist and support as necessary.

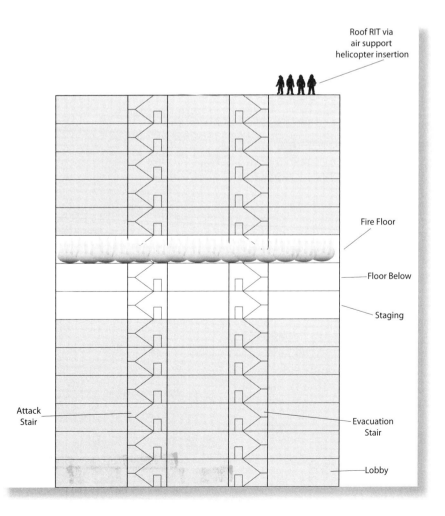

Fig. 16–12 Placing a RIT on the roof, if possible, could be very beneficial at a high-rise fire.

Fig. 16–13 Low-rise buildings, specifically large warehouses, require a significant RIT complement.

Low-rise and other standpipe-equipped buildings

Generally, the initial RIT strategies for low-rise buildings can follow the same basic recommendations outlined for commercial and residential high-rise buildings, respectively. Extremely large low-rise buildings may require a significant resource commitment to RIT operations.

Another very complicated fireground operation, especially from the perspective of RIT, is the extremely large buildings equipped with horizontal standpipes. For these buildings, simply envision them as a high-rise building lying on its side or *wide-rise* (fig. 16–13). You should work to provide a RIT resource as close to the fire as possible, along with RITs positioned on all sides. For example, at a serious fire in a large warehouse, positioning the first RIT close to the main fire area with immediate access is desirable, and then expanding to include RITs on at least two sides, preferably on all four sides, of a typical building, depending on fire and smoke involvement, as well as access (fig. 16–14).

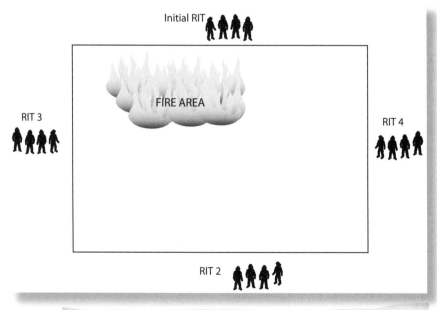

Fig. 16–14 Placing a RIT a close to the fire area as possible, along with a RIT on all four sides, may be necessary at a large warehouse fire.

Remember, unlike our high-rise standpipe operations, these buildings with horizontal standpipes don't allow for firefighters to begin operations from a safe area below the fire. Once inside these horizontal standpipe buildings, and depending on fire involvement, you are likely in an IDLH atmosphere. Therefore, RIT placement will likely be outside the building, but strategically placed to allow for the best, safest, and fastest access to areas where firefighters are operating.

Summary

The RIT operation at high-rise and other standpipe-equipped buildings is extremely complicated. Fire departments should not rely on and attempt to apply single family-dwelling RIT procedures to fireground operations in these much larger and more complicated buildings. The commercial high-rise is a particularly dangerous foe, and should not be underestimated. If we want to achieve *rapid intervention* to assist the injured, trapped, or incapacitated firefighter, we must place multiple RITs in strategic positions, so that rapid intervention can be a reality. This means early and proactive calls for help and additional resources. It also means utilizing companies for dual roles, including RITs supporting various operations, while maintaining their primary assignment of rapid intervention team.

CHAPTER 17

Command and Control

Up to this point in the book, I have addressed numerous different topics related to the overall subject of firefighting operations in high-rise and standpipe-equipped buildings. I have placed a significant priority and importance on specific operations, including getting the first handline in operation quickly, and completing a primary search of the most critical areas. However, every single component of a fireground operation plays a very important part in achieving the overall mission, and none of these components should be understated. Ultimately, there is a synergistic effect, resulting in operational success and safety, when all of the various fireground components are accomplished in a timely and efficient manner.

The subject of this chapter is arguably the most critical, and, in many respects, the most important of all the fireground operations. Those leading the overall battle are the choreographers, responsible for prioritizing the myriad of critical tasks and ensuring that they are accomplished as quickly and safely as possible. There is a huge difference between having an experienced and strong leader in command of a serious fire versus a weak and tentative individual (fig. 17–1). The engine company fireground saying, "As the first hand line goes, so goes the fire," can be given a command and control equivalent of "Operational success and safety goes, as the incident commander goes."

Of course, all of the numerous command staff positions are also of paramount importance. The IC must rely on his various incident command system (ICS) section chiefs, branch directors, division and group supervisors, area managers, and unit leaders, as well as the countless company officers, to make good decisions, based on sound judgment, and to specifically communicate conditions, actions, and needs (CAN report), in a calm, proactive, and timely manner. But it all still boils down

Fig. 17–1. A strong and experienced incident commander is the single most important component to a safe and successful fireground operation.

to one person, one individual who is in charge, in command, and ultimately responsible for the success or failure of an operation. That person is the *incident commander* (IC). As the saying goes, "The buck stops here," and that certainly applies to the incident commander.

ICS/IMS

As a young man growing up in the fire service, I was taught the incident command system (ICS). Over the years there have been various other versions out there, most notably Phoenix Fire Department Chief Alan Brunacini's *Fire Command*, and of course various hybrids that took what some believed were the best of different models and developed a system that worked best within a specific organization.

The most recent trend has been to move away from the term *incident command* and instead refer to it as *incident management*, something I consider to be new age, politically correct terminology. I am a firm believer that the fire service has suffered greatly over the past several years because of the growing number of "managers" and emphasis on management, while placing little or no emphasis on the importance of leadership and developing future leaders. We've got plenty of *managers* in the fire service: we need more *leaders*—good, strong leaders, who are willing to take a few risks, step up to the plate, and lead.

This philosophy goes completely against the grain and is contrary to the all-too-common contemporary management style that has unfortunately inundated the fire service, and society as a whole. Good leaders are hard to come by, and only include a small number of individuals who are willing to stick their necks out, often times only to be chastised, and, in some really sad cases, even formally punished because they stood up and did the right thing.

I am a firm believer in the formal system used to help leaders command and control major emergency events such as a serious high-rise fire. In this book (and on the fireground), I have and do utilize terms adopted and recommend by the National Incident Management System (NIMS). My only disagreement is that it is not management, it's leadership. The good people in leadership positions of the federal government and various national fire service bodies would make tremendous strides toward reestablishing a badly needed leadership mind-set in our fire service by utilizing the term *command* in our system, and leaving management to the numerous functions in society where lives don't hang in the balance.

Managers should be found at your local fast food restaurant. They should continually strive to ensure that there are enough hamburger patties and french fries on hand for the lunch crowd. Leaders should be found in front of and inside fire buildings. They should continually provide a confident and strong level of command and control to ensure operational success and safety for all members, along with good service to our many constituents, providing for their life safety and the conservation of their property. Fire department managers perform white-gloved inspections of firehouses, looking for serious violations, like crumbs in the toaster. Fire department leaders lead men into battle. There's a huge difference.

Yes, we do manage in the fire service, and there are numerous management functions that take place daily in the process of running a fire department. However, those management functions are generally not based on critical time constraints. We usually have the necessary time to deliberate, call our friends down at the human resource bureau, and bring in a variety of resources to assist in our decisions related to management issues. That's not the case when operating at a serious emergency incident, specifically fires in high-rise and standpipe-equipped buildings. Decisions typically have to be made quickly, with little time to deliberate or query several colleagues for advice and guidance. Furthermore, those decisions have to be accurate, because the decisions will affect the lives of civilians and firefighters alike. It is only through strong leadership that these critical decisions can be made, and made with confidence.

I will not use the term *management* in my book because this is not a management book. It is a book designed to provide guidance to fire service leaders, as they lead firefighters into battle at fires in high-rise and standpipe-equipped buildings, where the lives of firefighters and civilians hang in the balance. The new age management style of *"I'm OK, you're OK,"* and *"could you please, if you don't mind, search the attack stairwell, that is, if it's not too much trouble,"* is not leadership. Success at all fires and emergency incidents, especially serious fires in high-rise and standpipe-equipped buildings, hinges on the strength and leadership provided by a good, strong incident commander. Let's put the leadership back in the fire service. We can start by making a clear differentiation between *fire service management* and *fireground incident command and control*, which is true leadership.

The Initial Incident Commander

The first fire officer to arrive at the scene of an emergency incident should become the initial incident commander. At a working incident, this officer will operate in a fast attack command mode. At a serious high-rise fire, this initial IC will be inundated with many problems and will have to make accurate decisions in a very compressed time frame, all of which can either positively or negatively effect the overall operation.

This initial IC should keep it simple, focus on the known facts, and give a thorough but concise initial radio report that truly paints a picture for the other fire companies and chief officers responding to the fire. The *CAN report* is an excellent model, communicating the conditions, actions, and needs.

Most importantly, this initial IC must take action. As I said in chapter 11, placing the first handline in position quickly is a critical determining factor of the overall operational success. This initial IC will likely be in command for only a short period of time, and therefore should provide that initial command from a mobile position, while directing and coordinating the initial fire suppression operations.

Formal Command

Like many systems, on the Denver Fire Department (DFD), in the absence of a chief officer at the outset of an event, the command process will continue as the company officer of the second fire company to arrive, typically a truck company, assumes command, and establishes a formal command. In many situations, especially in larger cities, two or more companies may in fact arrive simultaneously, and thus the first command will actually be a formal command, where the IC actually operates in the command mode. For firefighting operations in high-rise and standpipe-equipped buildings, the first formal command, provided by the company officer of the second arriving fire company, should be established inside the fire building. This will typically be in the lobby, unless the fire happens to be somewhere on the lobby level, or if there are smoke conditions at the lobby level from a fire that is below that level.

The officer establishing this first formal command will announce that he is assuming command, and specifically communicate this to the initial IC, either via radio or face-to-face. At this point, the IC will have to wear many hats. Numerous critical fireground command and control functions and components will need to be implemented and maintained. Specifically, in addition to overall command, the IC will have to establish and maintain lobby control, as well as the position of systems control or simply "*systems*," responsible for evaluating and monitoring critical building systems, including the fire detection and protection systems, elevators, and HVAC.

Once a chief officer arrives, he should assume command, with a *formal* transfer of command, generally via radio communications. Upon assuming command, the chief officer should assign the former IC to his new position that will likely be the dual role of lobby control and systems, until relieved of one or both assignments (fig. 17–2). I provide more specific explanation regarding the duties of those positions later in this chapter.

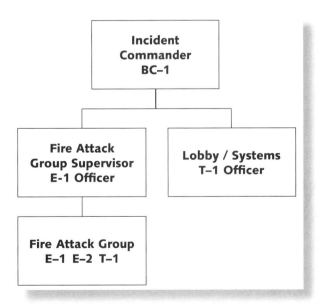

Fig. 17–2 The first chief officer to arrive will assume command, with the previous IC continuing with lobby control/systems.

Incident Commander

Of course, the most important of all the ICS positions will be that of incident commander (IC). The overall operation and all assignments given to fire companies will be the ultimate responsibility of the IC. All decisions, good or bad, regardless of whether the decision was made by the IC or one of his subordinates, will in fact be the responsibility of the IC. This position should not be taken lightly, as it truly is the most important position on the fireground, with the responsibility for everything that occurs.

Location of the Command Post

A topic that certainly generates a good deal of discussion, debate, and perhaps even some controversy, is the location of the command post (CP) at a high-rise fire event. There have historically been two distinct schools of thought regarding this: one of those is that the CP should be established, and maintained in the lobby of the fire building, and the other is that the CP should be established and maintained outside of the fire building, located a safe distance from the building. In reality, both of these CP positions have various pros and cons.

Interior command post

Establishing the CP inside the building, typically at the lobby level, has many advantages, most notably the fact that there is immediate access to critical information regarding the building and its systems, at least in many buildings, specifically modern commercial high-rises. Setting up the CP at or near the building's fire command center will provide the IC with critical information that would otherwise have to be communicated via radio from companies, teams, or units operating inside. Also, the IC has immediate access to building security and management personnel, as well as members of the building engineer's staff, who are vitally important to the success of the operation.

The other consideration that makes an interior CP more desirable is a simple matter of space. In many of the very large and densely populated cities such as New York, Chicago, Philadelphia, and San Francisco, buildings are built and located very close to one another, based solely on the reality of a limited, finite amount of real estate space available. Because of that, in many cases there may not be a very convenient, advantageous, and safe location on the outside perimeter of the building in which to establish a CP. In fact, in some situations, an exterior CP located in a relatively safe location might not allow for an unobstructed view of the fire building, if other structures and large buildings are in the way.

Exterior command post

On the other hand, the exterior CP provides many benefits that the interior CP does not. First and foremost is the opportunity to see at least two exterior sides of the fire building. For a serious fire in a high-rise or standpipe-equipped building, being able to evaluate and continually monitor the outside of the fire building is of critical importance. The events of 9/11, specifically the fires and subsequent collapses of the World Trade Center Towers are a very significant historical lesson underscoring the importance of this exterior viewpoint.

Ultimately, I believe that most incident commanders would like to have the ability to watch and continually evaluate their fire building. There are a significant number of cues, based on visual and audible senses, that can help guide the IC in his overall decision making.

Another very important element and benefit to utilizing an exterior CP is that it facilitates interagency communication. Having an easily identifiable, exterior fire department CP, placed in as safe a position as possible, enables other agency heads to report to that location and provide assistance to the fire department. Police, EMS, and the utility companies, are just a few agencies on the long list of outside agencies that the fire department must have immediate access to, and direct assistance from, in order to successfully and safely combat a serious fire in a high-rise or standpipe-equipped building. If the CP is inside the building, especially if falling glass and debris is an issue, safely getting outside agency representatives to an interior fire department CP will be difficult, and in most cases, will not likely occur.

So the question remains, CP inside or outside for a high-rise fire event. Generally, I prefer to establish and maintain an exterior CP because this typically works well in most areas of Denver. However, I also agree with and certainly recognize the many benefits of an interior CP. Bottom line, depending on the circumstances and specific situation, you may have to use one or the other. The good news is I believe you can, and should actually have both.

I believe that most fire department systems are well suited to accommodate a command and control position both on the interior and exterior of the building. Specifically, for the first or initial CP, when formal command is established by a company officer, the best location is clearly inside the building, at the lobby level near the security desk and fire command center, if possible. This IC will have to perform many tasks and wear many hats, initially. This can only be done from a position inside the building.

Next, upon arrival of that first chief officer, typically a battalion or district chief, he can locate and establish an exterior CP, if possible. By delegating specific responsibilities, and reassigning the first IC to lobby control/systems (all of which should be previously determined and written into a comprehensive operating procedure), you now have the best of both worlds. That is, a formal exterior CP with all its advantages, along with the establishment of an ICS position in the lobby of the fire building in charge of lobby control (firefighter and occupant accountability), as well as systems control, monitoring the critical building systems.

Clearly lobby control and systems is not a one-man job. However, in most organizations, with our typically limited resources, one person will have to perform dual roles and wear many hats initially. I am not a proponent of using an entire company from the first alarm assignment for lobby control, as there are too many other critical life safety and fire control tasks that must be completed and that I consider to be a higher priority.

As event unfolds, and more resources arrive, ideally a chief officer should be assigned to supervise all of the various support functions associated with lobby control/systems. That officer should have a support staff that includes the previous lobby control/systems officer assigned as necessary, and any number of available personnel to assist. This could be one or several later arriving companies. The ICS should expand to include a logistics section chief, with lobby control, systems, and numerous other support functions under his direct supervision.

Lobby Control Unit

The lobby control unit leader is an ICS role specific to high-rise operations. The primary responsibility of lobby control is that of accountability, specifically firefighter accountability, but also the accountability of all civilian occupants of the fire building. In addition, lobby control has the responsibility for numerous other critical areas and items at a serious high-rise fire event. Because of that, the IC should establish and maintain a logistics section chief as soon as possible, in order to help coordinate and supervise all of the various support functions.

The specific position of lobby control is one of several functions under the overall command of a logistics section chief. Once again, lobby control is responsible for firefighter and civilian accountability. This accountability is particularly important at a high-rise event, because numerous companies and firefighters will be operating throughout the tens of thousands of square feet within a very large and complicated building. It would be very easy to lose one or several members in such a large building, especially under heavy smoke conditions. Therefore, in order to ensure firefighter safety and survival, proactive firefighter accountability from the outset, is of paramount importance.

There is a wide variety of accountability systems used throughout the American fire service. Some of these systems are very basic, with little or no hardware involved, while others are quite comprehensive, with various components. The passport system is one example, using company and member identification to serve as a passport leading into the fire building.

I believe the very best accountability system, is also the oldest. That system is the incident command system (ICS) itself. This may sound simplistic, and is certainly not meant to discount some very good accountability systems. However, if ICS is used properly, and all members operate in a disciplined and professional manner, accountability will be achieved through the ICS.

For example, establishing command and control at the outset and maintaining it throughout the emergency event, with appropriate expansion of the system as necessary, is the first step toward a comprehensive and successful accountability system. The overall incident commander gives the broad strategic assignments and establishes the necessary division and group supervisors, and ICS section chiefs, as appropriate. The IC is responsible for the safety and accountability of these sections, divisions, and groups. Those specific section chiefs have divisions and/or groups under their command and are responsible for their safety and accountability. In turn, the various division and group supervisors have various companies, units, and personnel under their command and are responsible for their safety. Lastly, individual company officers, operating within a specific division or group, are responsible for the safety and accountability of their respective members (fig. 17–3).

If each of the commanders and supervisors at these various levels does their job in a disciplined manner, with no freelancing or independent contracting, then accountability is established and can be maintained throughout an emergency event. So regardless of what specific accountability system is used by a fire department, good accountability begins with a strong system of command and control, and that system is the ICS.

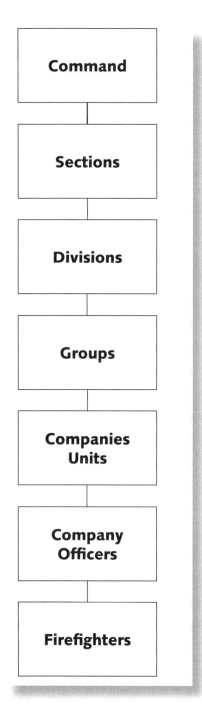

Fig. 17–3 The very best accountability system is the incident command system.

Fig. 17–4 As part of our overall accountability system, the DFD uses simple identification cards for each member, attached to a large ring, for the company that these members are assigned.

This is not meant to diminish the importance or value of any specific accountability system; but passports, Velcro, and name tags are of absolutely no use if the operation is without good discipline and command and control. Immediately knowing the name of an injured, trapped, missing, or dead firefighter is meaningless, if the system of accountability does not include good, disciplined operating procedures, within the scope of good command and control.

Now, with all that said, I prefer to follow the concept of keeping it simple, especially in regard to a formal accountability system. The more hardware associated with a system, the more cumbersome I believe it becomes. Also, the more complex systems are simply subject to a greater potential of failure. Most departments include some form of identification for the members riding on a specific company, typically referred to as *riding lists*. This is very important, and some form of riding list, identifying the names and any critical information for each member, should be utilized by all fire departments (fig. 17–4).

However, the lobby control officer does not need to know the age, blood type, and the mother's maiden name of the nozzleman from Engine Co. 3. He simply needs to know what Engine Co. 3's assignment is, where they are going and how they are getting there, how many members they have operating inside the building, and what time they checked in at lobby control. Remember the four *W*s: *who*, *what*, *where*, and *when*. *Who* is Engine Co. 3 with four members. *What* is their assignment, for example, to report to staging. *Where* is via the fire attack stairway. *When* is 1945 hours.

The accountability at a high-rise event is very dynamic and continues throughout the event. When Engine Co. 3 arrives at staging, they are accounted for by the staging area manager, who is aware that they have been assigned, and is awaiting their arrival. Depending on their specific assignment, they will eventually be passed on to another division or group supervisor. For example, if they are assigned to relieve engine companies operating on the fire floor, they will be passed from staging to the fire attack group supervisor.

In other words, there is a built-in tracking system, within ICS, that maintains accountability of a given company (fig. 17–5). When this company returns back to lobby, they must check out with the lobby control unit leader, or his designee. In the event that a member leaves without his entire crew, such as because he's injured, the lobby control unit leader will note this specifically.

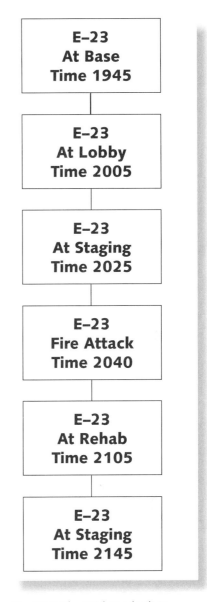

Fig. 17–5 The ICS has a built in accountability component because companies can be tracked as they are assigned and move from one location to another.

Systems control unit

The overall logistics section is also responsible for information with regard to critical building systems. Within the logistics section is the position of systems control, which may become a unit, depending on needs and available resources. Early in the event, systems control is one of many dual roles being performed by one individual. As the event evolves, this will be one individual, referred to as "systems," who could become a systems unit leader, if it becomes necessary.

The systems control unit ("systems") is responsible for monitoring and providing information about the critical building systems, specifically the fire detection and protection systems. Much of this can be monitored via a fire alarm enunciator panel, typically found at the lobby level and/or located inside a fire command center or control room in most modern high-rise buildings, especially commercial high-rise buildings (fig. 17–6).

The fire protection systems must be evaluated, and monitored. What type of standpipe system does the building have? Is there a sprinkler system? Are there any pressure-regulating devices used in conjunction with the standpipe system? If so, are they fireground adjustable? If they are adjustable, what tools are needed to make adjustments, and are any of these tools located inside the building, and/or does the fire department have any of these specific tools? Does the fire protection system include one or more fire pumps? Are they operating properly? Are there any on-site water tanks that are part of the fire protection system?

In addition, "systems" monitors the building's elevators, specifically identifying how many there are, if they are operating properly, their specific locations, and whether or not they are under fire department control. Any elevators that appear to not be operating properly, such as an elevator stalled at a specific location, must be searched as soon as possible. Any elevators located above the fire floor should be the highest priority. Each must be located and searched.

Even if a remote panel indicates that an elevator has been properly recalled in fire service Phase I to a specific location such as a sky lobby, it still needs to be located and searched. It may have recalled to the sky lobby, but the doors may not have opened; therefore, it must be opened and searched. However, the first priority is any stalled elevators that are not in an appropriate Phase I fire service recall location, such as stalled between floors at a location above the fire floor.

Fig. 17–6 A systems control officer located inside the fire command center, monitoring the many building systems.

Most high-rise buildings have a remote panel that indicates the current location of all elevators in the building. This panel is usually located at the lobby level, often inside the fire command center. However, the panel may be located at a lobby security desk, or there may be a simple indicator above the elevator door.

Stairway support unit

A stairway support unit (SSU) is an essential part of any long term, fireground operation in a high-rise building. The concept of and tactical use an SSU is addressed in chapter 7.

For purposes of command and control, the SSU is a unit placed under the logistics section chief. This SSU has a unit leader, and that leader should be a company officer located close to the lobby level, preferably at the lobby level. In this position, he can communicate directly to the logistics section chief, face-to-face, thus reducing unnecessary radio traffic. In fact, ideally, the SSU should operate on a separate radio channel altogether, if possible. The SSU leader reports directly to the logistics section chief.

Fire Attack Group

The fire attack group should be responsible for operations on the fire floor (or floors, if more than one floor is involved), and one floor above and below. This will typically be three floors, the fire floor, the floor above, and the floor below. However, at major fires where the fire has extended upward, this group could include several more floors. Because more than one floor is part of this group's responsibility, the term *fire attack* is most appropriate, rather than using the fire floor number, such as the 22nd floor, and calling this Division 22.

All of the floors and operations associated with fire attack is led by a fire attack group supervisor. The first or initial fire attack group supervisor is the officer in charge of the fire attack/investigation team that first operates on the fire floor. The DFD uses the company officer from the first arriving engine company as the first or initial fire attack group supervisor.

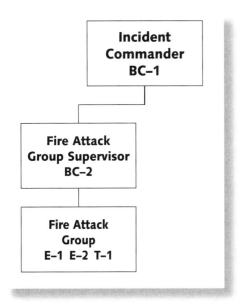

Fig. 17–7 The fire attack group supervisor operates under the incident commander or operations section chief.

A comprehensive operating procedure should include a significant resource allocation for a fire in a high-rise or standpipe-equipped building. These resources should include several chief officers, with a minimum of at least two dispatched on the first alarm assignment. With the first arriving chief assuming command and operating as the incident commander, the second arriving chief should be assigned to operate as the fire attack group Supervisor. He will communicate with, and assume this position from the officer of the first engine company (fig. 17–7).

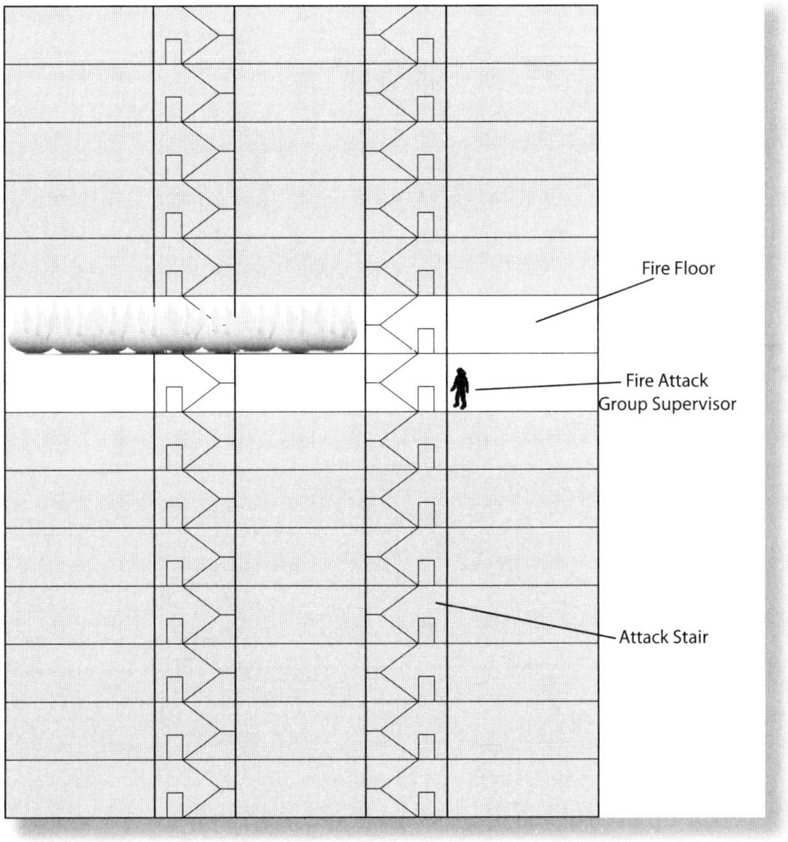

Fig. 17–8 The fire attack group supervisor should be located at or near the fire floor.

The fire attack group supervisor should operate in the most advantageous position possible, because he is responsible for the area that includes the entire fire floor, the floor above, and the floor below (fig. 17–8). All fire companies and units operating on the fire floor, floor above, and floor below are under the command of this fire attack group supervisor (fig. 17–9). A very good operating position is in the fire attack stairway, at or near the fire floor, when there is a significant portion of the fire floor involved. In situations where the fire is confined to a small portion of the fire floor, such as a fire involving one apartment of a high-rise multiple dwelling, the fire attack group supervisor might be able to operate out on the fire floor.

The fire attack group supervisor will likely not operate in a static position, but in most cases will have to be quite dynamic. He will have to move from one location to another in order to effectively evaluate conditions and communicate face-to-face, as necessary, with company officers and unit leaders operating within the fire attack group. This includes evaluating the fire floor, the floor above, and the floor below. Depending on conditions in the fire attack stairway, he may have to utilize the evacuation stairway, from the floor below to access the floor above the fire for evaluation. The fire attack group supervisor will communicate directly with the IC or the operations section chief when and if it is established, and all the company officers and unit leaders assigned to his group.

Upper Search and Evacuation Group

Once sufficient resources have been assigned to the fire attack group, at least to effectively initiate attack and provide for relief, the next priority should be the critical areas above the three floors encompassing the fire attack group. In many cases, this could be a very substantial area, including many floors of perhaps tens of thousands of square feet. In an already labor intensive operation with limited resources, searching and evacuating all of the floors above, will, in most cases, be a very difficult and dangerous operation requiring a significant number of resources. Therefore, this operation, like all others during high-rise fires, will have to be broken down into specific priorities.

The first priorities in this upper search and evacuation (USE) group include the top floor, the fire attack stairway, the evacuation stairway, and all elevators, first searching those that may be stalled in a hoist-way and have not successfully been recalled to a pre-designated location, such as a sky lobby.

There will be a need to assign multiple companies and/or teams to accomplish these tasks. The upper floors will likely have to be accessed using the evacuation stairway. Conditions in the fire attack stairway will have to be evaluated as it is searched. However, heat and smoke conditions inside the fire attack stairway may be so extreme

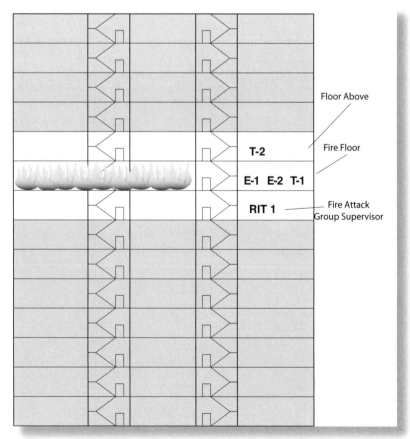

Fig. 17–9 All companies and units operating on the fire floor, floor above, and floor below report to the fire attack group supervisor.

that firefighters will not be able to operate at all inside this stairway above the fire floor, making it difficult, if not impossible to search in the pre-fire control phase of the operation.

The company or companies, assigned to reach the top floor and search this area will also need to evaluate the fire attack stairway at the top, and determine whether or not there is a roof access door, and if that door can be opened. The officer in charge of this search team should communicate this information to the upper search and evacuation group supervisor, who will in turn communicate with the IC or operations section chief, and the fire attack group supervisor. These commanders will determine if the roof access door above the fire attack stairway should be opened or not, based on conditions, the need for vertical ventilation, and whether or not this could negatively affect operating units on the fire floor. If the decision is made to open this door, especially if the door must be forced open, the company that completes this operation should make certain that they maintain the integrity of the door so that it can be immediately closed if necessary.

Once these most critical areas have been addressed, the next priority for the USE Group is a complete primary search of the entire floor area of all floors above the fire attack group. Smoke and heat conditions on these floors above will determine how quickly searches can be completed, or in extreme situations, whether or not a search can be initiated and/or completed at all. To accomplish this and the numerous other functions, the Upper Search and Evacuation Group Supervisor will likely have many companies, units, and teams assigned under his supervision. For example, for a fire on the twenty-ninth floor of a 43-story commercial high-rise building, the USE Group supervisor may have one search team for floors 31 thru 36, and a second search team for floors 37 thru 43 (fig. 17–10).

Like the fire attack group supervisor, the USE Group supervisor will likely not operate in one, stationary, static position. He too will have to be very dynamic, choosing to operate in the most advantageous position. However, the smoke and heat conditions on the floors above will determine where this group supervisor can safely operate.

A good location would be at or near the fire floor, specifically, close to the fire attack group supervisor to allow for periodic face-to-face communications. It is important that both of these Groups operate on separate radio channels, if possible, in order to facilitate more effective communications. In addition, both of these group supervisors, in the absence of a chief's aid, should take one member from a company or unit under their command to operate as their partner. This is critical for these group supervisor positions, or any others that might have to operate inside an area of heavy smoke conditions that is in an immediately dangerous to life and health (IDLH) atmosphere.

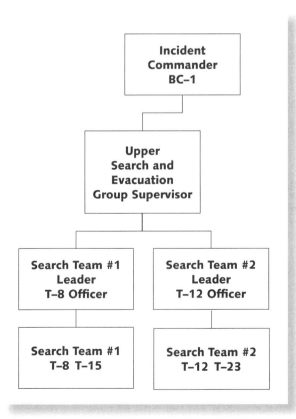

Fig. 17–10 Search teams operating under the Upper Search and Evacuation Group supervisor

Staging Area Manager

There is a large number of support components associated with a successful fireground operation in a high-rise or standpipe-equipped building. In fact, for a serious fire in a high-rise building, specifically those that occur in large commercial high-rise buildings, the resources needed for support functions will usually be much greater than that which is actually needed for the fire attack and rescue operations. Key to success is calling for help early and often. Falling behind the resource curve at a serious high-rise fire can be devastating and can quickly lead to operational failure and potential tragedy, including injuries and deaths to both civilians and operating firefighters.

At a serious high-rise fire event, there will be a need to establish an interior staging area, which I will refer to simply as *staging*. An appropriate location for staging is at least two floors below the fire floor, on a floor that is not contaminated by smoke or heat conditions, and one that provides adequate space for manpower and equipment. The staging area location should be far enough away from the fire floor to provide for safety of members, but close enough to the fire floor so that the fire floor can be accessed quickly in order to provide additional backup or relief resources for the fire attack group as immediately as possible after they have been requested (fig. 17–11).

There will need to be a staging area manager assigned as soon as possible. This manager will operate under the command of the IC or operations section chief. It is recommended that a latter arriving company, preferably an engine company from the second alarm, be assigned to operate as the staging support unit, with the company officer as the staging manager. This unit will organize the staging area, establishing appropriate areas for specific equipment, and communicating by writing directly on the walls, or making signs and then attaching them to the walls, in order to label equipment locations, as appropriate.

Expended equipment, specifically empty air cylinders should be separated from full air cylinders by a significant distance. Ideally, empty air cylinders need to be placed as close as possible to the entrance/exit to and from the staging area, so access and removal to an air filling location can be accomplished quickly. Any broken equipment, such as burst hose lengths, should also be separated from serviceable equipment items. This staging support unit should assist in receiving, organizing, and deploying all equipment and resources.

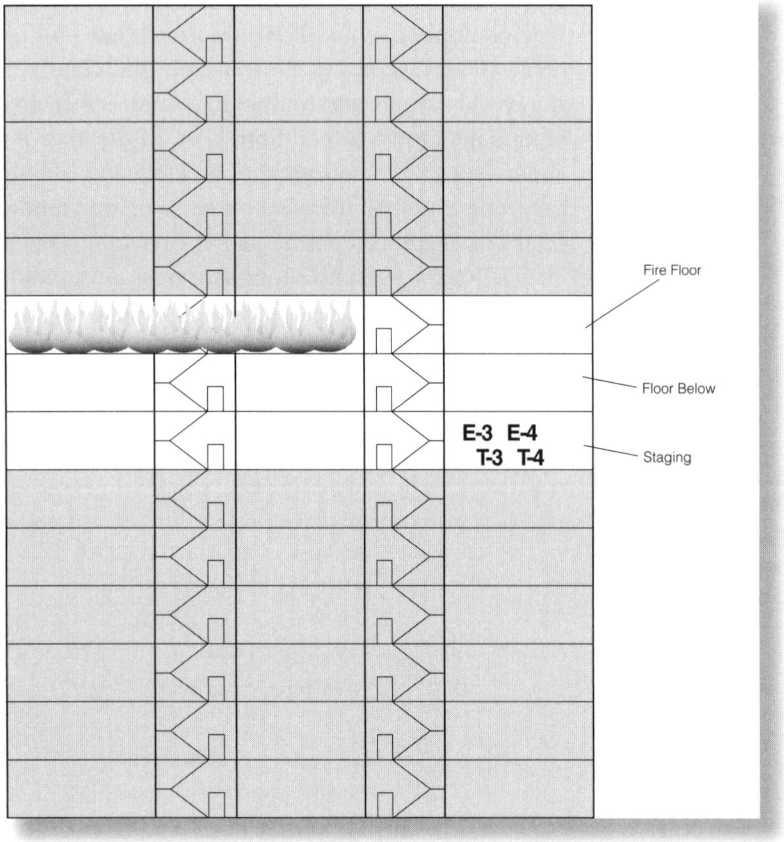

Fig. 17–11 Ideally, staging should be located two floors below the fire floor.

BASE

There are many different terms used to define the area where additional resources dispatched to an emergency event will go first. The term *exterior staging* may be used and is self explanatory. In some systems, the term *Level II staging* is used. The term *BASE* comes from ICS and was designed specifically for large campaign operations, such as large wildland fires. BASE, in conjunction with a major wildland operation, includes numerous resources, from staged fire attack equipment and personnel to mobile kitchens and portable toilets. BASE is primarily designed for extremely large operations that will likely go on for several days. Any of these terms can be used for high-rise operations, however, I recommend and am going to use the term BASE in this book. I believe this makes it easier, as there is then only one staging area, which is located inside the fire building; a BASE area is located outside.

A good BASE location should follow the same basic criteria as that for the staging area inside the building. It should be far enough away to provide for the safety of personnel and equipment awaiting deployment, yet close enough so that requested resources can get to the fire building in a timely and efficient manner.

Generally, the first unit to arrive at the BASE location, after a specific location has been designated, should become the BASE support unit, with the officer in charge of that unit becoming the BASE manager. He will operate under and communicate with the IC, until and if a logistics section chief is established. The BASE manager will be called "BASE," and is responsible for the organization and deployment of resources from the BASE location. Initially, with limited resources, one member, such as an engine company engineer/pump operator, can be assigned to BASE and serve as the BASE manager, thus allowing the IC to utilize the remaining members of that crew as an additional resource for firefighting and/or support operations inside the fire building.

In addition, the BASE manager must keep the IC or logistics section chief informed as to the amount of resources in BASE. For example, it is a good idea to have a predetermined minimum number of resources, which can serve as a benchmark for when additional resources should be requested. It certainly depends on the overall resource needs of the specific incident, and how quickly resources are being deployed and depleted, but limiting the BASE minimum to one engine company resource and one truck company resource is a good starting point. This minimum can also be two engine companies, specifically for those fire departments with a limited number of truck companies.

The BASE officer should contact the IC or logistics section chief when resources fall to a point at or below the minimum, at which time additional resources should be requested if the incident is not yet under control. This request will typically be a call for an additional alarm assignment, third alarm, fourth alarm, and so on, or perhaps a request for five additional task forces or strike teams. The specific terminology may vary from one fire department to another. The bottom line is this: stay well ahead of the resource curve, and call for additional resources in a proactive manner.

The BASE manager also serves an important part in the overall firefighter accountability process, by literally initiating the accountability process. BASE keeps an accurate and up-to-date account of all companies in BASE, along with those that have been assigned to the incident and who are responding to BASE. The BASE manager should maintain communication with the lobby control unit leader in regard to companies assigned to the fire building and who have been sent to the lobby from BASE (fig. 17–12). Lobby should communicate back to BASE regarding the arrival of said companies, thus eliminating any gaps in accountability. A

BASE

Co.'s in BASE (3rd Alarm)	Co.'s Assigned to Fire (1st and 2nd Alarm)	Co.'s Responding (4th Alarm)
E-7, E-8, E-9	E-1, E-2, E-3, E-4, E-5, E-6	E-10, E-11, E-12
T-5, T-6	T-1, T-2, T-3, T-4	T-7, T-8
R-2	R-1, HM-1	
BC-4	BC-1, BC-2, BC-3 DC-1	BC-5

Fig. 17–12 Firefighter and fire company accountability begins at BASE.

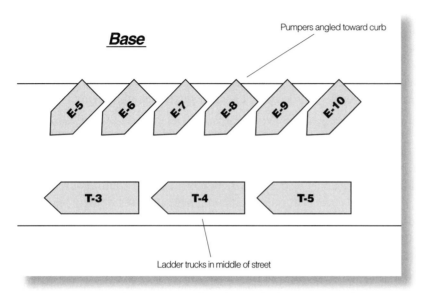

Fig. 17–13 At BASE, fire apparatus should be parked in a manner that will facilitate rapid and independent movement of each unit as necessary.

resource status and situation status (re-stat/sit-stat) officer should also be established by the IC as early as possible to help coordinate and facilitate this overall accountability.

At the BASE location, it is very important to assemble the various resources in an organized fashion. Numerous pieces of fire apparatus will remain at BASE, as the high-rise fire event requires a lot more manpower than it does actual pumper and aerial ladder apparatus and other vehicles.

If possible, large parking lots or other areas with lots of open space are well suited for a BASE location. However, often, especially in large, congested cities, BASE will have to be located on a street. Good organization and discipline from the outset should include parking all of the fire apparatus in such a manner so that each unit can be moved and relocated independently, and placed back in service as quickly and efficiently as possible. Generally, parking aerial apparatus down the middle of a street in a position parallel to the curb, and parking pumper apparatus at an angle, backed into the curb, works well to facilitate an organized re-deployment of apparatus (fig. 17–13).

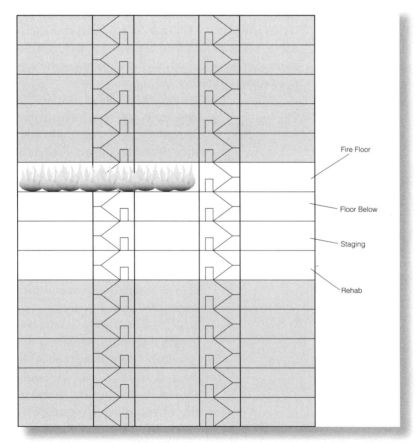

Fig. 17–14 REHAB should be located close to staging, preferably one floor below.

Other ICS Operational Components

There will be numerous other groups and units established to support the overall operation at a serious high-rise fire event. Each will have a specific function, typically defined within the title of the group. Some examples are listed here.

Rehabilitation support unit (REHAB)

There are only a few extremely large fire departments, with enough internal resources to handle most high-rise events without outside help. The average-sized fire department, including the numerous medium and large departments, will be stretched thin from the outset of a major high-rise fire event. The firefighters, who comprise these limited resources, will likely be called upon to work beyond the typically accepted limit. Those in top physical condition may in fact be reassigned numerous times to complete several different assignments, one after another, with only short periods of rest.

At our more frequent operations, I prefer to send firefighters to REHAB after one assignment, and/or after depleting one air cylinder. At significant fire events that are not brought under control quickly, members may have to be quickly reassigned after only a short period of rest, and upon receipt of a full air cylinder. As a chief officer, I attempt to strictly adhere to a *one air cylinder* rule for all members when the temperature is above 80° Fahrenheit or below 40° Fahrenheit, based on the typically moderate climate in Denver. Chief officers who work in more extreme climates, such as Buffalo, New York in the winter, or Phoenix, Arizona, in the summer, must adjust their REHAB benchmarks accordingly.

However, at a major high-rise fire event, there are so many problems that must be addressed immediately, and countless critical operations that must be completed, that firefighters will be called upon to work to extreme limits. A great example of this was the First Interstate Bank Building fire, in Los Angeles in 1988. The LAFD is a very large fire department with significant resources. Even with those resources, several firefighters were assigned to numerous assignments, one after another, with many members going through five or more air cylinders before being sent to long-term REHAB.

We must anticipate and prepare to deal with these extremes. However, establishing a REHAB area early in the event, and conducting proactive REHAB of all operating members, is of paramount importance. The preferred location for REHAB is at or below staging and as close as possible, ideally one floor below (fig. 17–14). This location will provide reasonable access to REHAB from the fire floor, and immediate access to staging, so as members complete their REHAB, they can be sent immediately back to staging for reassignment.

A support unit of at least one company should be assigned to establish and maintain a REHAB area. One of the most critical items that must be delivered to REHAB is a large supply of bottled water. The company assigned to establish REHAB should bring a supply of water in with them, if possible. The bottled water should be transported up to REHAB via elevators, if they are operable and safe to use. If not, hopefully a stairway support unit (SSU) has been established and is in place. Several cases of bottled water, one case at a time, can be quickly shuttled up to REHAB from the ground level by an efficient SSU. In addition, for long-term operations, such as serious campaign fires, food and medical supplies may also need to be shuttled up to REHAB.

In conjunction with the rehabilitation of firefighters, proactive medical evaluations should also be conducted. Initially, members of the fire company that established the REHAB area will serve dual roles, including that of assessing members for any medical problems, and providing first aid and basic life support as necessary. Most fire departments have their own EMS Division, with firefighter/paramedics and advanced life support capabilities. Ideally, a fire company or unit with paramedics should be assigned to the REHAB.

Medical unit

It is important to also establish a medical unit, which should have a primary area located close to, and working in conjunction with, REHAB. The location of this primary medical unit should be as close to REHAB as possible, ideally on the same floor, or on the floor below if necessary (fig. 17–15). This area should be staffed, ideally with several firefighter paramedics. Injured firefighters or civilians should be brought to this location first for medical assessment, triage, and treatment, as appropriate. Firefighters in REHAB should be evaluated by medical unit personnel, and if any medical problems or concerns are identified, they should be addressed immediately.

In addition, there should be a secondary medical unit area, preferably at or near the ground of lobby level (fig. 17–16). This location should serve as a staging area for injured firefighters and civilians who need to be transported to a medical facility. It can also serve as an overflow triage and treatment area in the event that there are multiple victims, or for assessment of injured firefighters or civilians who are located closer to this area to begin with.

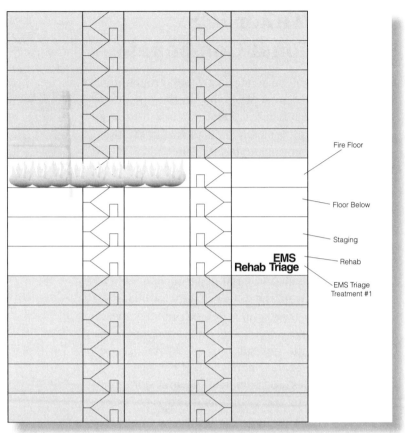

Fig. 17–15 An EMS triage (medical unit) located on the same floor as REHAB.

Ventilation group

Just as the name implies, the ventilation group will be responsible for completing the necessary tasks and tactics to achieve the overall fireground strategy of ventilation. This will likely involve several companies, units, and/or teams within the ventilation group, as the ventilation of smoke and fire gases within a high-rise building is typically a very complicated operation. Like all other groups, the ventilation group will have a group supervisor. The ventilation group should maintain communication with the fire attack group, and the upper search and evacuation group in order to evaluate the effectiveness of ventilation and accomplish the goal as safely as possible (fig. 17–17).

Utilities control unit

Controlling the utilities at any structural fire event is of critical importance, especially at a high-rise fire, or a fire in a large standpipe-equipped building. However, this will be a much more complicated and difficult task, than it is at our more frequent operations in smaller buildings.

Once again, this operation will likely require several companies and teams of firefighters. We are concerned with the control of all utilities, specifically electric, gas, and water. Depending on the specific building, it may only have electric and water, or it may also have and utilize natural gas. Controlling these utilities as necessary, and/or providing the necessary protection for these utilities falls under the responsibility of the utilities control unit. This unit typically operates under the logistics section chief and in conjunction with the systems control unit.

Salvage group

For most fireground operations, salvage procedures are typically considered post-control operations. However, at a serious high-rise fire event, Salvage procedures may need to be implemented and utilized in a proactive manner, during the pre-control phase of the operation.

Specifically, the salvage group may be called upon to initiate protection for areas related to critical building services. For example, elevator hoist-ways, elevator machine rooms, and electrical rooms are all very vulnerable to collateral damage from fireground operations, primarily water damage. In addition, there are countless other mechanical rooms and entire mechanical levels that are integral parts of the overall building systems. These areas may have to be protected, in a proactive manner, and as early as possible, if these building systems are to remain in service, specifically for fire department use.

The salvage group works directly with the systems control and utilities control units, with regard to protecting building systems. Simple procedures can be used, such as using salvage tarps to build dikes and to direct the flow of runoff water away from critical service areas.

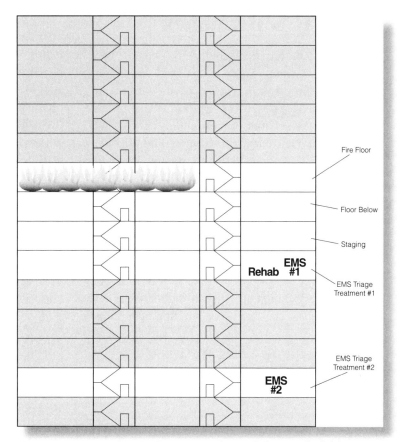

Fig. 17–16 A secondary EMS triage area located close to the lobby.

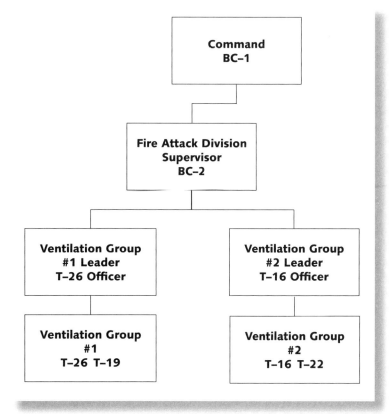

Fig. 17–17 A ventilation group within the ICS.

ICS Branches

The ICS has the ability to expand and compensate when span of control becomes excessive. The use of *branches* allows for two or more different types of operations to be under the operations section. For example, there might be a fire branch, a haz. mat. Branch, and an EMS or Medical Branch at a major emergency incident (fig. 17–18). Probably the best example of a branch component at a serious high-rise fire would be the establishment of an air operations or air support branch.

At a serious high-rise fire event, there will likely be a need for some sort of air operations. Anything from aerial observation, to the insertion of firefighting companies, teams and/or units on the roof, and/or the rescue of building occupants from the roof via helicopter, if this procedure is possible. The specifics of an air support operation are comprehensively discussed in chapter 18.

Needless to say, an air support operation at a serious high-rise fire is a very complex and major operation. Strong command and control will be essential, in order to complete the operation successfully and as safely as possible. Under the air operations branch there will likely be several other ICS components, including, but not limited to, Air Support and Helibase. Also, the establishment of a roof division may fall under the air operations branch or directly under the operations section.

ICS Organization Sections

Within the overall incident command system, there are several different specific areas which allow for the ICS to expand as necessary. Those areas are called Organizational Sections, and include:

- Operations
- Planning
- Logistics
- Finance/administration

Because of the size and complexity of a serious fire in a high-rise or large standpipe-equipped building, several or perhaps all of the specific ICS organizational sections may be implemented.

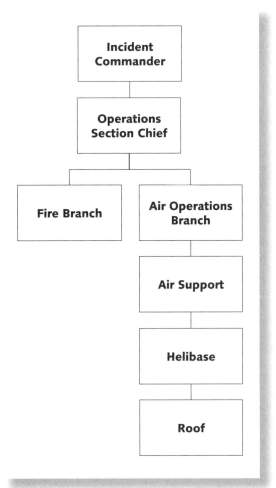

Fig. 17–18 An air support branch within the ICS.

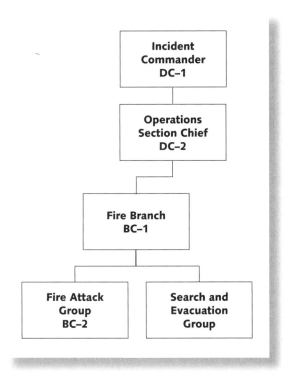

Fig. 17–19 Operations section within the ICS.

Operations section

It is almost certain that an operations section will be established with a chief officer assigned to lead this section at a serious high-rise fire. Once the operations section has been established, most of the tactical operations and various fireground divisions, groups, and units will operate under and answer to the operations section chief. Areas such as the fire attack group and the upper search and evacuation group will be under the direction and supervision of the operations section chief (figs 17–19).

In most cases, this will likely be the first ICS organization section established. The operations section chief could operate at the command post, however, I believe that once established, and if it is safe to do so, the operations section chief should set up and operate inside the lobby of the fire building.

Logistics section

Although the expansion of the ICS has no specific order, generally the operations section would be the first to be established. Listed in order, planning, logistics, and finance/administration would follow. However, I believe that the logistics section is probably going to be the highest priority at a serious high-rise fire event, in many cases, maybe even being established prior to the operations section. That is, if there are already chief officers in position and supervising the most critical tasks/groups, specifically the fire attack and upper search and evacuation groups.

Once again, the support network and functions at a serious high-rise fire will, in many cases, be much more complex and require a greater commitment of resources than the actual firefighting operations. One of those support areas is that of *logistics*. Therefore, it makes sense that a logistics section be one of the first, if not the first, organizational expansions of the ICS. Logistics, as previously addressed, includes many areas, such as BASE, lobby control, systems control, stairway support, and so on (fig. 17–20). The logistics section should certainly be established at a serious high-rise fire event. This section will focus on directing many of the operations that support the actual firefight, and its importance cannot be understated.

Planning Section

As the event unfolds and it becomes clear that it is not going to be brought under control in a reasonable period of time, specifically if control

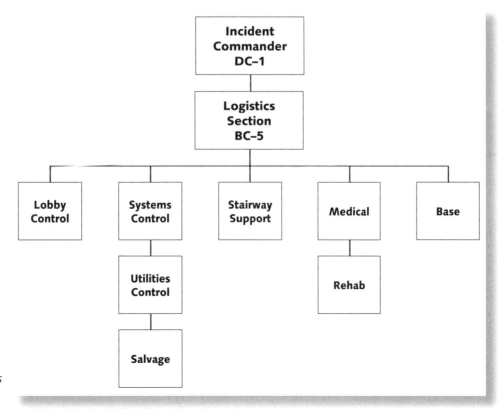

Fig. 17–20 The Logistics Section within the ICS.

is doubtful at all, then it has become a long-term, campaign fire. It is obvious that the expansion of the ICS to include the organizational component of planning will be essential if the fire is going to be brought under control. A chief officer should be assigned to the position of planning section chief. He will lead a team of fire department members, with assistance from various representatives of outside agencies, to gather information and develop an overall plan to bring the incident under control. In essence, planning basically serves in an advisory role for the incident commander, evaluating information, providing advice, assistance, and an overall strategic and tactical plan to bring the incident under control.

Finance/Administration Section

Establishing a finance/administration section, may or may not be necessary at a high-rise event. If it does, the IC can expand the ICS to include this organizational section. This section can assist with various financial services from documentation of costs to cost recovery.

Summary

The incident command system has been compared to a large box of tools. Fireground commanders can utilize whatever tools are necessary to form an effective and efficient system of command and control. At major events, such as fires in high-rise buildings, there will likely be a need for more ICS tools than those that are typically used at our more frequent operations, such as fires in single family dwellings and small commercial buildings.

The essential ICS components at a serious high-rise fire include the incident commander, a fire attack group, an upper search and evacuation group, a logistics section, with BASE and lobby control, staging, and an operations section. A chief officer should be placed in the leadership position of the major ICS components, including sections, branches, divisions and groups.

I consider the most important part of successful command and control on the fireground to be good strong leadership. Because of that, I refer to this as incident command rather than incident management. Success at a serious high-rise fire event hinges on strong leadership at the top, providing confident direction to bring one of our most complex and serious emergency events to a safe and successful conclusion.

CHAPTER 18

Air Support Operations

Most fire departments, including even many of the largest, don't have their own air support division. The cost of acquiring and maintaining one or several aircraft is significant. There is also the additional cost of establishing, training, and paying for a staff to fly the aircraft. In the continued battle to hold on to the limited budget that most fire departments currently have, the addition of an air support division/unit, or even just one aircraft, is highly unlikely for most. Simply keeping the basic firefighting units of engine and truck companies in service with the appropriate manpower levels continues to be a significant budgetary challenge.

Although many fire departments, especially the larger ones, would like to have a helicopter or two, it is a very unrealistic expectation for most of us. Therefore, barring a gift from a kind philanthropist or the acquisition of a healthy federal grant, most fire departments will have to resort to borrowing this resource from someone else when it is truly needed. In this chapter, I focus on how to accomplish air support operations at high-rise fires and emergencies for the vast majority of fire departments that don't have their own internal air support resource.

Outside Resources

For the majority of fire departments that don't have their own air support resource, one of the first steps in establishing some sort of air support capability for high-rise operations is to identify the available outside resources. For high-rise operations, the type of aircraft that would be necessary for air support is a helicopter. Therefore, I will focus on that one type of aircraft in the information to follow.

In larger metropolitan areas, where the potential for a serious high-rise incident is greatest, there are a number of potential outside resources. Fire departments should ask the following questions in an attempt to identify available air support resources:

- What agencies, public, private, and/or military, have a helicopter(s)?
- What is the size and capability of their helicopter(s)?
- Could this helicopter(s) be used for fire department air support operations?
- Would the owner/operator be willing to share it with us?
- Would the owner/operator be willing to participate in regular training drills?
- Would there be a monetary cost to the fire department?
- Can we trust the owner/operator's pilots with our people on board their aircraft?

Fire departments should start by contacting the various agencies and/organizations that might be willing to assist. Establishing and maintaining a positive relationship with these various outside agencies will pay dividends when air support is needed.

Police department aircraft/helicopter

Many police agencies have at least one aircraft in the form of a helicopter. In some cases, one helicopter is shared by many smaller jurisdictions. Law enforcement and EMS are the two agencies that fire departments work with most closely and frequently. You must attempt to establish and maintain a good relationship with these agencies. In some cases, that's easier said than done, but it's critical in order for you to have success at large scale operations, such as high-rise incidents.

Of all the outside agencies that have air support capability, the police will likely be in the best position to provide air support to us in the fastest, most efficient manner. In many cases, their aircraft may already be in service and airborne when we need it. Some larger cities and metropolitan areas have police departments with significant air support resources. In some of these cases, there is at least one helicopter in the air, 24 hours a day. However, that is likely more the exception than the rule in most jurisdictions.

In many jurisdictions with a law enforcement aircraft resource, the significant high-rise fire or emergency that occurs in the late night or early morning hours may require a callback of police personnel to respond and get their helicopter in service quickly. As with all of our operations, once a need, or even a potential need, for air support has been established, an early call for this resource is critical. In some cases, it might take several hours to actually get a helicopter in the air and operational. Even with the best-case scenario, the reflex time to get in service might be at least an hour before we can use this resource.

In Denver, the police department is one of the best agencies to call for a fast and reliable response to assist the fire department with air support operations. Even after normal work shift hours, the Denver Police have the ability to get their helicopter airborne and to any city location within an hour. The speed associated with the police operation is directly related to the size of their helicopter. The Denver Police Department (DPD) helicopter that would be used for air support operations is a Bell Model 407. It is a lighter aircraft, making it easier to place in service and operate; however, there is the disadvantage of the limited payload and carrying capacity.

As part of the operational procedures, the police personnel will first remove the rear doors and seats of their helicopter in order to lighten the aircraft and provide maximum space for firefighters and equipment. Furthermore, they will plan to carry less fuel on board than normal, which will also help increase its carrying capability. This procedure can be completed quickly as the helicopter is being prepared for flight. However, the issue remains that this and most police helicopters are lightweight aircraft, designed primarily for patrol and pursuit operations, not transporting firefighters and heavy equipment.

Interagency training drills conducted by the Denver Fire Department (DFD) and DPD have concluded that a maximum of three, possibly four average-sized firefighters with their associated equipment can be transported from a ground helibase location to the rooftop of a high-rise building (fig. 18–1). Firefighters will be equipped with full personal protective equipment (PPE) and self-contained breathing apparatus (SCBA), with each firefighter typically carrying a spare one-hour air cylinder, one section of 2½-inch hose (DFD high-rise hose pack), and a forcible entry/exit tool. One member will be assigned to bring a high-rise/standpipe equipment kit. Additional equipment, specifically truck company equipment, will be addressed in the operations portion later in the chapter.

Medical aircraft/helicopter (air ambulance)

Another potential air support option for fire departments is that of medical aircraft/helicopter or air ambulance. I don't know of any metropolitan area that doesn't have at least one, or more often several helicopters, used by medical agencies for the purpose of transporting the sick and injured to medical facilities. Because of their widespread use and immediate availability, this can be an excellent resource.

Specific advantages associated with an air ambulance include the fact that the reflex time needed to place one of these helicopters in service is usually very minimal. There is typically a staff on-site 24 hours a day at their base location, typically a hospital where the helicopter is stationed. Because this flight crew is on call and usually set up for a fast response, they could potentially be able to respond to and assist with fire department operations at a high-rise incident in a short period of time. Furthermore, because there are usually multiple air ambulance helicopters operating in many areas, it may be possible to utilize multiple aircraft for your operations. Of course, this would have to be carefully coordinated, and the overall logistics of the operation would have to be supervised closely.

Fig. 18–1 The Denver Police Department's helicopter during an air support operations training drill with the Denver Fire Department.

The disadvantage of using such aircraft once again lies with the area of payload and carrying capacity. The typical air ambulance helicopter has a significant amount of equipment on board for their primary function, emergency medical service (EMS). Items such as heart monitors, defibrillators, oxygen tanks, and so on, all add to the overall weight of the aircraft. Some of these items may be removed, but others are permanently anchored in the helicopter. The helicopter itself is usually similar in size to those used for law enforcement. However, due to the medical equipment on board, the overall capability of the medical helicopter for fire department operations may be less than that of the typical law enforcement helicopters. Nevertheless, this is an outside resource that should be identified and a determination made to its viability for assisting with fire department operations. At the One Meridian Plaza fire in Philadelphia, a medical helicopter was used successfully to transport firefighters to and from the roof for various operations during that incident.

News media aircraft/helicopter

One particular agency that should not be overlooked with regard to their potential for assisting with fire department operations is the news media. These organizations typically have healthy budgets and are able to buy the best equipment, including aircraft. They are constantly monitoring our communications via scanner or radio, and when a working fire or other emergency event occurs in a high-rise building, they will likely be there, above you in their helicopter, arriving almost as fast as you do.

I operated at a mid-air plane crash in January of 2003. I was the second due district chief on the first alarm assignment. Because there were two separate crash sites, I ended up being the initial incident commander at one of the two sites. I remember early in the incident seeing numerous helicopters hovering overhead, at what felt like only a few hundred feet. The noise was significant, and I was actually concerned that we might have another mid-air collision, perhaps between channels 7 and 9. We needed the aircraft removed from that airspace so we could work as safely as possible and hear one another to communicate.

Furthermore, my crash site also involved a collapsed building, which collapsed after a natural gas explosion due to a major gas leak caused when one of the planes involved in the mid-air collision crash landed into the rear of a house. We needed to control the ambient noise in order for my crews to listen for and hear the sounds of any possible victims trapped in the collapse. This incident underscores the fact that, often, the news media can be much more annoying than helpful. However, they can potentially be a valuable asset at high-rise operations, one that you should at least investigate.

News media aircraft is typically similar in size to those used by law enforcement and the medical community. These helicopters will also have significant equipment on board adding to the overall weight, and potentially limiting its capabilities for transporting firefighters and equipment. Like the air ambulance helicopter, some of this equipment can likely be removed temporarily, but certainly not all of it.

The obvious advantages associated with news media aircraft include the fact that there are typically many of them in most of the larger cities and metropolitan areas. In Denver, like many cities, there are three network stations, along with Fox Broadcasting. Each has its own helicopter that it owns and operates daily. Furthermore, their response time is unmatched. They're in the business of getting the story first. They make our business, their business, and when we are responding, if it sounds good, you know they're on the way. If our initial radio report includes the words "working fire" the news media will likely be on scene quickly, especially if it involves a high-rise building.

At the very least, perhaps a news media helicopter could be utilized to carry a fire department chief officer, charged with command and control of the overall air support operations. From a vantage point high above the incident, with the ability to see the entire operation (at least during daylight hours), including the fire building, and all other support aircraft, he could effectively coordinate the air operations. Of course, many potential problems and issues would need to be worked out well ahead of time, such as establishing radio communication between all agencies involved.

The news media will likely embrace the idea of assisting us in this manner because they will have an excellent opportunity to achieve their mission, and collect exclusive footage of the incident as it unfolds. I am sure they won't hesitate to pat themselves on the back as part of their lead story during the five o'clock news either. If they help us achieve our mission, then more power to them.

Military aircraft/helicopter

The obvious advantage to using military aircraft is their payload and carrying capacity. The military has larger, more powerful helicopters designed specifically for military operations. They are thus well suited for use to support the fire department during high-rise operations. Whereas we might only be able to transport three or four firefighters with equipment on the smaller helicopters used by police, medical, and news media, military helicopters can potentially deliver one or several companies of firefighters with their equipment in one trip. These aircraft are also much more suitable for operations involving rope, such as rappelling, as well as hoisting and carrying equipment nets attached to the belly of the aircraft (fig. 18–2).

Unfortunately, there are also some significant disadvantages to the use of military aircraft. Most notably is of course that of time. The reflex time needed to make ready and place in service a military helicopter is, in most cases, significantly greater than that for the other aircraft identified. Unless the military actually has some of these helicopters in the air, their response time could potentially be several hours.

Many cities and metropolitan areas don't have an active military installation located close by that conducts daily air operations. Therefore, military response to an incident within a typical metropolitan area will likely take an extended period of time. In many situations, the military aircraft used will come from the National Guard. To become comfortable with an operation that involves transporting firefighters and equipment in

Fig. 18–2 Military aircraft/helicopters have the added benefit of a significant payload and carrying capability.

an urban setting from ground to rooftop requires a regular schedule of training to increase everyone's efficiency (including the fire department's) to ensure as safe an operation as possible. The DFD conducts quarterly training drills with the Army National Guard, using their aircraft, pilots, and support personnel who are located at Buckley Air National Guard Base, which is about 20 miles east of downtown Denver in Aurora, Colorado.

Training is a critical component for anyone associated with this type of operation, including pilots from law enforcement, medical, and the news media. The pilots from these agencies are typically flying their respective helicopters on a daily basis. They are usually very comfortable with their aircraft and the urban environment.

While researching the operations of several other fire departments, a close friend and brother firefighter pointed out some important information to me regarding military aircraft as a resource. The world that I'm used to, Denver is landlocked. There are no large waterways, rivers, oceans, bays, and so on. In a conversation with Ted Corporandy, Battalion Chief (Retired) with the San Francisco Fire Department, he spoke of using the Coast Guard as an air support resource. When I look to the west, all I can see is big mountains and no water. Having no coast or ocean in Denver, I only see the Coast Guard on TV. For those fire departments located on the coast and near other large bodies of water, the Coast Guard could be an excellent resource.

Being a branch of the military, the Coast Guard operates a wide variety of aircraft, including large helicopters with the capability to lift and transport heavy payloads. Furthermore, the Coast Guards operation is similar to that of a fire department, in that they do have resources that are on call and available 24 hours a day. If the Coast Guard is located in or near your jurisdiction, I would encourage those fire departments that don't have an internal air support resource to research the possibility of asking the Coast Guard to assist with air support operations.

Forest service aircraft

For fire departments located in areas that deal with a wildland firefighting and urban interface, there may be an additional aircraft option. Personnel who utilize aircraft, specifically helicopters, and are used to combating wildland fires can potentially be used in an urban setting to assist with high-rise operations. This resource might be particularly valuable due to the experience that pilots of these aircraft have flying in situations with heavy smoke and thermal columns.

Agencies such as the Bureau of Land Management (BLM) and the Department of Forestry, and, in California, the California Department of Forestry (CDF) play a very important role in wildland firefighting. These agencies represent one more resource that some fire departments might have access to for air support operations. Once again, the key here is to first identify all potential resources, and, from there, fire departments can narrow it down.

Operational Plan

I would recommend that fire departments with high-rise buildings, and a potential need for air support operations at a major incident, begin to develop an operational procedure that involves public, private, and military air support. Look at the potential resources available, with a strong emphasis on using law enforcement (in most communities) as the first phase of an overall air support operation. Certainly medical helicopters and those from the news media could potentially be used to bolster the logistical operation, either through direct involvement or logistical coordination in Phase II. For serious incidents that will be long-term events (campaign fires), mobilizing a military aircraft response in a Phase III of the operation would likely be beneficial.

This operation must be designed a lot like all of our other emergency operations, that is, developed with several different options and redundancies built into the system. We certainly don't want to have all our eggs in one basket, so to speak. The idea of establishing a system that includes public (law enforcement), private, and military aircraft will provide the speed and finesse of law enforcement and private agencies, as well as the more powerful, long-term logistical capability of the military.

Training

Regardless of whether the aircraft used for air support at a major high-rise fire is from law enforcement, private agencies, or the military, the most essential component associated with success and firefighter safety is *training*. There must be a comprehensive, consistent, and ongoing training program established and maintained in order for any air support operation to be successful (fig. 18–3). Loading firefighters into the back of a helicopter and dropping them off on top a fire training tower once every five years is not an effective program and would do little to ensure competency, as well as operational safety and success, the night it's needed.

First, start with identifying the resources available. From there, identify who you want to use as part of your program. Let them know that they must be willing to commit the time necessary to participate in regular training. Also, unless your budget allows for it, make it clear to them that there is no money available to compensate them for the use of their staff and aircraft, including fuel and maintenance. This will likely be a significant hurdle to overcome, because most agencies will want some form of compensation, as there is a significant expense associated with operating aircraft, even for only a few hours. One possible alternative is to seek funding from federal grants, specifically those associated with homeland security. This is what the DFD has done in order to pay for air support training. Once the various agencies have made a commitment, fire departments can begin to establish and implement a comprehensive training program.

Helicopter High-Rise Teams (HHRT)

The term *helicopter high-rise team* (HHRT) is used by the Los Angeles County Fire Department (LACoFD). Captain Larry Collins of the LACoFD is a national expert on urban search and rescue (USAR), and is highly experienced in the use of helicopters for a wide range of emergency operations. In his March 2003 article in *Fire Engineering* magazine, "Helicopter Operations For High-Rise Emergencies," Captain Collins gave the fire service an excellent introduction to the rapidly developing concept of air support for fireground operations at high-rise buildings.[1] The following is based largely on information presented by Captain Collins in his article, with adaptation by the author.

Fig. 18–3 A comprehensive and ongoing air support training program is essential for operational success and safety.

There are several advantages to the use of helicopters for air support operations at high-rise incidents, specifically high-rise fires. Most notably, the aerial view that a helicopter can provide makes it ideal for reconnaissance and critical information gathering (fig. 18–4). Specifically, the four parameters of *lookouts*, *communications*, *escape routes*, and *safety zones* (LCES) typically used by firefighters for wildland fire operations, can be applied to the arena of high-rise operations.

From the bird's eye view, above a high-rise building, the pilot, observer, or an onboard fire department member, preferably a chief officer, can serve as an aerial lookout for the incident commander (IC). Critical information can be communicated to the IC regarding the conditions that are visible from this aerial view. Furthermore, all fire department personnel, command officers, and division and group supervisors operating at the incident and monitoring the radio channel being used by the helicopter team, are made aware of conditions that, in most cases, will be completely hidden from their view as they operate inside the building. Questions still remain regarding what was seen by law enforcement personnel in helicopters above the World Trade Center Towers on 9/11. Information regarding the potential collapse of the WTC towers was never received by fire department personnel operating inside the buildings.

In addition, the helicopter team can potentially provide emergency escape for firefighters who become trapped or otherwise cut off from their escape route while operating above the fire. Civilian occupants of the building cut off by fire, heat, and smoke conditions might also attempt to escape by going up to roof, although most fire departments advise against this. The helicopter team might be able to remove civilians and firefighters to safe zones, away from the high-rise building.

There is a wide range of critically important tactical operations that can be performed by firefighters inserted onto the roof of a burning high-rise building. Keep in mind, the emergency is not limited to fire, but could include buildings with severely damaged stairwells and elevator systems, possibly caused by earthquakes or explosions that are either accidental or intentional acts of terrorism.

Size-up

Specifically, firefighters from HHRT can conduct a much more in-depth reconnaissance, answering numerous questions and conducting ongoing size-up for the IC. As with any emergency operation, specifically fireground operations, the size-up of a high-rise building fire or emergency is ongoing and cumulative. Your size-up truly begins long before the emergency incident, in the form of pre-fire planning and ongoing training and development. The answers to many critical questions can be obtained long before an incident occurs, but only through a continued dedication to excellence.

Formal size-up begins at the time of the alarm, when you start to put together the numerous pieces of variable and dynamic information. Rarely, if ever, will you have all of the critical information necessary to comfortably make split second decisions on the fireground. That's why the area of ongoing size-up is so critical. Conditions change; we are dealing with an ever-evolving situation with countless dynamic pieces of information.

The events of 9/11 are the greatest example of an incredible amount of rapidly changing information being thrown at fireground commanders in a very short period of time.

Critical size-up information will be obtained once the fire building is in view and as personnel arrive on scene from the outside perimeter. Once inside the building, the size-up continues, as the operation unfolds. Important pieces of the puzzle are connected with information from the building occupants, management, and security personnel. Fire department personnel assigned to the fire

Fig. 18–4 The aerial view provided by a helicopter during high-rise operations is extremely valuable.

command center, specifically the systems control officer, will also be able to provide the IC with information regarding building systems, such as the fire alarm, alarm location(s), status of the heating, ventilation, and air conditioning (HVAC) system, location and status of elevators, sprinkler, standpipe systems, and so on. As the initial attack teams make their way to the suspected fire floor or location of the emergency, the IC begins to get a more precise picture of conditions above, and what will be needed to bring this under control. All of these will collectively add to developing a clear, overall size-up of the problem.

However, there will continue to be more questions than answers for the IC. Many areas of a large building take time to identify and evaluate, especially if fire department personnel must climb many flights of stairs to access these areas. The proactive use of an HHRT will definitely enhance the overall operation, including that of ongoing size-up. Some of the most critical areas, many of which might not be easily addressed until well into the operation, can be quickly surveyed by an HHRT. Arriving via the roof places firefighters in an immediate position to evaluate stairs, mechanical areas, and elevator machine rooms, just to name a few. All of these areas are of critical concern to the IC, and the information gained by the HHRT can be quickly communicated to the IC, filling many gaps in the currently available size-up information.

Fig. 18–5 A team of HHRT firefighters prepare for high-rise, air support operations

In addition to the reconnaissance and ongoing size-up, the HHRT can perform numerous fireground operations, including vertical ventilation, search and rescue, and fire attack, and they can serve as a rapid intervention team (RIT). All of these functions are discussed in greater detail later in this chapter.

Fire department operations

Once an air support resource has been established, either with internal resources for a few fire departments, or more likely from the outside for the vast majority of fire departments, the fire department must then begin to focus on the specifics of their air support operation. One of the first steps internally will be to identify the firefighters and fire companies that will be utilized as the members of the department's air support teams. In the Los Angeles City Fire Department, certain companies have been designated as airborne engine companies, and so on. Remember, the terminology and names we use are not nearly as important as establishing and maintaining an air support program in the first place. In this book, I have adopted the term *Helicopter High-Rise Teams* (HHRT) that has been recommended by Captain Larry Collins of the LACoFD.

The members of a fire department chosen to be part of a HHRT should be that fire department's very best. In most cases, the training and assignment will be given to the fire department's special operations companies, specifically, heavy rescue companies, squad companies, and urban search and rescue (USAR) units. In the process of establishing a program on the DFD, I made the recommendation to the special operations assistant chief, and the division chief of operations, that, in addition to our special operations companies, two companies from my district, an engine and truck company, should be utilized for this operation. This was based on the fact that at many high-rise incidents in our city, all of the special operations companies and units will likely already be involved in the operation and committed when a need for air support operations is recognized. Therefore, fire departments should also have other fire department units trained to perform the HHRT functions. Companies that are not due on the first alarm in high-rise areas would be ideal.

After I made the recommendation to use some of my companies for the HHRT operations, I also had a long conversation with the company officers, and specifically the company captains of those two companies. I made it very clear that all the members of their companies must understand that this is an extremely hazardous and demanding operation. It should not be taken lightly. I specifically wanted highly motivated, physically fit, and mentally prepared members to be a part of this team.

Members selected to be a part of an HHRT should be in top physical condition, with a keen mind-set focused on preparation. It must be made clear to every single member of the HHRT that this operation will be implemented in the worst-case scenarios. Members will be exposed to a wide variety of hazardous situations and no one should accept an assignment to a HHRT unless they fully understand and can accept the risks involved. Everything from flying in the helicopter to off-loading onto a roof, the actual operations on the roof and inside a burning high-rise building all present some of the most dangerous operations firefighters will ever face. This is a high-risk low-frequency operation with a significant potential for injury.

HHRT equipment

The following list includes the minimum suggested equipment that members assigned to an HHRT should have upon initial insertion onto the roof of a burning high-rise building (fig. 18–5):

- Full personal protective equipment (PPE) including self-contained breathing apparatus (SCBA). The preferred SCBA is a long-duration, closed-circuit, rebreather-type SCBA, rated for at least four hours work time. If a fire department is not equipped with these types of SCBAs, at the very least, members must be equipped with one-hour-rated air cylinders for use with a standard SCBA.
- A minimum of a Class III rescue harness
- Forcible entry/exit hand tools (set of irons)
- High-rise hose packs, 2½-inch
- High-rise/standpipe tool kit with nozzles, appliances, adapters, etc.
- Rescue rope
- Tagline rope
- Thermal imaging camera(s)

En route

All members of the HHRT, especially the team leader must monitor the fireground radio traffic en route to the high-rise incident. Valuable information regarding the incident, operations under way, progress or lack thereof, can be ascertained via fireground radio communication. Most importantly, the team leader must make contact with the incident commander (IC) and provide the IC with an estimated time of arrival for the HHRT. The mission or missions to be completed by the HHRT must be confirmed, along with any specific operations.

In addition, while en route, members of the HHRT should be taking this opportunity to check one another's PPE, harness, rope, equipment, and so on. Pre-established check lists will make this much more effective with less chance of missing critical items. Any crew members onboard the aircraft should be included in this safety check, with a focus on ensuring that their safety tethers are secure and in place prior to deploying members onto the rope, especially for rappelling operations.

On arrival

The following are specific tasks that can be completed by the HHRT upon their arrival at a high-rise fire:

1. ***Aerial survey:*** Prior to inserting members onto the roof, the helicopter can circle the building to conduct a thorough aerial survey from above. Hazards on the roof should be identified, and a determination made as to whether a roof insertion can actually take place and how to complete it in as safe a manner as possible. Locate all areas of fire and evaluate the smoke and fire conditions, especially any at roof level. Look for any victims located at windows or on the roof. The use of night-vision and thermal imaging equipment can be very beneficial, especially for nighttime operations, or when a clear view is obscured by heavy smoke conditions.

2. ***Structural instability:*** Especially in light of the collapse of the World Trade Center Towers, fire departments must be particularly cognizant of the potential for structural failure and collapse. From the aerial view, members of the HHRT should survey the building and identify any areas of potential structural instability that might be a sign of potential collapse.

3. ***Feasibility of HHRT insertion on the roof:*** A determination must be made as to whether members of the HHRT can in fact be inserted onto the roof. What is the feasibility of landing the helicopter directly on the roof, allowing HHRT members to rappel onto the roof from the helicopter, or the ability to lower members and equipment down onto the roof via a hoist or a tethered basket.

4. ***Communicate with the incident commander:*** Any and all information obtained during this aerial survey, as well as the specifics associated with the deployment of HHRT members must be communicated to the IC. As with overall fireground size-up, the specific information of the aerial operations is also dynamic, and, as conditions change, these changes must be communicated to the IC.

HHRT deployment onto the roof

There are several potential methods for the deploying HHRT members onto the roof of a high-rise building. The method used depends on several different factors, including the capability of the helicopter being used, the building rooftop and obstructions, and the overall conditions, as well as what methods the HHRT members have been trained to complete. The following methods are listed in order of preference:

1. ***Helicopter landing insertion:*** For this method of HHRT insertion, the helicopter actually lands on the roof of the building. This can only be done in situations where there is a helipad on the roof, or the roof is certified to hold the weight of the helicopter plus the firefighters and equipment. That information must be previously determined through a comprehensive pre-fire plan and HHRT training drills. Furthermore, in situations where there is not a dedicated helipad on the roof, even if the roof can hold the weight of the helicopter, other items such as obstructions could still prevent a rooftop landing. If this type of landing is possible, once the helicopter is safely on the roof, members of the HHRT step out of the helicopter with their respective equipment and move away from the helicopter in a low, crouched position (fig. 18–6).

2. ***One-skid insertion:*** This operation requires a very skilled pilot, especially if smoke and heat conditions exist. The pilot places one skid lightly on the edge of the roof, typically at a corner, maintaining the necessary power to hold the helicopter in position, similar to a hover (fig. 18–7). Members of the HHRT step off the helicopter onto the roof in a controlled fashion, one member at a time. With the weight of each member and their respective equipment exiting the helicopter, the pilot must compensate by adjusting power to keep the helicopter in position, with one skid placed lightly on the roof's edge. Control of the helicopter would be much more difficult if HHRT members exited rapidly, one right after another. Control and finesse are critical components to the success and safety of this operation. Once again, as members exit they should do so in a low profile, crouched position, moving away from the aircraft as quickly as possible. If heavy smoke conditions exist at the roof level, members should move slowly, and crawl, if necessary, to avoid falling off the roof.

3. ***Rappel insertion:*** This operation requires a very high skill level of HHRT members. Where the one-skid insertion method is highly dependent on the skill of the pilot, the rappel

Fig. 18–6 HHRT members exiting the helicopter in a low, crouched position

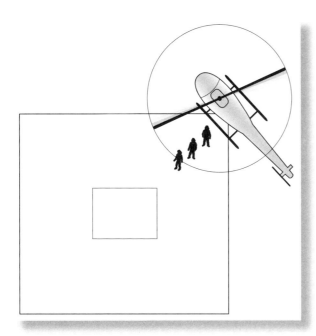

Fig. 18–7 This diagram illustrates a one-skid landing on the corner of a roof.

insertion method is highly dependent on the skill and training of HHRT members. The pilot positions the helicopter above the building inline with the roof, holding the helicopter in a hover. Members of the HHRT rappel onto the roof. Members should be trained to complete this rappel without shock loading the rope or anchor system, and to avoid creating unnecessary movement of the helicopter during their exit and descent. In fact, members of the LACoFD HHRT are trained to keep their feet placed on the skid until they are completely inverted, at which point their feet come off the skid and they complete their rappel to the roof (fig. 18–8). With full PPE and SCBA, a firefighter's center of gravity is up higher, making this evolution very difficult and uncomfortable to complete without extensive training. A comprehensive and continued training program will be essential to maintain the necessary skill level to safely complete this on the fireground. After members of the HHRT have rappelled to the roof, helicopter crew members can then lower equipment to the roof using the same rappel rope. Once all the equipment and HHRT members are safely on the roof, the rope can be released and dropped to the roof, giving the HHRT members an additional rope for rescue operations, or to be used as a search or tagline.

4. **Hoist insertion:** This type of insertion is dependent on the capabilities of the helicopter. Obviously, only larger helicopters equipped with a hoisting system will be capable of this type of insertion. Like the rappel insertion, the pilot once again holds power to the helicopter hovering safely above the building inline with the roof. HHRT members are lowered to the roof via the hoist system, followed by their respective equipment. The hoist is typically a strong and very stable method of lowering firefighters and equipment.

5. **Tethered basket insertion:** This method of insertion is used by the DFD when we work with military aircraft. A large, very strong metal basket, the *Heli-basket*, built by Precision Lift, Inc., can carry several firefighters and their equipment (fig. 18–9). The Heli-basket is tethered to the belly of the aircraft and is lowered into place on the roof. Communication between the lead member in the Heli-basket and

Fig. 18–8 Rappelling onto the roof of a high-rise building requires comprehensive and ongoing training.

Fig. 18–9 A large metal basket, the Heli-basket, can be used to lower HHRT members to the roof of a high-rise building.

the helicopter load master is critical for success and safety. Once members are on the roof, taglines can be used to stabilize the Heli-basket when it is being raised off the roof, and then lowered to the roof on a return trip (fig. 18–10).

HHRT fireground operations

Once the HHRT members are placed safely on the roof with their respective equipment, they can begin to carry out their primary missions. There are a number of very important operations that the HHRT can perform during the course of a serious high-rise building fire or other emergency. The following items represent a partial list:

1. **Remove rooftop obstructions:** During the aerial survey, any obvious rooftop obstructions will be identified. One of the first functions that the HHRT must perform upon arrival onto the roof is to locate and remove any and all of these rooftop obstructions that might create a flight hazard for the helicopter (fig. 18–11). This should be done immediately, by the first HHRT members that arrive at the roof level, unless a serious life safety issue exists and must be dealt with first. The specific tools and equipment brought to the roof by the HHRT must include some substantial forcible entry/exit tools that can be used to remove roof obstructions. A complement of these type tools should be included with the first delivery of firefighters and equipment to the roof. The clearing of said roof obstructions will allow for better helicopter access to the roof, making for safer flight operations, and, in some cases, might even allow for a full rooftop landing. This will also facilitate faster evacuation of any civilians and firefighters from the roof. HHRT members must be careful to identify and watch out for any electrical or other hazards during the removal of such obstructions. There is also one other specific note of caution when removing rooftop obstructions. Many of these will be antennas; in some cases they might actually be part of the fire departments communications system. For example, there are some buildings in downtown Denver that have rooftop repeaters to assist with the DFD's daily communications. In some cities, internal repeaters have been added to some high-rise buildings to assist with fire department communications within the specific building. HHRT members should attempt to determine (assisted by the incident command team) if there are any such communications systems on the rooftop that, if removed, could negatively effect fire department communications. In some cases, there will be no choice, and removal will have to be completed if it is determined that the air operations are the higher priority.

2. **Vertical ventilation via stairwells:** After the rooftop is cleared of obstructions, the next operation to be completed by the HHRT will be to access the interior of the building via doors at the roof level. In most cases, these doors will have to be forced to gain entry. It is critical that the HHRT use finesse when forcing these doors, as the need to maintain the integrity of the door or doors will be of paramount importance to control air currents and the strong draft of stack effect. Access to the interior will likely be from a bulkhead door, or some other door that leads to the interior of the building. The door or doors at the roof level may provide direct access to a stairwell, typical of many residential high-rise buildings, however, in the case of many

commercial high-rise buildings; roof doors will generally lead to a mechanical room, from which the HHRT firefighters will have to search for and locate the access door to the stairs. In either case, one of the most significant operations that can now be performed by the HHRT is that of vertical ventilation. This is where things get a little bit tricky. Obviously, ventilation is an important component associated with any interior fire operation. However, ventilation at a high-rise must be completed carefully, and with continuous communications. This HHRT must communicate with the IC, and with the various division and group supervisors, specifically the fire attack chief located near the fire floor and attack location before they initiate any vertical ventilation. If the door from the roof to the interior or any other doors that access the interior stairs must be forced open, members of the HHRT must be particularly careful to maintain the integrity of those doors, so if they need to be closed, they can be. Because of stack effect and other various ventilation phenomena in high-rise buildings, opening the roof door at the top of an attack stairwell could help the interior teams, but it could also prove devastating, especially if windows on the windward side of the fire floor have broken out and a strong wind condition exists. The specifics associated with ventilation are addressed in chapter 15, but for the purposes of this topic, the HHRT can quickly provide access and vertical ventilation at the roof level. Properly executed vertical ventilation can significantly improve the atmosphere above the fire, increasing the chances of survival for trapped occupants and making fire department operations easier with improved visibility.

Fig. 18–10 Once the operation is underway, tethered ropes can be used to control the descent and ascent of the Heli-basket,.

Fig. 18–11 The first HHRT to be inserted onto a roof must remove any rooftop obstructions, such as tall antennas, that might prohibit safe helicopter operations.

3. **Search and rescue:** Once access to the interior is gained, the HHRT can potentially initiate interior operations. The officer in charge of the HHRT will have to evaluate conditions and determine whether or not the HHRT can effectively operate on the interior with a reasonable degree of safety. In most situations, this will be like descending into a very deep

cellar fire. Heat and smoke conditions will be very punishing, and in some cases, the HHRT firefighters might not be able to actually enter the building at all. If entry is possible, the first priority will be the same at this fire as it is at all others; that is, life safety and the search for any savable human life. As the HHRT enters and begins to complete various tactical operations, a primary search will be ongoing. In fact, the HHRT might find occupants of the building immediately upon opening the access door to the stairs. Upon entry, the HHRT will have to immediately deal with any occupants located in the stairwell that became trapped attempting to escape the fire, heat, and smoke by moving upward. If this is the case, the HHRT will have their hands full from the beginning, especially if numerous people are at the top of that stairwell. Depending on heat and smoke conditions, these victims might be alive and conscious, unconscious, or dead. Regardless, the HHRT will have to make a fast determination as to what actions will be appropriate if victims are found. Any savable human life will have to be removed from this chimney to the open air of the roof. And, depending on conditions, specifically weather conditions, including wind, temperature inversions, and countless other factors, the conditions on the roof might not be all that good either. Ultimately, any savable victims removed will have to be assessed, basic life support provided as necessary, and removed from the roof via helicopter. One person or several people found at the top of the stairs, equates to an extremely demanding and logistically difficult operation. The HHRT is in the best position to complete such an operation. Once any obvious victims have been rescued and the victims removed to a safe zone, the HHRT will likely begin a thorough primary search from the top down, with a primary focus on the attack stairwell, the evacuation stairwell, and any elevator cars located above the fire floor. After the stairwells and elevators have been searched, all areas above the fire must eventually be searched, including all common areas, tenant spaces, mechanical rooms, restrooms, and janitorial closets. The use of thermal imaging cameras, search rope, and tag lines will greatly increase the efficiency and effectiveness of this search operation, as well as firefighter safety, especially in heavy smoke conditions.

4. **Fire attack:** The one thing that has to happen at any high-rise fire, in order for the myriad of other tactical operations to be completed successfully, is that the fire must be confined and extinguished. This will not be an easy task at a serious high-rise fire, and with the need to insert firefighters onto the roof, this is obviously a serious fire and might in fact be out of control at this time. The HHRT, while performing their primary search, is not only looking for victims, but, like a good truck company operation, they are also completing a thorough reconnaissance for the IC, with a specific focus on looking for and identifying any signs of vertical fire extension. The smoke and heat conditions should be evaluated at every level, and this information communicated back to the IC, operations section chief, or appropriate division or group supervisor. Those HHRT firefighters at the roof level may in fact be in the best position to stop the fire. Depending on where the fire is located within the building, the HHRT may have the best access to it. Furthermore, because they are going down to access the other areas of the building, including the fire floor, they will likely be less fatigued than the firefighters coming up several flights of stairs from below. If the stairwell is intact and fire has not penetrated the enclosure, the HHRT may be able to access the fire floor from above. For example, let's say the fire building is 50 stories tall, with a fire on the 35th floor. Fire companies at ground level have to climb up 35 flights to access the fire floor, but the HHRT will only have to walk down 15 flights. Once again, this will not be an easy operation. It will in fact be extremely dangerous and will seriously test the mental and physical preparedness of even the very best firefighters. Depending on conditions, it might not even be possible to attempt such an operation. However, the HHRT may be able to quickly access the fire floor(s) and, once below, regroup, changing to fresh air cylinders brought in with them, and then attaching a handline to the standpipe and initiating an attack on the fire. The HHRT must be equipped with 2½-inch hose packs (three to four 50-foot sections, 150 to 200 feet) and the appropriate nozzles and appliances to initiate

fire attack as necessary should they find an area involved with fire. Furthermore, depending on the circumstances of the fire, the HHRT may be in the best position to access and begin fire attack in the original or main fire area. In any event, the HHRT must make every effort to initiate their fire attack from as safe an area as possible, preferably below the fire. This might not be possible in many circumstances, but at the very least, the HHRT must coordinate any fire attack from the stairwell. Any fire attack is made in an effort to stop or reduce vertical and horizontal extension of fire, as well as protecting the means of egress, specifically the stairwells, for any victims. The tactical operation of fire suppression is one more of the many possible operations an HHRT can perform at a high-rise building fire.

5. **Rapid intervention:** A question that will certainly stimulate a great deal of conversation, debate, and a wide range of opinions is that of rapid intervention teams for high-rise operations. Specifically, where do you put your RIT when you have a serious fire in a high-rise building? The fact is, we must begin by understanding that it will not be a RIT, singular, but more likely several RITs—multiple companies, under the command of a chief officer, perhaps called the RIT group. Obviously, a lot depends on the size of the building, the location of the fire floor or floors, and where other companies are operating in the building. Once again, for a 50-story building, with a fire on the 35th floor, there needs to be a RIT close to that fire floor, preferably close to the attack stairwell and just below the fire floor. What about the firefighter from a search and evacuation team that becomes disoriented while searching the 47th floor? The RIT is 14 floors away from that firefighter. If an HHRT was inserted on the roof, the trapped firefighter on 47 would have help only three floors away, and they would be walking *down* to get to him, rather than *up*, making it much less fatiguing. A comprehensive discussion of RIT procedures for high-rise fires is in chapter 16. Suffice to say, the HHRT may also be in a very good position to provide RIT operations from above.

Extended roof operations

It may be that once the HHRT reaches the roof level via helicopter, they will have their hands full with operations on the roof. This would likely occur if there were numerous building occupants that fled the fire area and floors above by moving upward and were able to access the roof. Most high-rise buildings advise their occupants to not go to the roof in the event of a fire, but people will still do it, primarily because of the assumptions many people make. They've been watching TV and movies for too long, and believe that the roof is their best option. Members of the HHRT might encounter several occupants on the roof, with many more coming onto the roof from inside the building via the stairwell. In this situation, a determination would have to be made as to whether or not these occupants could be sheltered in place on perhaps an upper floor, conditions permitting, or if they would have to be evacuated from the roof via helicopter.

Some of the occupants might be injured, and may require medical attention, therefore requiring immediate removal from the roof. Furthermore, the fire may be significant and extremely serious, where control is doubtful. Therefore, an evacuation of all occupants from the roof would be mandatory.

If there are large numbers of people on the roof, a triage-type operation will have to be conducted by the HHRT in order to prioritize people being evacuated. Obviously, those with serious medical conditions, and those who have suffered traumatic injuries will have the highest priority. Having one or several members of the HHRT trained as medical providers, preferably fire department paramedics, will greatly enhance the effectiveness of the HHRT and the overall operation. Of the people that are not injured, women and any children should be the next evacuees. After that, all remaining civilian people should be evacuated from the roof, followed by fire department personnel.

One of the most important functions the HHRT will provide at roof level will be to control the crowd of people, along with the associated chaos and panic that will likely occur in this type situation. If members of the HHRT can access the upper floors, a controlled evacuation and rescue effort can begin inside the building, coordinating a smooth and regulated movement of victims to the roof level. If possible, keeping victims inside the building and only moving those people who will be immediately evacuated on to the roof will enhance the safety for both victims and rescuers alike. If stairwells are intact, with at least one stairwell that is free of smoke (evacuation stairwell), victims should be evacuated downward via that stairwell. That is, unless there is additional size-up information that indicates this would not be a safe means of evacuation, such as a fire that is out of control and rapidly spreading, or perhaps that the original problem was likely caused by a terrorist attack and there may be secondary attacks.

Rooftop Evacuation of Victims By Helicopter

Once the decision has been made that victims on the roof will in fact be evacuated from the roof via helicopter, there are certain procedures that should be followed. First, the methods of evacuating victims from the roof are identical, or, at the very least, similar to the methods used by the HHRT to get to and from the roof. These include the roof landing method and the one skid method as previously explained. However, rather than an insertion, these would be evacuations. Members of the HHRT on the roof would coordinate from the roof level, constantly communicating with the helicopter crew, and with one or more members of that crew coordinating the evacuation from inside the helicopter and assisting victims on board.

In addition to the roof landing and one skid methods, there are:

1. **Short-haul extraction:** This method depends on the availability of a safe zone where the victims can be transported. One possible location is the roof of an adjoining building. However, as we all know, the roof of an adjoining building could be a very dangerous location, especially if the event involves terrorist activities. The short-haul destination could also be to a ground location, but once again, it would have to be determined with certainty that whatever location is selected, it is in fact a safe zone. In order to use this method, the HHRT must be equipped with an evacuation harness for ambulatory people and a rescue litter for any injured and nonambulatory people. The helicopter must have some sort of life line, usually a hoist, to which the harnessed victims are attached for the short-haul. This method leaves the hoist or life line extended out, remaining in a static position. Constant communication must be maintained between the helicopter crew and the HHRT members on the roof in order to safely coordinate the rescue of any people from the roof.

2. **Hoist extraction:** This method will obviously require the use of a hoist-equipped helicopter. Furthermore, like the short-haul method, a rescue harness will be needed to evacuate people from the roof and a rescue litter for any injured victims. The difference between this and the short-haul method is that the hoist is actually retracted, lifting the victim up to the helicopter, at which point crew members pull the victim into the cabin of the helicopter. This is usually more time consuming that the short-haul method. When the hoist method is used, like the short-haul method, constant communication between the helicopter crew and HHRT members at the roof level must be maintained in order to ensure as safe an operation as possible. In the hoist method, after the victim is placed in the harness or rescue litter and that carrying device has been hooked up to the hoisting cable, rather than using the hoist to pull the victim up off the roof initially, the helicopter pilot first lifts with the helicopter until the victim is off the roof and the helicopter is holding the entire load. At this point, the hoist is activated with the victim in the air, and the victim is pulled up to the helicopter and brought onboard. This is done in order to give the pilot better control of the aircraft, as he first lifts the entire load off the ground and then applies the appropriate power to maintain a hovering position. If the hoist were used first, it would be very difficult for the pilot to gauge the weight of the load and maintain the proper control with a smooth, hovering position.

3. **Tethered basket:** This method of extraction, used by the DFD, uses a large, very strong metal basket—the Heli-basket—that can carry several victims. The basket is tethered to the belly of the aircraft and is lowered into place on the roof. Communication between a firefighter in the Heli-basket and the helicopter loadmaster is critical for success and safety. Members on the roof should use taglines to stabilize the Heli-basket when it is being lowered to the roof, and raised off the roof. Two firefighters should be

assigned to remain in the Heli-basket, one on each end, to protect civilians being rescued from falling out and to coordinate loading and removal.

Summary

Most fire departments don't have their own air support division/unit. Only some of the largest and best-equipped fire departments have an internal resource. Most fire departments have to look to outside agencies in the public, private, and military sectors for assistance. Fire departments that respond to and have high-rise buildings in their jurisdictions should develop some sort of air support procedures using available resources. A combination of resources from law enforcement and the military can form the basis of a solid program. A comprehensive training program must be established and maintained. Assembling specially trained teams of firefighters called helicopter high-rise teams (HHRT) and having them prepared for worst-case scenario situations will give the incident commander of high-rise fires and emergencies an invaluable strategic advantage. The HHRT can perform a wide range of tactical operations, including comprehensive size-up, rooftop evacuation, vertical ventilation, search and rescue, rapid intervention, and fire attack, just to name a few. In the fire service of the 21st century, where we face terrorism and other complicated emergency operations, specifically high-rise building fires and emergencies, an air support resource and operational program is essential.

References

1. Collins, L. 2003. Helicopter operations for high-rise emergencies. *Fire Engineering, March 2003*, pp. 105–128.

CHAPTER 19

Communications

Fireground communication has always been an imperfect science. Without fail, there will likely be some communications problems, to a greater or hopefully lesser extent, at many, if not most, fireground operations. When reviewing various fire service case studies, most will have the subject of *communications* high on the list of operational problems encountered.

Sometimes the communications problem is directly related to equipment issues, but often times it's operational. Some equipment issues may require millions of dollars to correct, including the acquisition and implementation of significant infrastructure. For most fire departments, continued budgetary constraints can stall an extremely important communications upgrade for years, and, in some cases, eliminate it altogether.

As for the operational issues, we should be capable of solving these through ongoing organizational training programs, but that too is an easier-said-than-done proposition. It's important that fire departments proactively establish comprehensive operating procedures with regard to emergency communications. In addition, a continuous program of training and development must be in place to teach and reinforce the fundamentals of effective, efficient emergency communications.

Fig. 19–1 The remote microphone allows for efficient communication

We've certainly come a long way toward improving fireground communications, but there is still plenty of work to be done. Modern technology has provided us with numerous pieces of valuable equipment that assist and improve our fireground communications. Rather than having to go through the cumbersome process of pulling a portable radio out of a pocket with gloved hands to transmit a message, the remote microphone allows for a much more efficient operation (fig 19–1). In addition, the use of SCBA face-piece-mounted voice amplifiers has given us a much better ability to communicate in the noisy, hostile environment typical of an interior structural fire (fig 19–2).

Just like everything else associated with fireground operations in high-rise and large standpipe-equipped buildings, communication is also made much more difficult. We frequently see some form of communications problem at most fires, including fires at small single-family dwellings but it is almost certain that communications will be difficult, cumbersome, and problematic at high-rise fires, and fires in the numerous other large, standpipe-equipped low-rise buildings.

There are several reasons why these buildings present many more potential and certain communications problems. The large size of these buildings sets the stage for communications problems, along with construction features that further exacerbate it. Specifically, the vast amount of concrete and steel literally creates a cocoon and radio waves typically bounce off these structural components, oftentimes never making it from sender to receiver.

Fire Department Radios

For all practical purposes, experience has proven that standard issue fire department radios do not consistently work in most high-rise buildings. Specifically, radios that rely on a repeater system coving a large geographic area frequently fail in fireground communications because the signal from a small, portable radio is generally not strong enough to get out to the repeater. Therefore, fire departments that operate in these large, complicated, concrete and steel buildings must use a radio system that includes specific components that can overcome and bypass the barriers associated with communications in these buildings.

Non-repeated radio channels

A good starting point is to specify a radio system that includes several non-repeated *simplex* radio channels. This allows for communication between operating units within a small geographic area that is typical of fireground operations. Furthermore, this does not require the use of a repeater to complete the communications. However, the disadvantage is that the fire dispatchers at the communication center, and any incoming units, at least those that are not within a few blocks of the fire building, will not receive any of the radio transmissions. The incident commander (IC) or any operating units needing to communicate with dispatch (communications center) would have to switch to a repeated channel or use a different radio all together.

On the Denver Fire Department (DFD), we use a radio system that includes a couple of these non-repeated simplex channels, referred to as *radio-to-radio channels* that usually operate effectively in most high-rise buildings. There is however a tremendous need for several more non-repeated simplex channels, and the DFD is currently working to acquire at least six more. When specifying a new radio system or completing an upgrade, fire departments that operate at a significant number of high-rise and standpipe-equipped buildings should request that several of these dedicated, simplex radio channels be included in the system.

Mobile repeaters

The DFD has taken proactive steps with the implementation of several mobile repeaters. These units have been placed on specific apparatus strategically located throughout the city to ensure that at least one mobile repeater will be on scene at all working fire incidents and major emergencies. There is a dedicated repeater channel on all DFD radios. It works by using the on-scene mobile repeater to boost the radio signal. When the repeater is used at an incident, the communications are linked to another radio channel, which allows the fire dispatcher and incoming units to monitor the fireground tactical radio traffic. The mobile repeater works well for many situations.

Internal Building Communication Systems

For fireground operations in these large high-rise and standpipe-equipped buildings, there may be an on-site radio and communications system that could be beneficial for fire department operations. Many of the large commercial buildings, especially high-rises, have a large management, engineering, and security force with several personnel on site, especially during normal business hours. They usually have a built-in communications system that includes several portable radios and often a dedicated radio repeater of their own inside the building.

These radios, in a lot of buildings, will actually work better than ours. Therefore, fire companies should first determine if such an internal communications system exists, and, if it does, we should attempt to utilize the building's radios in conjunction with ours. Initially, the radios may be difficult to obtain, as they will likely be spread throughout the building in the hands of various building personnel. We should attempt to round up as many of these radios as possible, as quickly as possible.

Keep in mind, there will certainly not be enough radios to equip all of our operating members, but in most large buildings, there will probably be enough radios to equip members in key command positions such as the incident commander or command assistant, ICS section chiefs, and division and group supervisors. In one particular commercial high-rise building in downtown Denver, there are fifteen portable radios and a built in repeater (fig. 19–3). This is probably a good example of what communications you could expect to find in a typical, medium-sized, commercial high-rise building.

Phones

Telephones are another good example of a communications device that can potentially be utilized to communicate inside a large high-rise and/or standpipe-equipped building. To begin with, in most of the commercial occupancies, there is on-site security, oftentimes located at a security desk in the lobby and/or a fire command center. These locations typically have telephones (landlines) that can provide reliable communications. Division and group supervisors operating on the upper floors should obtain the telephone number to the lobby

Fig. 19–2 The SCBA face-piece-mounted voice amplifier

Fig. 19–3 Many large high-rise buildings have their own built-in radio system, including several portable radios.

security desk and the fire command center. There will likely be numerous offices equipped with telephones throughout the upper floors of a typical building, and, barring a complete system failure related to the fire, these landlines can be used for backup communications. It is a good idea to try to locate a telephone equipped with a hands-free speaker. This will help facilitate communications from one point to another, as the sender and receiver will not be constrained by having to keep a telephone handset constantly up to the ear.

Cell phones

In the 21st century, almost everywhere you look, someone is talking on a cell phone. This can be annoying to many, but we should seize the opportunity to capitalize on this widely available resource. Within most fire department organizations, typically at least the chief officers, and sometimes the individual companies themselves are equipped with cell phones. Any that are available should be brought into the fire building and kept available as a backup or secondary communications medium. Don't forget to bring a phone battery charger, which, given an available power source, can be used to maintain a fully-charged battery during long-term operations.

We may encounter similar problems with cell phones that we do with radios, such as dead areas, dropped calls, and poor reception, but, once again, it's worth a try. Whether using radios or cell phones, attempting to get close to the outside perimeter of the building, near windows, will sometimes help with reception. Or, taking a position close to internal vertical channels, such as elevator hoistways, will sometimes help with communications from both radios and cell phones.

Sound powered phones

Many of the larger commercial buildings and some residential buildings, specifically high-rises, will have a system of sound-powered phones for internal communications. There is typically a small supply of telephone handsets located in the lobby, usually in a cabinet in the fire command center (fig. 19–4). As companies arrive and are assigned to positions on the upper floors, these handsets should be distributed accordingly. For example, the first companies to go aloft and that form the initial fire attack group should take one handset, as

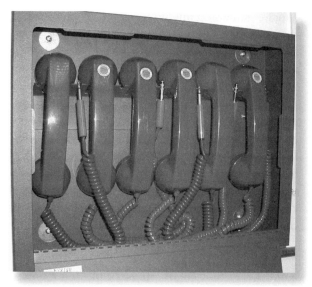

Fig. 19–4 A typical supply of telephone handsets in a cabinet located inside the fire command center of a high-rise building.

Fig. 19–5 Typical wall phone jack located near a pull station inside a high-rise building.

should all division and group supervisors.

These sound-powered phones work by plugging the handsets into convenient phone jacks, which are located throughout the building, specifically at locations such as pull stations and elevator vestibules (fig. 19–5). To initiate communication, the phone handset is plugged into a wall phone jack, which initiates a call, to generally, the fire command center (fig. 19–6). Someone must be in the fire command center to receive the call, so the systems officer or a member of the systems control unit should be in the FCC ready to answer the call.

Although these sound-powered phones are generally reliable, they are not without failure, and sometimes the reception is marginal. Once again, focusing on developing and maintaining good habits, this system should be checked whenever companies are inside the building on frequent incidents such as automatic fire alarms. Getting into the habit of sending phone handsets up with operating units will allow them to test the system and identify any problems before a major fire event occurs.

Public address (PA) system

Most of our fireground communication will be internal fire department communication between various fire department units. However, a very important component of fireground communication, specifically at high-rise buildings and some large low-rise buildings, will be fire department to civilian/occupant communication. This is a very important aspect of our overall operations and can truly save countless lives when factual information and instructions are communicated to building occupants, and then followed by those occupants.

Many high-rise and standpipe-equipped buildings will be equipped with a public address (PA) system (fig. 19–7). A well-designed PA system allows for communication with occupants throughout the building. During fire department operations, this tool can be an invaluable resource to provide instructions for occupants, such as telling occupants to stay in place (protect in place), evacuate to a lower floor, or completely evacuate the building. PA systems also allow firefighters to

Fig. 19–6 A telephone handset is plugged into a phone jack to initiate communication with personnel in the fire command center.

Fig. 19–7 A typical public address (PA) system located in the fire command center of a commercial high-rise building

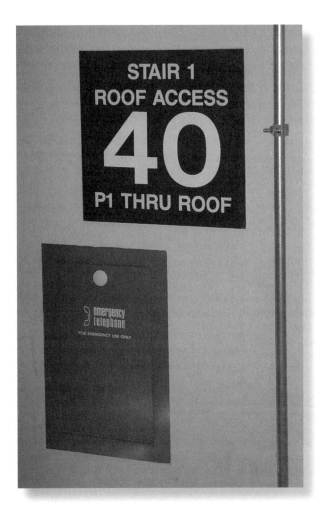

Fig. 19–8 Many high-rise buildings have emergency telephones inside the stairwell at designated floors.

provide guidance to the occupants as to the location of the evacuation stair, and emphasize to occupants to stay out of the attack stair.

Keep in mind that, as stated in chapter 14, not all occupants will follow the given instructions, especially after the events of 9/11, and many will hastily evacuate regardless. Nevertheless, a building PA system gives you an opportunity to provide important information and attempt to establish and maintain an orderly, systematic, and appropriate movement of building occupants.

It can also be used as an additional communications tool for fire department operations. If, for example, a specific company, team, group, or unit cannot be contacted via fire department radios, the systems control officer or his designee can use the PA system to attempt to make contact with them. In addition, instructions can be given over the PA asking that company, team, group, or unit to attempt other various methods to re-establish communication, confirm their location, and verify their well-being.

As part of the overall fireground operation in these buildings, the PA system should be used to provide periodic announcements, made at specific timed intervals, with critical information. This should include repeating various pieces of information, especially the location of the evacuation stair and attack stair, with an emphasis on instructing occupants to stay out of the attack stair. Also, early communication should be provided, when possible, to communicate with any occupants that are inside the attack stairwell telling them to get out and stay out, prior to companies commencing with fire attack. Obviously, the announcement would identify the stairwell being used for attack by the term the occupants are familiar with, such as Stairwell A, or the South Stairwell and so on. In addition, if a particular building has emergency telephones in the stairwell, as many do, occupants can be directed to the locations of these emergency telephones, and instructed to call for help (fig. 19–8). This will establish that they are inside a specific stairwell and their location. If it is determined that occupants are trapped inside the attack stairwell, above the fire floor, the fire attack will have to be delayed until all occupants can be safely removed from the attack stairwell.

Reverse 9-1-1

In chapter 14, I addressed the option of using reverse 9-1-1 to communicate with building occupants. Once again, in many communities, modern technology allows for the 9-1-1 emergency telephone system to be used in reverse, which allows operators to call occupants of a specific building and give them guidance pertaining to the emergency. This is another communication option that members should be aware of, and one that might prove beneficial at a serious high-rise fire. We must continually look for innovative ways to solve our problems. Reverse 9-1-1 could be used to contact a much greater number of occupants in a much shorter period of time than would be possible using typical fireground resources. This is an excellent option in those older buildings that are not equipped with a PA System.

Runners

This will remind you of years past, when chief officers used speaking trumpets to communicate orders on the fireground, along with runners to accomplish longer distance communication. The reality is that we have found ourselves in numerous situations before where we simply could not communicate when operating deep inside these large buildings. We will certainly encounter these same problems again in the future, regardless of how much high tech equipment is available to us.

A good example, or perhaps I should say a sad example, is that of a brand new high-rise hotel in downtown Denver. In the winter of 2005, the newest, most modern high-rise building in the city of Denver opened for business. However, fire companies were literally unable to effectively communicate inside many areas of this building.

When a fire or emergency occurs in this and many other buildings, various division and group supervisors operating in different locations of the building, may have to resort to using runners to communicate critical messages. This is not ideal, but something that we must be prepared for in order to accomplish the mission. This will not be a fun or glamorous assignment. Designated runners will need to be firefighters, preferably young members, in top physical condition.

Summary

Good communications are essential to achieve fireground success. At almost every fireground operation, you will encounter some communications difficulties. At fires in large high-rise, low-rise, and other standpipe-equipped buildings, you will definitely face communications problems. You must be prepared for that and proactively develop several backup plans in order to accomplish necessary communications. From radios to runners, you might have to use everything in your tool box.

CHAPTER 20

Preparing for Battle

As I write the final chapter of this book, I do so using the same central theme with which I started. In chapter 1, I laid out what I consider to be the most important characteristic of any firefighter—his *mind-set*. That critical mind-set includes a compilation of several essential factors, including a positive attitude, overall preparedness, and a belief that serious emergency events, including fires, will happen on your watch. Those top-shelf firefighters who possess a positive, professional attitude, believe that serious events *will* happen, including high-rise fires. With that attitude and belief, they will take the appropriate, proactive, and continual steps to get prepared and stay prepared.

Preparing your Firefighters for the Worst

There is no fire service crystal ball to predict when or where our next serious fire in a high-rise or other standpipe-equipped building will occur. Therefore, we are left with only one thing: *preparation!* We must always prepare for the fire in a large, low-rise standpipe-equipped building or high-rise building, because it may occur anytime and anywhere. In the fire service, the word *preparation* is synonymous with *safety*.

The most critical components of any fireground operation are proper training and preparation. High-rise firefighting and, specifically, standpipe operations, whether they occur in true high-rises, or in the myriad of other standpipe-equipped buildings, are no different. In fact, these operations truly require a higher level of training and development than our more high frequency events, those that occur every day, such as fires in single-family dwellings and small commercial buildings. For firefighters to operate effectively, efficiently, and, most importantly, as safely as possible in high-rise and standpipe equipped buildings, there must be a proactive, ongoing, and dedicated training program that focuses specifically on the skills associated with these operations.

The Denver Fire Department Hands-On Training Program

The Denver Fire Department (DFD) has used a unique, hands-on training program for the past 15 years to introduce recruits—probationary firefighters—to the extreme physical and mental demands of firefighting operations in high-rise and standpipe-equipped buildings. First developed in 1992 by several members of the DFD and me, this program has stood the test of time, and today, the majority of firefighters on the DFD have completed the training. Like many fire departments, DFD recruits are trained during an intensive, 20–week fire academy, which starts with the basics, and concludes with much more complex, multitask operations. One of these operations is a daylong high-rise standpipe training drill, conducted at a large, commercial high-rise building located in downtown Denver.

The first component of a quality high-rise standpipe training program is a very critical and valuable resource: a large standpipe-equipped building, preferably a true high-rise. I am not talking about a five-story drill tower, either. There is probably no fire department in the country, at least that I know of, that can afford to include the construction of a high-rise structure in its annual budget. Therefore, we must be creative and ambitious in our quest to find a high-rise building we can borrow from time to time.

The DFD is very fortunate: since 1993, we have been conducting our high-rise and standpipe training in a large, 44-story, commercial high-rise building located in downtown Denver. Our department has developed and fostered an excellent relationship with the owners, management, and security personnel of this building. They generously welcome us into their building on a daily basis; it is not unusual to see several DFD companies stop by throughout the day, every day, to climb stairs as part of an ongoing physical fitness training program (fig. 20–1). In addition, various training drills, from brief, company-level training, to large scale, multi-unit drills and academy recruit training, are conducted in the building throughout the year. In return, the DFD provides uniformed personnel to assist during the building's biannual fire drills, presents fire and life-safety demonstrations and lectures during fire prevention week, and provides whatever assistance and guidance is necessary, whenever possible.

Large-scale academy recruit drill

The large-scale training drill for recruit firefighters involves several evolutions. The academy class of recruits, generally 24, is broken down into two groups of 12. During the morning

Fig. 20–1 Firefighters from DFD Engine Co. 20 completing their daily physical fitness program of climbing stairs.

session, one group tours the building and is acquainted with its various systems, including HVAC, fire protection and life safety systems, and elevators. Special emphasis is given to elevator rescue operations. The other group of 12 is divided into two groups of six. They complete a fire-attack scenario, rescue scenario, and a water supply evolution. After a short lunch break, the groups switch training assignments for the afternoon session, which is a repeat of the morning drills. The entire operation takes approximately nine to 10 hours, including set-up time, and cleanup.

Fire-attack scenario

The fire attack scenario is the most physically demanding drill in the daylong session, and, many believe, of the entire fire academy training curriculum. The scenario is based on the potential failure of several critical high-rise building systems, including elevators. The hands-on training is accompanied by the presentation of facts and statistics that constitute the tragic history of system failures at high-rise building fires that have occurred throughout the country. Problems associated with failures of standpipe pressure-regulating valves (PRVs), elevators, and ventilation difficulties are continually emphasized.

The recruits operate as two engine companies, paired up to stretch and advance the initial attack line. The simulated fire is on the 29th floor, of the 44-story, commercial high-rise building. The water pressure at all standpipe outlets is reduced and regulated to 70 to 90 psi, via factory preset, non–fireground adjustable, pressure-regulating valves (PRVs). The initial assignment for the recruits is to report to the 27th-floor staging location via the fire attack stairwell.

Initial assignment

- Report to the 27th-floor staging area using the interior stairs, stairwell #1, the fire attack stair.

  Rational provided: The elevator system has completely failed or the officer in charge has determined that the elevators are not safe to use.

- Recruits are instructed to bring with them all of the basic equipment needed to initiate an effective, solid fire attack.

  Rational provided: The fire apparatus, pumpers, and aerial ladders, are parked on the street, 27 floors below the staging area; the reflex time needed to retrieve any additional equipment will be significant and impractical. An uncontrolled fire will grow exponentially and the window of opportunity to stop the fire via manual firefighting might be lost.

- Each recruit must transport the following equipment:
  1. Full personal protective equipment (PPE), including SCBA
  2. One spare SCBA air cylinder
  3. One 50-foot section (length) of 2½-inch hose carried in a horseshoe-type load (DFD high-rise/standpipe hose pack) over the SCBA cylinder
  4. One forcible entry/exit tool (each member carries a different type of tool so that collectively they will arrive with a striking, prying, and pulling tool available for their operation)
  5. Two DFD standpipe equipment kits (one per engine company, carried by one firefighter).

- With the aforementioned equipment loaded up, the recruits proceed into the building by way of the parking garage ramp, located remote from the perimeter of the building.

  Rational Provided: This is done to illustrate the importance of accessing a building remote from the perimeter, if possible, to avoid any falling debris, glass, and so on, which is frequently encountered at major high-rise fire operations.

  Note: For operations where the fire floor is above the 10th floor, our firefighters are trained to remove their turnout coats and hang them over the SCBA air cylinder before the hose is loaded on, along with removing and carrying gloves, hood, and helmet, for the climb up to staging. This is done to reduce the extreme heat buildup that occurs during this arduous evolution, and allows for the dissipation of more heat via exposed body surfaces, especially the head.

  This equipment adjustment usually takes place at the lobby level, as firefighters will typically arrive at a fire building with all PPE in place. However, this can and should take place at the BASE location for later arriving companies, if a safe access into the building has been established. In the event that members must approach the building from the perimeter and enter via the main entrance doors, all PPE, specifically and most importantly, the helmet must be left in place to give members some protection (although limited) from any falling debris. Obviously, if any of this (PPE) safety equipment is needed to protect the firefighters, they should leave it in place. Common sense should dictate what constitutes proper PPE in a given situation.

1. Once inside the building, the recruits proceed up to the lobby level by way of an interior stairway—the fire attack stair.

2. At the lobby level, the recruits check in with the lobby control officer (LCO) and are given specific instructions on the critical importance of maintaining a strict accountability process at all high-rise operations (fig. 20–2).

3. In addition to checking in with the LCO, the recruits are instructed to always know their exact location within the building, for example inside stairwell #1, fire attack stair, or on the 27th-floor (staging), and so on.

   Rationale Provided: Should operating members become disoriented or completely lost, become trapped, start to run out of air, or encounter any number of potential problems, they will be able to communicate their exact location, which will hopefully help facilitate a more rapid response and rescue by a rapid intervention team (RIT).

4. At this point, the recruits re-enter the stairway and proceed as rapidly as possible up to the 27th-floor staging location.

Needless to say, recruits must be in top physical condition for this segment of the training drill. Because they have been involved in a very comprehensive daily physical fitness program at the DFD Fire Academy for the past several months, and have had the opportunity to train by climbing, walking, and running the stairs in this very building, most are well prepared for the physical demands of the drill.

During a classroom session the day before the hands-on drill, recruits are given comprehensive training and instructions on various methods they should use to carry their equipment for a long stair climb, specifically, how they can make the task of carrying approximately 100-pounds of equipment up several flights of stairs easier and safer. For example, the recruits are taught to remove their turnout coat and carry it over the air cylinder of the SCBA hose pack. They are taught to carry their helmet by hooking it into the SCBA harness using a lightweight carabineer. Once again, this helps to reduce heat buildup. Also, methods on how to properly carry hand tools, use tool belts, and

Fig. 20–2 Denver Fire Department, recruit firefighters, checking in with the lobby control officer prior to clinbing 27 flights of stairs to staging in full gear with equipment.

Fig. 20–3 Recruit firefighters initiating the stretch and hookup from the floor below during a high-rise firefighting/standpipe training drill.

load the hose on the SCBA air cylinder in a well-balanced and secure fashion, will all help lead to a safer, more efficient operation.

The recruits are encouraged to strive for a pace of 20 seconds per flight, which puts them at the staging level within approximately eight to nine minutes from the time they leave the lobby. If you have not had the opportunity to try this, it may appear to be almost impossible. However, time and time again since 1993, our recruits and fire companies throughout the city have proven that it is not only possible, but also very realistic, that is, if the firefighters train for the occasion.

You may be asking, "Okay, so they made it up there, but do they have anything left to fight a fire?" The answer is definitely yes. These recruits are trained to a physical level that will allow them not only to climb up several flights of stairs quickly and efficiently with a heavy load of firefighting equipment, but also to have the physical stamina and strength left to do battle once they arrive. Besides, with the elevators out of service, what other choice is there?

At the 27th-floor staging level

At the staging level (floor 27), the recruits unload the extra hose, spare air cylinders, and any tools and equipment that will not be immediately needed during the initial fire attack operation. They are instructed to organize their spare equipment and store it against walls out of the way on the staging floor.

From this point, they are assigned to fire attack, which involves advancing a 150-foot long, 2½-inch attack handline, out onto the 29th-floor level (fire floor). They proceed up to the 28th floor with three sections (lengths) of 2½-inch hose (DFD high-rise/standpipe hose packs), and the DFD standpipe equipment kit, where they will hook up to the standpipe outlet. Once again, recruits have been given comprehensive instruction and training on all of the related tools and equipment that they are now using. This includes, specifically, how to properly hook up to the standpipe outlet, and how to properly stretch the attack hoseline and advance it out onto the fire floor (fig. 20–3).

The recruits are instructed on how to properly complete the following tasks:

1. Flush the standpipe outlet by opening the outlet valve a couple of turns to flow water before attaching the hoseline in order to clear debris.

2. Attach the DFD standpipe in-line pressure gauge directly to the standpipe outlet, in order to regulate pressure at the outlet once water is flowing and the fire attack gets underway

3. Attach one or two lightweight, 60-degree angle elbows as necessary, to prevent kinks created in the hose when the standpipe outlet is in tight quarters, such as facing straight up inside the standpipe cabinet.

4. Complete a *stairwell stretch* by laying three hose packs on the floor near the standpipe outlet, attaching them together, and stretching the line up the stairs toward the fire floor.

5. Lay out as much hose as possible up past the fire floor landing to the next landing above prior to charging the line with water. This makes the initial advance of line out onto the fire floor much easier, as several feet of hose is pulled down from above, assisted by gravity.

6. Lay out the hose so that the nozzle ends up approximately two to four steps above the fire floor landing. Therefore, when the hoseline is charged with water, and naturally stretches out, the nozzle team will not be pinned up against the door to the fire floor, but will instead have plenty of room to work.

7. Take time to make time, by properly laying out the hoseline and fully stretching the entire hoseline, leaving no piles of hose. This will prevent kinks from occurring in the line when it's charged with water, and will ultimately lead to a safer, more effective advance out onto the fire floor.

Note: Fortunately, the management of this high-rise building allows us to actually charge the hoseline with water from the standpipe system. This creates a much more realistic training drill for the recruits, as they must advance a charged 2½-inch handline out onto the simulated fire floor. However, in order to prevent water damage during the training drill, the recruits only simulate flushing the standpipe outlet prior to hooking up the hoseline, and the nozzle is not opened during the drill except to drain the hoseline at the conclusion of the evolution. All nozzle operations are simulated. Also, support personnel have hose clamps immediately available should a section (length) of hose burst, and one member of the support staff is assigned to remain at the standpipe hose valve to immediately shut it off, if necessary.

Members of the building engineer's staff are present in the building throughout the duration of the drill. Before the drill begins, they disconnect the jockey pumps to the main fire pump, and the water flow alarms at the 28th floor. This prevents the fire pump from starting and/or a false activation of the building's fire alarm system. Of course, before DFD personnel leave the building at the end of the day, we ensure that building engineers have returned their entire fire protection system to normal.

With the hoseline properly laid out and charged with water, the recruits are now ready to advance out onto the simulated fire floor. The basics are always emphasized, including the specific fundamentals associated with advancing a 2½-inch handline. These procedures include the following:

- The attack team should have master keys, previously retrieved from the lobby control officer or the lock box inside the fire command center.
- The floor plan on the floor below the fire should be reviewed prior to commencing with attack. Because this building has scissor stairs, the layout two floors below the fire floor is reviewed by attack team leaders.
- Feel the door for heat, and open it slowly to determine if advancement out onto the fire floor is possible based on the simulated volume of fire and the attack team's ability to suppress it.
- The attack team brings one set of irons with them, in case the master keys don't work or the heat of the fire has warped the door, making forcible entry necessary.
- If the door must be forced, recruits are taught to maintain the integrity of the door so that it can be closed again in order to keep the heat and smoke out of the attack stairwell.

The fire attack

The team of recruits advances the hoseline out onto the simulated fire floor wearing full PPE including SCBA, and all recruits are working *on air*. Six recruit firefighters, simulating two DFD engine companies, advance the hoseline out onto the fire

Fig. 20–4 Recruit firefighters positioned in the "unglamorous positions" during advancement of the attack line out onto the fire floor

floor. This simulation represents the officer and two firefighters from two engine companies, with the engineers/pump operators outside the building completing the water supply. (DFD minimum staffing is four members per company).

The attack team assignments include Engine Co. 1 as the nozzle team, with Engine Co. 2 members providing support. This support includes assisting with advancement of the attack line by positioning members at friction points, corners, stairs, and the floor below—the "unglamorous positions" (fig. 20–4). The Engine Co. 2 members' positions do not remain static, but are in fact dynamic, as they move forward and backward on the line to help feed additional hose for advancement to the simulated seat of the fire. Numerous aspects of the high-rise fire attack are reviewed with the recruits during the training drill. This includes addressing the issue of backup lines and protection of the egress.

Because of their top physical condition and excellent basic training, the recruit firefighters are consistently able to get water on the simulated fire in about 15 minutes from the time they left the lobby. Time is stopped when the charged hoseline enters the fire floor from the stairwell. This is an outstanding time for a fire on the 29th floor, especially when it happened without the logistical benefit of elevator use.

After the handline has been fully advanced out onto the fire floor and fire attack simulated for several minutes, the evolution is stopped, and the recruits are instructed to go *off air*. Training officers review the evolution with the recruits, and provide feedback on their performance. The recruits are given the opportunity to ask questions, and each recruit is evaluated for any physical problems, medical concerns, injuries, and so on. If any such problems are identified, the recruit is immediately pulled out of the training drill and evaluated by members of the medical support unit. Having barely caught their breath, the recruits are sent back to staging to prepare for the next segment of this training drill.

Note: Keep in mind, the DFD trains our recruits to operate as safely as possible. In an ideal world, a team of firefighters who has just completed the above assignment would be sent to REHAB for a long period of rest and recovery. However, the DFD, like most fire departments, has limited resources, and is stretched thin during major high-rise operations. Therefore, the firefighters involved in such an operation must be prepared to, and will probably have to, complete numerous assignments one after another during the overall operation. We are literally preparing our recruit firefighters for the worst-case scenario.

After the fire attack scenario has been completed, the team of recruits is sent back to the staging area on the 27th floor. They are instructed to work together, as a team, and change out empty air cylinders for full ones. Once each recruit has a full air cylinder, they are given a new assignment.

While the recruits are preparing for their next assignment, members of the support staff drain the attack hoseline on the simulated fire floor, pick up all equipment, and transport it to the lobby for the next evolution. The simulated fire floor is actually a mechanical level in this building. This level has floor drains that are used to drain the hoselines. Furthermore, because this is a mechanical level, it is much less susceptible to damage during this training exercise.

Truck company assignment

The recruits are given the assignment to search the attack stairwell above the fire floor. Because of a simulated smoke condition, the recruits must use their respective SCBAs and go on air just below the 29th floor (simulated fire floor) level. This is also a very physically and mentally demanding task.

Fig. 20–5 Recruit firefighters completing the rescue of a downed firefighter from the top of the attack stairwell

When the recruits reach the top floor, they locate a downed firefighter, who is unconscious and out of air. They must provide air to the downed firefighter via DFD rapid intervention team (RIT) kit, which contains a one-hour air cylinder and associated RIT equipment. The recruits must then rescue the victim firefighter by carrying him down 15 flights of stairs to a location that is below the simulated smoke condition. This evolution must be completed before any member of the rescue team runs out of air (fig. 20–5).

The recruits are taught several options that may be applied during a real incident. These include potentially removing the victim firefighter to the roof level, if it is safe to do so, and if air support to evacuate the member is a viable option. Also, they may potentially use the evacuation stairwell for ascent and descent, given that it is clear of smoke.

Support staff members are used to protect the recruits from slips, trips, and falls as they are descending the stairs with the firefighter victim. In addition, because we use a live firefighter staff member to serve as the victim, he is constantly monitored to ensure that he is not being beat up (too badly) or injured during the evolution. If any injuries occur, the evolution is stopped, and a medical support unit is immediately assigned to assist the injured member and provide first aid as necessary. Very few injuries have occurred over 15 years of completing this evolution. Most of the injuries are minor, and related to dehydration and or exhaustion. Once again, the recruits and all support staff members are evaluated throughout the day, and given plenty of fluids to maintain appropriate hydration.

Water supply evolution

While the fire attack scenario is being completed by one group of six members, the other group of six is outside completing a water supply training drill. This involves the use of two pumpers; the recruits complete all hose layouts necessary to simulate water supply via tandem pumping into the buildings standpipe system (fig. 20–6).

The Engine Co. 1 group places their apparatus at the FDC; supply lines are stretched to the FDC and attached. The Engine Co. 2 group lays out supply lines from the Engine 1 apparatus to a fire hydrant. All connections are made, and supply lines charged up to the Engine 1 pumper apparatus. Supply lines to the buildings FDC remain dry.

In addition, recruits are taught methods to overcome various problems encountered, including frozen female swivels on the FDC, broken clapper valves, debris inside the FDC, and so on. Furthermore, they are taught how to back feed the standpipe via hookup to a standpipe outlet on the first floor, if possible. A review of pressure-regulating devices is presented, along with clarification that a pressure-regulating valve (PRV) is also a check valve, and, if one exists at the standpipe outlet, water cannot be back fed into the system to supply the standpipe.

Inevitably, whenever I speak about this recruit training drill, many firefighters from other fire departments comment that it might be possible for recruits, most of whom are younger and in exceptional physical condition, but they are skeptical that the more senior members of an organization could do it. My experience has been that the ability to complete this evolution has little to do with one's age, and much more to do with one's attitude. I like to refer to these examples as *models of excellence*.

Models of excellence

There are countless examples of excellence that occur on every fire department every day. My good friend Sean Roeper is a young firefighter on the DFD in his late 20s. One would certainly expect that someone his age could excel during physically demanding operations. He certainly does, in fact he excels at everything he does. Whenever I climb stairs with him, he rarely gets winded and can literally run me into the ground. I like to use the excuse that he's half my age, but, in reality, he's a top-shelf firefighter in superior physical condition. But once again, it isn't his age, but his attitude. Sean trains daily, and he trains hard.

I unfortunately can cite other examples of individuals younger than Sean Roeper who have literally let themselves go since graduating from the academy. Being overweight, out of shape, and disinterested in developing as a professional firefighter is not a function of age, but attitude; it is in fact a choice, albeit a very bad choice.

There are many top-shelf company officers on the DFD who set high standards at their respective companies, including the implementation of a daily physical fitness program. One of the very best is Lieutenant Mike Shepherd. Mike leads by example as he is the first one to the top of the building every shift. The members under Mike's command are well trained and prepared for battle.

The examples that I am most impressed with are the very senior members of the fire department, who, regardless of their chronological age, are still able to complete the most physically demanding of tasks. It is their attitude that drives them to maintain physical excellence. One such example of excellence is Firefighter Mike Young.

I had the honor and privilege to work with Mike Young when I was the captain of Rescue Co. 1. Mike was my senior man, and if I were to use one phrase to describe Mike, it would be that he's 100% dependable, 100% of the time. This top-shelf firefighter with over 30 years on the DFD is definitely no spring chicken. I see Mike with the other members of his crew climbing stairs almost every work shift. The other day I was climbing stairs at the same time as them, and I saw that Mike had them doing a new routine. As they rapidly climbed the stairs, they stopped at every tenth landing and completed 25 push-ups and 25 sit-ups. They usually complete three sets of 44 flights, unless they are interrupted by an emergency call.

Another exceptional example for the rest of us to follow is my good friend Phil Miller. Phil is a lieutenant on Engine Co. 3, and an invaluable member of my FDIC Standpipe HOT Team. Phil's company has implemented several different initiatives during 2006, including their goal to have each man on each shift at Engine Co. 3 climb at least one million stairs during the year. By midsummer they were well ahead with over 500,000 stairs under their belt. Whenever I see Phil and the members of Engine Co. 3 climbing stairs, they are always in full gear and loaded up with equipment. This company is ready to do battle with a high-rise building fire.

One day about a year ago, Phil called me up and invited me to an Engine Co. 3 training evolution. He realizes from experience that standpipe systems can and do fail. He wants himself and his crew to be prepared. Phil wanted to see how long it would take two engine company crews to stretch 800 feet of 5-inch supply line up 38 flights of stairs. The off-going shift volunteered to stay late and participate in the drill as the second engine company. I watched as eight top-shelf firefighters from Engine Co. 3 stretched the line up to the 38th floor. Each man started out with one 100-foot section of 5-inch hose on his shoulder. That's a lot of weight. They took the hose out of the main supply hose bed and completed the stretch in less than 20 minutes. Outstanding! (fig. 20–7).

It is truly gratifying when models of excellence exist at the top. The district commander in my district is Chief Kevin McDonnell. He holds the number two seniority number of all district chiefs on the DFD. He's been around for a long time and has many years of life experience under his belt. In fact, his belt buckle is not under stress, but actually serves to hold Kevin's pants up, as his stomach is

Fig. 20–6 Recruit firefighters completing the water supply segment of the training drill

Fig. 20–7 DFD Engine Co. 3 completing a very demanding physical training drill, stretching 800 feet of 5-inch hose, up 38 flights of stairs, in less than 20 minutes.

professionally flat. He's in top physical condition and remains that way due to a consistent diet of physical exercise, including daily stair climbing. One day, Kevin asked some of the other chiefs on his shift if they'd like to join him climbing stairs. One of them abruptly said, "I don't have anything to prove." Kevin replied, "I do. I want to prove to myself that I can still climb 44 flights of stairs in full gear, in a reasonable period of time, so when I am assigned to be the fire attack chief, I can do the job." Kevin further stated, "When the day comes that I can no longer do it, then it's time for me to go downtown and retire."

I am very proud to be associated with these men and many others like them. They are models of excellence and their leadership by example will serve as the foundation for the future of the fire service. Once again, when someone asks, "Can the older senior firefighters perform as well as the younger recruits?" the answer is, "yes, absolutely!" Certainly not all of them, but the vast majority can, specifically those with professional, can-do attitudes the ones that truly walk the walk. I am very proud of my fire department, as physical fitness has been given a very high priority and the vast majority of department members are dedicated to physical excellence.

Fig. 20–8 DFD special operations units at a recent, quarterly air support operations training drill

Quarterly District/Battalion Training

Good training programs are the foundation of long-term, firefighter operational safety and survival. Unfortunately, one of the things that happens all too often is that good training programs don't always last. They are usually started by ambitious people with great intentions, but without critical support from above, they can slowly be placed on the back burner, and eventually end altogether.

It is absolutely critical that all training, but especially training for high-risk events, specifically high-rise firefighting, become part of the organizational culture. Starting with introductory classes for recruit firefighters, the training should continue long after these new members complete their probation. All training programs should be ongoing, cumulative, and include all members. How often do we have to re-certify on cardiopulmonary resuscitation (CPR). We should take a cue from our EMS training programs and ensure that our firefighters are given refresher courses and recertified on the essential fire-related topics just as often.

A good place to start is at the district/battalion level. Ideally, fire departments should strive to complete a multi-unit high-rise/standpipe training drill at least on a quarterly basis. These drills should include all of the companies from a respective district or battalion, and should be repeated as necessary, so that members from all work shifts or platoons can complete the training. The logistics associated with such a program are not always easy to accomplish, but with a strong commitment from the top down, it can be done.

On the DFD, each district is assigned specific months during the year for training purposes. The district training chiefs work together to schedule the training and provide fire companies to cover the unprotected areas of one another's district as necessary. These more involved training drills are usually scheduled on weekends, when traffic is lighter and call volume is lower. There have also been successful high-rise drills conducted in the past that included fire companies from adjoining districts that frequently operate together at emergency incidents.

Smaller- and medium-sized fire departments should look to other jurisdictions to provide assistance. Mutual aid and automatic aid agreements can be expanded to include providing mutual assistance on a quarterly basis to one another in order to complete comprehensive training drills. There will be naysayers and those who will say that it can't be done. I completely disagree. Once again, there must be a commitment from the top if any of this is going to be achieved.

Let me give you an example. Unfortunately, we recently had another line-of-duty death on the DFD. In the spring of 2006, Lieutenant Rich Montoya died of injuries sustained while operating at a dwelling fire on Denver's northeast side. The department organized and provided a beautiful funeral for Rich. There were hundreds of firefighters in attendance from all over the Denver metropolitan area, and many from other cities and states. The funeral procession had hundreds of fire apparatus, with personnel on board each rig.

My point is this, we always show tremendous respect and support for our brother and sister firefighters who die in the line of duty, as it should be. I am merely proposing that we show them the same respect while they are alive. We can do that be providing the same level of support, in terms of apparatus and manpower, to assist one another in achieving large-scale, quarterly training programs. Specifically, neighboring fire departments should commit to sending apparatus and manpower to cover another's district for a few hours while they complete some quality training. In doing so, we might save the life of a firefighter or two, and then we won't have to get together for another funeral.

Quarterly Air Support Operations Drills

Recently, the DFD initiated an air support operations program to be used in conjunction with our high-rise firefighting operations. So far, this has been an excellent program and has every indication that it will continue. It has been set up as a quarterly training program, as well, and includes the DFD's special operation companies working in conjunction with the Denver Police Air Support Division, and the Colorado Army National Guard (fig. 20–8).

Tenacious efforts by the DFD special operations chief managed to achieve funding for the program from Homeland Security grants. This is just another example of how quality training programs can become a reality. It all starts with an attitude that it can be done. If someone proposes an idea, and the first comments from other members, especially chief officers are, "that won't ever happen," then the idea is doomed for failure. You can't make the people around you have positive attitudes, but you can make sure your attitude is positive. I believe in the power of one. Anything can be done, and I mean anything. It requires hard work, someone to get it started, and support from above and below.

Do We Really Need to Do That Every Time?

Ponder the question *"Do we really need to do that every time?"* as it relates to your daily operations.

Several years ago, the DFD adopted four-way hydrant valves as part of our water supply operations. The DFD uses 3-inch hose as our primary supply line, and we generally employ a forward hydrant-to-fire hose lay for most of our fireground operations. Our operating procedures mandate that the first-arriving engine company lay a supply line from a hydrant and thus establish a sustained water supply to all reports of fire or smoke in any structure.

At working fires, prior to the implementation of hydrant valves, the second-due engine company gets a separate, secondary water supply as a backup to provide a redundancy if a system fails, be it a hydrant, a supply line, or the pumper. At those fires that are not quickly controlled, the next-due engine companies go to work at the first- and second-due engine companies' hydrants, thus establishing relay pumping to increase volume and pressure. However, this practice of "going to work" on a hydrant that is already supplying water to another pumper requires shutting down the hydrant, thus interrupting the water supply to complete the necessary hookups.

With the four-way hydrant valve, a later arriving engine company can go to work on another engine company's hydrant without shutting down the hydrant. Therefore, there is no interruption of water supply. It's a very simple concept. Although new to the DFD, our research indicated that numerous fire departments across the country have been successfully using hydrant valves for many years.

I consider my fire department to be a fairly progressive organization. Over the years, many positive changes have led to a safer work environment for our members and much more effective operational procedures. The four-way hydrant valve is a prime example.

As you may have guessed, I am a proponent of the hydrant valve. I was involved in the research and development, implementation, and training related to this new water appliance. Although the DFD is a progressive organization, we still have a nearly 150-year history and close to 1,000 members. Needless to say, change is not always achieved easily in older, larger organizations. That's especially true when it involves not just the equipment, but also major changes in our operational procedures as well.

As an outspoken proponent of the four-way hydrant valve, oftentimes I found myself in the middle of discussions, frequently trying to open the minds of some who had not seen the light of day for years. On one particular day, the members of an engine company were having a friendly discussion behind their pumper. As I passed by, one of the members stopped me and asked if I could answer a couple of questions. Several of the members asked me questions, mostly just to clarify operational issues they had encountered. Then it happened. I heard those now unforgettable words, "Hey Chief," one of the men asked, *"Do we really need to do that every time?"*

The easy answer to that question would have been, "Yes, you do need to do that, every time, so just do it." But words like that are tough to overcome once they are out of your mouth. The old saying, "You get more bees with honey than with vinegar," goes a long way when dealing with firefighters. Over the years, as both a company officer and a chief officer, I have found that the members under your command will do anything you ask, especially if they have a complete understanding of the reason behind the request. However, you must take the time to educate and train your members long before the commands are given.

A motivational factor

"Do we really need to do that every time?" That seemingly benign question has evolved into a significant motivational factor for me and countless other firefighters that I teach, mentor, and work with every day. I realized that at the root of that question was one of the deadliest aspects of human nature: complacency. The question truly yields not just one clear-cut answer, but several answers, at least on a theoretical and philosophical level.

To begin with, we all know that within the so-called "routine" daily operations of all fire departments, both large and small, we respond to countless events, many of which turn out to be something less than a major emergency. Those annoying automatic alarms in large commercial buildings that frequently prove to be system malfunctions, unintentional false alarms, smoke scares, and a long list of other categories are far from the classification of building fire.

The reported building fire turned out to be smoke from a barbecue grill. The parties trapped in an auto accident ended up being a minor fender bender with no injuries. The odor investigation turned up a small plastic spoon that was resting on the heating element inside a dishwasher. Most fire departments have had these statements on more than one incident report narrative. It's a fact of life in our business. Unfortunately, as human beings, we can quickly be lured into a false sense of security, and that deadly disease of complacency can start to take hold.

It is easy to see how anyone could ask the question when our experience has proven many times that our good habits and disciplined actions frequently are not needed on a response. On a daily basis, DFD engine companies working in districts across the city lay out supply lines in response to reports of smoke or fire, only a portion of which turn out to be bona fide working fires. As for the four-way hydrant valves, they certainly will not be needed at the nonworking incidents and perhaps even on some of the working incidents. The DFD, like most fire departments, will quickly stop most working fires with the first handline, generally a 1¾-

inch preconnect. At those incidents, the demand for water typically will not require relay pumping and the use of the four-way hydrant valve.

So once again, the question remains, *"Do we really need to do that every time?"* Before we jump to a hasty conclusion and answer no, we must first examine the facts and figures on the other side of the ledger. For every full first-alarm assignment that was cancelled because of smoke from a barbecue grill, we must recall those significant events that started with a single engine company response to perhaps an investigation of outside smoke.

Brother Firefighter Brett Tarver of the Phoenix (AZ) Fire Department was lost at an event that began as a report of rubbish burning outside a building. I will always remember a good fire I had a few years ago. It was about 42 minutes past midnight, and I responded with a first-alarm assignment to a report of smoke on the first floor of a high-rise multiple dwelling. That building, like many others, has a strong and well-deserved reputation for late-night burnt food incidents. There were no second-source calls, and my first-arriving companies reported nothing showing. I struck the second alarm four minutes later and the third shortly after that. What early on showed indications of being a routine nuisance call proved to be a heavy volume of fire on the lower level that connected two large, high-rise multiple dwellings, each containing three separate wings with nearly 1,000 occupants. And, there were no sprinklers!

If we could accurately predict the future, it would be simple to answer our question. However, I am still searching for that magic "fire service crystal ball," and Nostradamus didn't include addresses or building locations in his prophecies.

Prediction versus preparation

We can't predict, but we can prepare. Preparation is a lot like size-up: it's ongoing and cumulative. It begins with the physical and mental preparedness of all firefighters. It rests with a strong mind-set and belief that it could happen today. What is *it*? *It* could be just about anything. The greatest prophecy in the fire service today is our own history. We continually hear about how firefighters have been killed in the same ways as in years past. Fire Department of New York Brother Vincent Dunn has told us on many occasions, "Learn from your history, or you're doomed to repeat it."

As a young man growing up, I wasn't the best student. I especially disliked history class. After all, at 16 years old, what's the point? I had a lot of things on my mind, and history wasn't one of them. Now well into my 40s, I am driven to learn as much history as I can, especially as it relates to our profession. As I aggressively study history, I continue to ponder the question, *"Do we really need to do that every time?"* My research took me outside the fire service, and I discovered some very interesting things.

Lessons learned from American history

There have in fact been countless occasions in American history that began with complacency, neglecting to take one or several actions that should have been accomplished. Some classic examples of these failures relate to our question and illustrate that, yes, we should have done that this time, but didn't do it. In many of these examples, the people involved knew better but gave in to complacency that resulted in tragedy.

My first example occurred several generations ago: "A day that will live in infamy," the famous words of then president Franklin D. Roosevelt. Countless books and movies recall the events related to the Pearl Harbor attack that occurred more than 65 years ago. Sure, hindsight is 20/20, but perhaps we sometimes see only the things we want to see and ignore the real important stuff when making decisions. In the case of Pearl Harbor, historical facts indicate that plenty of intelligence pointing to a possible attack by Japan existed long before it ever took place. In fact, on the day of the attack, several young GIs assigned to radar screens saw the enemy aircraft long before the first strike. Their pleas to take action fell on deaf ears, with comments such as, "It's probably just some birds," or "It has to be some of our aircraft." *"Do we really need to do that every time?"*

I was a "new boy" with only three years experience on the job in 1985. It was the morning of January 28, 1986, and I was on duty in the fire station that morning. I was busy mopping the kitchen floor. The TV was blaring in the background. Suddenly, the TV screen was filled with a fireball in the skies over the Kennedy Space Center in Florida. The Space Shuttle Challenger had just blown apart in midair. The lives of seven very talented people were lost in a matter of seconds. What went wrong?

The subsequent investigation uncovered several disturbing items. During what appeared to be a political battle, rocket engineers from Morton Thiokol, who had designed parts of the aircraft, had literally pleaded with NASA managers not to go forward with the launch of Challenger. Their concern was that the cold weather would negatively affect the O-rings that were supposed to seal the joints of the solid rocket boosters. The engineers believed it was too cold and that the O-rings might fail and allow the extremely volatile fuel to escape and cause the spacecraft to catastrophically fail. Having had already postponed the launch because of cold weather three times, the NASA managers were growing increasingly impatient. Politics prevailed over safety. The launch went forward. The temperature was 36°F, lower than that for any other previous launch. The O-rings did, in fact, fail, and 73 seconds after the launch of Challenger, seven lives were lost forever. *"Do we really need to do that every time?"*

More than 17 years later, it wasn't O-rings, but foam insulation panels that led to the Columbia spacecraft disaster. Once again, history repeated itself. The post-investigation uncovered correspondence within NASA that warned that if foam insulation panels broke loose from the space shuttle's fuel tanks and struck the aircraft, catastrophic failure was highly probable. *"Do we really need to do that every time?"*

The events of 9/11 certainly have overshadowed the 1993 bombing of the World Trade Center. But as history lessons go, this was extremely significant. I'm not a big fan of politics and do not want my message to be distorted by political commentary. However, I have to believe that, whether a person is liberal, conservative, or somewhere in between, our country really missed the message in 1993. The malicious intentions of those involved were clear, and their tenacity obviously was underestimated. Had a vigorous campaign against terrorism started in 1993, those 343 Fire Department of New York brothers, who we all sorely miss, would still be with us today. *"Do we really need to do that every time?"*

History lessons within the fire service

Moving from those examples outside the fire service, we don't have to struggle to identify countless events that have ended in tragedy for us. Reviewing the fire service case studies in which firefighters were injured or killed is a significant learning tool. The sad part involves the discovery that much of what we learn often is nothing new at all. As I said in chapter 1, I believe that most of the tragic events that occur in the fire service are the result of several factors that set in motion a chain reaction, such as that of dominoes falling one after another until they're all down. During that process, individual decisions can literally stop or increase the speed of the fall of the dominoes.

It boils down to good habits, plain and simple. We don't have control over everything that can affect us, but we do have the ability to develop and maintain good habits and the basics, like stopping two floors below a reported fire floor and walking up the last two flights. Basic and elementary, sure. But failing to strictly follow that basic procedure has, on more than one occasion, been the first domino to fall on a path that resulted in the death of brother firefighters. *"Do we really need to do that every time?"*

Every single call to which we respond has to be taken seriously. We don't know for sure what it is until we get there, investigate, gather the facts, and take care of business. The chief officer who consistently fails to respond to automatic building alarms because "it's probably false" sends out a strong message to every one under his command. That message is one of laziness and complacency. Just like a dysfunctional family, soon, most of the members under his command will follow his lead and absorb the deadly habits of complacency. Many will get away with it for weeks, months, years, perhaps even an entire career. But it takes only one time to be wrong, and the dominoes will fall.

As my father has told me on many occasions, "The most important fire you will ever go to is the one you're at right now." Take every call seriously, conduct a comprehensive and professional operation every time, and consider it over only after you have covered all the bases and brought it to a safe and successful conclusion. If you think it's nothing before you even get there, the dominoes of complacency are falling; the result could be deadly for somebody you should be protecting, specifically the brothers around you.

The American Fire Service's "Dream Team"

During my 25 years as a firefighter, I have had a number of significant influences that I attribute to my passion for our profession. I have been very fortunate and blessed with opportunity and positive guidance throughout my career. As the son of the man who I consider to be the greatest firefighter that ever lived, I have received quality mentoring from my dad throughout my career, mentoring that continues today. In addition, I have worked with some of the best firefighters in the world, those being my many mentors from within the Denver Fire Department.

Unlike many firefighters, I have had the added opportunity to meet, become friends with, and receive continued mentoring from countless firefighters across the American fire service. As I mentioned in chapter 9, one of those mentors was the late Andrew Fredericks. Andy introduced me to a whole new world. As a member of Andy's FDIC Engine Company HOT Team, I was introduced by him to Jerry Tracy, Battalion Chief, FDNY. Jerry and I started the FDIC Engine Company Standpipe HOT Team, which has grown to become what I consider to be the American Fire Service *"Dream Team."* These are men who I have been lucky enough to know for over a decade now, and who continue to teach me the fine points of being a professional firefighter (fig. 20-9).

If I were to offer some advice to the young, up and coming firefighters that I mentor, I would say, make sure you get out of your world, and find out what some of the other guys are doing. The American fire service is a global place, with numerous talented individuals from big cities and small towns. I have seen far too firefighters get completely locked into *"their way"* of doing things, usually because *"that's the way we've always done it."*

The FDNY is probably the most recognized fire department in the world. I am fortunate to have many friends from that department. In the past, I have heard negative and condescending comments, like *"hey, this isn't New York, and we don't do it that way here."* There is no helping the people who

Fig. 20–9 With several hundred years of combined experience, the FDIC Engine Company Standpipe Operations Hands on Training (HOT) Team, ("American Fire Service Dream Team") set the standard of excellence and provide invaluable training to hundreds of FDIC participants each year.

make these statements, but you can help yourself. Start by opening your eyes and learning from other professional firefighters, including those from outside your own department. My exposure to countless great firefighters, including those who I consider to be part of the "Dream Team," has tremendously helped my professional growth.

Carpe Diem: Seize the Day

Today's fire service is filled with youth, and we must protect them. In my conversations with young firefighters, they all say the same thing: they want experience. They want the opportunity to prove themselves to the older firefighters. Many comment that they just want to be a great firefighter. But when will they get their chance?

To those young firefighters, I always say the same thing, "*Carpe diem*, brother." That's right, *seize the day!* Often the young, eager firefighter is waiting for his moment to shine. I remember my own thoughts and desires as a young firefighter. I, too, was waiting for that moment when I could step up to the plate and show the veterans I had what it took to do the job. As it turns out, they were watching me all the time, not just during the serious events.

I frequently think about the 343 brothers lost on 9/11. I was fortunate to have known a handful of them personally. They all demonstrated greatness through their incredible courage and actions that horrible day. We all hope that we would be as strong as they were if ever faced with something that enormous. But it wasn't that one event that defined those men and who they were. This is especially true with our brother, Andrew Fredericks, who I think of every day. Those who knew him fully understand that he wasn't waiting for his moment—he was seizing the day. The positive impact he had on the American fire service and countless brothers across the country, including me, was absolutely phenomenal. The man will not be defined solely by his heroic actions on 9/11 but, more specifically, by his daily actions and contributions to the American fire service and the FDNY (fig. 20–10).

As firefighters and human beings, we can make specific choices. There are two basic paths we can take daily on the job. The easier path is *complacency*—donuts in the morning and *Oprah* in the afternoon. It doesn't get any better than that. The second path is for those interested in hard work, study, dedication, and professional development. It is the path of *preparation*. It may lead you to

Fig. 20–10 Our brother, friend, and mentor, Lieutenant Andy Fredericks taught us so much during his unfortunately short life. His message will live on, forever.

climb the stairs at the local high-rise building after roll call or to make a quick stop at a construction site on the way to the store to size up lightweight construction—the latest firefighter killer—or to lead a drill on the essential components of an RIT pack and train on rapid intervention operations for commercial high-rise building fires. All this work is in between calls and other department activities. Yes, you can have your cake and eat it, too. There's nothing wrong with taking some time to relax, watch some TV, or do some personal stuff. Just remember, take care of business first. Get prepared, be prepared, and stay prepared, both mentally and physically.

In the movie *Dead Poets Society*, Robin Williams's character Mr. Keating, in a motivating dissertation, told his young students that *"the powerful play goes on and you may contribute a verse."* He then asked them, *"What will your verse be?"* That powerful play is life and our careers as firefighters. The verse we may contribute can be either positive or negative. What will your verse be? Have we answered the question yet? If not, remember this: when you walk through the doors of the firehouse to start your next work shift, you can be a truly great firefighter, or not. It will be easy to distinguish the real firefighters from those who consider our great profession to be just a part-time job. Simply put, the truly great firefighter never asks the question, *"Do we really need to do that every time?"* The truly great firefighter just does it, every time! *Carpe diem*, brothers.

Index

A

accountability, 181, 279–281, 287–288
adapters, 14
    for PRDs, 58, 59
    in standpipe kit, 166, 175–176, 205
    for supply lines, 214
ADULTS (Advanced fire upon arrival/ Defensive operations/Unable to determine the extent or location of the fire/Large, uncompartmentalized areas/Tons of water to cool the heat/ Standpipe operations), 130–131
aerial surveys, 300–301, 303
aid agreements, mutual and automatic, 33, 121, 264–265, 331
air ambulance helicopters, 296–297
air cylinders
    checking/maintaining, 14
    rating for/use of, 201
    REHAB and, 201, 289
    spare, 78, 117–120
    at staging area, 286, 327
air movement. *See also* Stack effect
    mechanical, 250–251
    natural, 245–249, 251, 257–258
    neutral pressure plane and, 246, 247–248
    operational, 251–257
    smoke and, 246, 248–249, 250
    unpredictable, 152
    ventilation and, 26, 30, 32, 224, 245–258
air support operations, 295–311. *See also* Helicopters
    branch, within ICS, 292
    drills, quarterly, 330, 331
    equipment for, 296, 297, 303, 305, 306, 308
    operational plan for, 299
    resources for, 22–23, 295–299, 300, 311
    from roof, 22–23, 28, 75, 76, 105, 264, 271
    training for, 296, 299, 300, 305, 327–328, 330, 331
alarms, false, 1–2, 11, 42
    complacency and, 1–3, 42–43, 332–333
    during training, preventing, 326
alarms, fire
    activated shutters, 112
    audible, 101, 185, 206, 233
    automatic, 204–205, 317, 332
    doors unlocked automatically with, 225
    elevator recall and, 90
    first, 200
    fourth, 201, 287
    multiple, 28
    panel, monitoring, 27, 96
    pressurization of stairwells and, 29, 112, 115
    second, 15, 120, 200, 232
    smoke management systems activated with, 112, 251–252, 254
    third, 15, 120, 201, 232, 287
Albinger, Paul, 68
Americans with Disabilities Act (ADA), 81
apartment fires, 8, 33, 48–49, 107, 152, 170–174, 183, 184–185, 187–188, 195
    flanking attack at, 248
    forcible entry/exit and, 225–226
    hose stretches with, 187–188
    number of firefighters in, 195
    precontrol overhaul with, 223–224
    protect in place strategy at, 49, 226, 230–231, 232–233

apparatus placement
   with BASE, 288
   with fire department connection, 209–210, 211–213, 216
   truck company operations and, 209, 221–222
arson fires, at schools, 42
attack. *See also* fire attack
   flanking, 248
   teams, 107, 109, 137–138
attacking and Extinguishing Interior Fires (Layman), 153
auto exposure, fire extension via, 32–33, 49, 222, 224
axe, 91, 118

B

balconies, on residential buildings, 32, 235
BASE, 286–288, 293, 323
below grade fires, 45, 186, 187
Borelli, Dave, 67
Bousman, Zach, 206
Brannigan, Francis, 2
bresnan distributor (cellar nozzle), 175
bridges
   standpipe systems and, 43–44
   transient shanty fire under, 146
   ventilation and, 40
Brunacini, Alan, 276
brushes, wire, 166, 177
btu's (British thermal units), 8, 124–125, 183
buffers, for hoist-ways, 85
building communication systems, internal, 314–319
building engineers, 35, 63–64, 94, 112, 252, 278, 326
building systems
   control of, 281–282, 291, 301
   failures in, 89, 91, 105, 106, 119, 204, 255, 316
burns
   nozzle selection and, 145, 146, 148
   propane Christmas tree, 145, 151

C

campaign fires, 201, 289, 294, 299
can (2 1/2-gallon water extinguisher), 48, 91, 92
canvas bags, for stairway operations, 117–118
carbon monoxide, in smoke, 249
cardiopulmonary resuscitation (CPR), re-certify on, 330
carpe diem (seize the day), 336–337
center-core high-rise buildings, 25–26, 28, 31, 149, 151–152, 195–197
chalk, 166, 179, 238–239
Chicago Housing Authority, 256
children, 235, 309
Cipri, Joe, 248
Clark, William E., 144
clothing
   for elevator operations, 96, 103
   stairway operations and, 117–121
codes, fire and building, xxi, 18, 27, 29
   on fire-stopping material, 26
   Hansen on, 63
   on PRVs, 60
   on smoke management systems, 251
   on sprinkler systems, 36
   on stairs, 109–110
   on standpipe systems, 50–51, 53, 57, 157, 170, 186
collapse rescue unit, wood from, 216
Collins, Larry, 300, 302
combined systems (standpipe/sprinkler), 47
   defined, 52
   pressure-regulating devices with, 57, 69
   zone/floor-control valves for, 57
combustibles
   class B, 8
   at sports venues, 40
combustion, process of, 78
command and control, 275–294. *See also* Fire attack group; Fire command center; Incident command system; Incident commander; Lobby control/systems
   accountability with, 279–281, 287–288
   BASE and, 286–288, 293, 323
   communication with, 277, 278, 283–287, 290
   lobby and, 277–283, 287, 290, 291, 293, 294
command, formal, 277, 279
command post (CP)
   location of, 278–279

RIT, 262
commercial buildings
    overhaul at, 222–223
    residential v., 19–21, 29, 33, 88–89, 94, 107, 193, 224, 230, 232, 262
    search and rescue in, 230, 233–234, 235, 239–243
commercial high-rise buildings
    backup lines in, 195–197, 200
    coordinated attack from opposite directions in, 196
    forcible entry/exit in, 224–225
    RIT at, 262, 267, 269–271
    two separate handlines operating side by side, 196
communication, 313–319. *See also* public address system; radios
    with command and control, 277, 278, 283–287, 290
    with fire command center, 316–317
    with HHRT, 300–302, 303, 304, 305–307, 310
    news media and, 297, 298
    during search and rescue, 230, 231–232, 242
    utensils for, 179
    verbal, 192–193
company officers, responsibilities of, 3, 12, 13, 121
complacency
    false alarms and, 1–3, 42–43, 332–333
    injuries and deaths from, 1–3, 5, 11
    mind-set and, 1–3, 4–5, 10–11, 336
conditions, actions, and needs (CAN report), 275, 277
construction, 23–28
    core, 25–26, 28, 31, 149, 151–152, 195–197
    first generation, 24–25, 30
    fourth generation, 26–27
    lightweight, 9, 261
    poke-through, 30–31, 222–223
    post-9/11, 26–27
    post-World War II, 25–26, 30
    pre-World War II, 25
    second generation, 25, 26
    specific concerns of, 27–28
    stacked floor layouts in, 107, 223
    third generation, 25–26, 30
    truss, 26, 27, 28, 261
    tubular, 26
    type, 261
    of World Trade Center buildings, 9, 23–24, 26, 28

Cook County Administration Building, Chicago, 236
Corbett, Glenn, 26
Corporandy, Ted, 185, 299
curtain wall gap, 26, 31–32
customer service, xx, 5, 7–8, 155

D

dangerous procedures, 194
dangers, of modern firefighting, xxi, 8–9
Davis, Harry Lee, 141, 197
Dead Poets Society (film), 337
deaths, xix, xx
    from complacency, 1–3, 5, 11
    elevators and, 82, 93, 95–96, 103
    from falling debris, 215–216
    from lightweight construction, 9
    line-of-duty, 1–3, 5, 8, 11, 151, 184–185, 266, 331, 336
    at One Meridian Plaza, 57
    physical fitness and, 11
    from smoke, 30, 249
    from steam, 152
debris
    falling, 210, 215–216, 230, 253, 278, 324
    in fire department connection, 54, 55, 207–208
    passing, with fog v. smoothbore nozzles, 160, 163, 170
    in standpipe systems, 159–160, 162, 176
Denver Fire Department (DFD), xix
    models of excellence in, 328–330
    training programs, 322–333
Denver Fire Department (DFD) hands-on training program, 322–330
    at 27th-floor staging level, 325–326
    fire attack with, 323, 326–327
    initial assignment with, 323–325
    large-scale academy recruit drill with, 322–323
    truck company assignment with, 327–328
    water supply evolution with, 328, 329
Denver hose packs, 118, 139, 172, 187
device, defined, 62
diet, importance of, 11
disabled persons
    buddy for, 231, 235
    evacuation of, 115, 235

removal of, 231
discipline
    elevators and, xxi, 4, 94, 95, 102
    mind-set and, 9, 10, 11, 12
    training as, 12, 14
disorientation, impacts of, 37, 185, 231
dispatching systems, computer-aided, 111
district chief, responsibilities of, 95, 116, 205
"Do we really need to do that every time?", 331–334, 337
Donovan, Regis, 214–215
donut effect, wind and, 152
door chocks (wedges), 49, 116, 166, 177–178
door straps/search markers, 166, 178, 236–238, 239, 266–267
doors
    automatically closing devices on, 233
    elevator, 85, 86, 91, 93, 97–100, 98–104
    fire, 48, 85, 112
    integrity of, 49, 174, 185, 226, 233, 255, 285, 307, 326
    roof, 255, 306–307
    unlocked automatically with fire alarms, 225
doors, open
    to public hallways, 150, 232–233
    stack effect and, 247
    to stairwells, 26, 49, 115–116, 117
Dream Team, American Fire Service, 335–336
drop point, 188
Dunn, Vincent, xix, 1, 9, 333

E

educational facilities, 42
egress
    backup line and, 195
    for firefighters, 261
    by following hoseline, 37–38, 186, 233, 241
    primary search for paths of, 234–235
    problems with, 26, 195
    windows for, 106, 235
elbows, 60-degree lightweight angle, 166, 168–169, 214
electricity, 30, 290
    electrical vaults/rooms, 226–227
    elevators and, 30, 83–87, 89, 95–99, 101, 103, 227
    hazards to, 226–227, 261
    water damage and, 226–227
elevator(s)
    air movement and, 254
    banks of, 88–89, 91, 98
    below grade, 93
    counterweights with, 87
    discipline and, xxi, 4, 94, 95, 102
    electric traction, 82–87, 88
    electricity and, 30, 83–87, 89, 95–99, 101, 103, 227
    emergency exits for, 102–104
    emergency stop for, 2, 99, 101, 104
    express, 86
    fire service recall and control of, 1, 29–30, 90, 91, 93, 97, 101, 103, 254, 282
    firefighters in, 1–3, 29, 89, 91–94
    in high-rise buildings, 1–3, 4, 17–18, 19, 29–30, 75, 76, 81–104
    hoisting ropes in, 83, 84–85, 87
    hoist-ways, 32, 83, 85–86, 89, 92, 93, 97–100, 102, 254
    hydraulic, 82, 87–88
    injuries/deaths and, 82, 93, 95–96, 103
    inspection of, 87, 92–93
    inspection station, 87
    invention/history of, 17, 29, 81–82
    keys for, 90, 91, 96, 98–100, 102, 103, 266
    lighting and ventilation for, 84, 102
    logistics and, 75, 76, 81
    in low-rise buildings, 86, 91, 95
    mechanics, 94–95
    monitoring, 27, 282
    multiple, 83, 86, 88–89, 91, 96, 98, 101, 103
    occupants trapped in, 30, 84, 95–104
    operator, designating, 91, 92
    overloading, 1, 91–92, 104
    panel, 98, 99, 282
    to parking garages, 88
    roller guides, 87
    safeties, 84–85, 87
    shafts/shaft-ways, 32, 83, 85, 108, 109, 223
    smoke, 89, 91, 92, 93, 95–96, 254
    stairs and, 3, 90–91, 92–93, 100, 105–106, 116, 119, 160
    stalled, 30, 84, 86, 89, 94–104, 232, 282
    taking elevator to floor of alarm, 1–3, 91, 93
    taking elevator to two floors below floor of alarm, 3, 91, 93, 94, 106
    testing, 91–94
    traction/drive sheave of, 84, 96
elevator cars, 87

hatchway door, 91, 93, 102–103
location of, 97–104
elevator machine room, 83–85, 88
controller in, 84
hoisting machine in, 84, 85
main disconnect in, 83–84
motor generator in, 83, 84
power control of, 95–96, 101
smoke or fire in, 89, 91, 95–96
speed governor in, 84–85
water in, 89, 119
elevator operations, 81–104
backup plans for, 89–90
door-locking mechanisms in, 91, 98–103
equipment/tools for, 91–92, 96, 101–103
during fires, 88–94
personnel assignment in, 91, 95–96
poling across in, 100–101
poling down in, 100, 101
recommended procedures for, 90–94
resetting system for, 97
roof hatch access in, 91, 102–103
size-up consideration for, 94, 97, 99, 102, 103
staging and, 91, 93
elevator rescue operations, 94–104, 101, 103–104, 323
methods of, 97–104
removal procedures for, 95–104
removal v., 95, 96, 102
roof hatch door in, 91, 93, 102–103
side emergency exit door in, 103–104
Emergency Medical Services (EMS), 7, 216, 289, 290, 291, 292
emergency service, customer service v., xx, 5, 7–8
Empire State Building, New York City, 21, 25
engine company operations, 181–202
connecting hose lengths, 187
dangerous procedures and, 194
elevators and, 91, 95–96
fire attack and, 195, 200–201
first handline and, 194, 220, 275, 277
fourth due company and, 200, 212–213
functions of, 182, 184
hooking up on floor below fires and, 184–187
offensive use of master stream appliances by, 197–199
overall strategic plan for, 200–202
priority one: stopping the fire, 181–182
proper positions for, 137–138
second due company and, 137, 171–172, 188, 191, 193, 194, 200, 203, 211–212, 277
stretching handline, 184
third due company and, 200, 212–213
truck companies and, 219–220
two engine companies, one handline, 183–184
water supply and, 182–194, 203–217
engineer/pump operators
BASE and, 287
protecting, 215–216
on SPG and, 166
standpipe kit, 205, 207–208, 211
water supply and, 47–49, 54, 61, 166, 204–216
equipment
for air support operations, 296, 297, 303, 305, 306, 308
checking/maintaining, 14
for elevator operations, 91–92, 96, 101–103
for search and rescue operations, 236–241
stair climbing with, 14, 76–78, 90, 91, 106, 116–121, 134, 160, 176, 323–325, 330
standpipe equipment kits, 64, 67, 165–179, 205, 207–208, 211
training with, 14, 139, 323–326
user-friendly apparatus and, 133
water supply, 29
equitable building, Denver, 25
evacuation, 18
communication for, 76–77, 230, 231, 317–318
of disabled persons, 115, 235
rooftop, by helicopter, 309–311
stairs, 22, 29, 76–77, 93, 113, 114, 115, 231, 235, 255, 264, 270, 284, 308, 318
exposures, 32–33
logistics and, 73–78, 182
in low-rise buildings, 36, 38
outside survey and, 261

F

families, xx
fans, ventilation, 84, 245, 255–257
FAST (firefighter assist search team), 259

FDIC (Fire Department Instructors Conference), xx, 12
   flow tests at, 158–159
   HOT and, 56–57, 141, 155–159, 192, 199, 329, 335–336
field adjustable
   defined, 64–65
   PRDs, 59
   PRVs, 57, 63–65, 66
finance/administration section, of ICS, 292, 294
fire(s)
   free-burning phase of, 159, 193
   growth of, 71, 183
   location of, 29, 39, 75, 114, 130, 184, 215
fire academy, 11, 12, 144–145, 147, 222, 322–328
fire attack
   coordinated from opposite directions, 196
   engine companies for, 195, 200–201
   HHRT and, 308–309
   REHAB, staging and, 201
   RIT, 262–263, 269
   stairs, 26, 29, 49, 76, 93, 113–115, 232, 234, 236, 254–255, 284–285, 307–308, 318
   three-dimensional, 194
   three-pronged approach to, 200–201
   training with, 323, 326–327
   two separate handlines operating side by side, 196
fire attack group, 91, 113, 195, 232, 283–284, 290, 292
   supervisor, 277, 283–284, 291
fire blanket, 253
fire branch, of ICS, 292
Fire Chief's Handbook, 144
Fire Command Center (FCC)
   communication with, 316–317
   elevator operations and, 90
   lock box with keys in, 224, 225
   PRV tools in, 63–64, 66–67
   responsibilities of, 27
   smoke control/air management panel in, 251
   systems control officer in, 282, 301
Fire Department Connection (FDC)
   apparatus placement with, 209–210, 211–213, 216
   for bridges, 44
   bypassing, 213–214
   clapper valves with, 54, 208–209, 214

   debris/trash in, 54, 55, 207–208
   locations of, 205–207
   priority for sprinkler and standpipe systems, 47–49
   Siamese connection, 53, 182, 208–209, 214
   standpipe systems and, 44, 47–49, 53–55
   Storz coupling for, 53, 54
   supply lines for, 53–54
   supply lines for, bypassing, 213–214
   supply lines to, securing, 210–211
   swivels on, 208–209
   water supply and, 29, 53–55, 205–215
fire department operations
   HHRT and, 302–303
   standpipe systems in, 38, 42, 50–52, 70
fire detection systems, 9, 28–29, 96, 282
fire doors, 48, 85, 112
Fire Engineering magazine, xx, 12, 56, 133, 145, 148, 300
fire escape, 22
fire extension
   via access stairs, 112
   via auto exposure, 32–33, 49, 222, 224
   to carpets, 23
   checking for, 106, 222–224, 308
   via curtain wall gap, 26, 31–32
   poke-through construction, 30–31, 222–223
   preventing, 199
   smoke spread and, 30–33
   through vents, 223
Firefighting Tactics (Layman), 153
fire floor
   hooking up on, 37, 184–185, 187
   hooking up on floor below, 37, 135, 184–188
   search and rescue operations and, 232–234
   standpipe cabinets on, 184–185
fire hydrants, 207
   four-way valves, 331–333
   tying into, 182, 211–213
fire investigations, 49
   of 1993 World Trade Center bombing, 18
   of Heenan's death, 151
   of PRVs, 57, 63
fire prevention, xxi, 63, 159–160
Fire Protection Handbook (Shapiro), 50, 52, 58
fire protection systems, xxi, 28–29, 69, 282
   water supply and, 29, 47–49
fire pumps
   electricity and, 30
   pressure from, 56, 57

room, PRV tools in, 55, 63, 67
water supply and, 29, 55, 204
fire service
    agencies assisting, 278
    daily activities in, 1
    expanded role of, 7–8
    history of, xix, xx, 6–8, 334
    learning curve, 12–13
    mission of, xx, 3
    on nozzle selection, 154
    reshaping, xx, 5
    Web site, 12
fire service recall and control, of elevators, 1, 29–30, 90, 93, 97
    Phase I, 90, 91, 97, 103, 254, 282
    Phase II, 90, 91, 101, 103
Fire Stream Management Handbook (Fornell), 132, 149, 151, 155–156
firefighters
    egress for, 261
    in elevators, 1–3, 29, 89, 91–94
    lost on September 11, 2001, 336
    required numbers of, 33, 195
    rescue of, 260–261, 266–269, 327–328
fireground operations
    HHRT and, 306–309
    water supply and, 29, 204–205
fireground priorities, 154–155
fireground support functions, 220–221
firenuggets, 162
fire-resistive components, 24, 28, 85
fire/smoke towers, 26, 114, 115, 171, 254–255
fire-stopping material, 26, 31–32
The Fire Officer's Handbook of Tactics (Norman), 181, 221
First Interstate Bank Building, Los Angeles, 23, 105, 223, 263, 289
"first-aid firefighting," 51
flanking attack, 248
flashovers, preventing, 9, 151, 193
Flatiron Building, New York City, 87
floors, of buildings. *See also* fire floor
    fires in lobbies v. upper, 74–75
    hooking up on floor below fires, 37, 135, 184–188
    hooking up on floor below standpipe systems, 37, 38, 40, 48, 55
    identification/labeling of, 27
    layouts of, 107–108
    number of, 27

pumping water to approximate range of, 215
spandrel distance between, 32
structural components of, 28
flow tests
    at FDIC, 158–159
    for SPG, 167
    for standpipe systems, 37, 68, 199
    valves and, 62, 68
    for weapons, 156–159
"flowing the floor," 199
force, disproportionate, 141
forcible entry/exit
    in commercial high-rise buildings, 224–225
    in residential high-rise buildings, 225–226
    tools, 91–92, 102, 198, 224–226, 233, 265, 266
    truck companies and, 224–226
forest service aircraft, 299
Fornell, Dave, 56, 132, 149, 151, 155–156
Frank, Dave, 102
Fredericks, Andy, xx, 148, 159, 335, 336
friction loss
    hoseline selection and, 125, 126, 141
    water pressure and, 37, 52, 57, 68, 125, 126, 141, 215
full alarm assignment, 120
full service fire department, 5

G

gated wye, 159, 175–176
generators
    emergency, 228
    motor, 83, 84
Girard, George, 6
GPM (gallons per minute), Btu's v., 124–125, 183
gravity, hose stretches and, 110, 189
gravity tanks, 29, 50, 182
grease pencils, 166, 179, 238–239

H

habits, good, importance of, 3, 16, 94
Halligan, Hugh, 102

hand wheel
    for standpipe hose valve, 40, 55–56, 166, 175, 176–177
    wrench, 166, 177
handlines. *See also* hoseline
    backup, 195–197, 200
    first, 194, 220, 275, 277
    one, two engine companies and, 183–184
    stretching, 184
    two separate, operating side by side, 196
Hansen, Thor, 63, 159–160
hazardous materials, 42, 43, 261–262
heat buildup, stair climbing, 117
Heenan, James, 151
heli-basket, 305–306, 310–311
Helicopter High-Rise Team (HHRT), 300–310
    deployment onto roof, 304–306
    equipment for, 303, 305, 306, 308
    extended roof operations with, 309–310
    fire department operations and, 302–303
    fireground operations and, 306–309
    size-up for, 301–302
helicopters, 292, 296–299, 310–311
    landing insertion, 304
HHRT (helicopter high-rise team), 300–310
high-rise buildings, 17–33. *See also* Commercial high-rise buildings; Construction; Residential high-rise buildings
    access to, 20–21, 23, 75–77
    aerial photograph of, 206
    age of, 27
    center-core, 25–26, 28, 31, 149, 151–152, 195–197
    collapse of, 2, 8, 23–24, 25, 26
    defined, 18–19, 21, 70
    elevators in, 1–3, 4, 17–18, 19, 29–30, 75, 76, 81–104
    features of, 27–30
    fuels in, 8, 18
    history of, 17–18
    jumping from, 22
    ladders and, 19–21, 22–23, 75, 105, 221–222
    lobbies of, 74–75, 86, 90, 97
    logistics in, 32–33, 73–78, 105–106, 119
    low-rise v., 17, 22, 35–36
    multiple dwellings, 32, 48–49, 115, 149–150, 152, 205
    practical definition of, 19
    practice spotting at, 21
    RIT at, 260, 262–271, 273

size of, 21, 33
steel frames of, 17–18
strategy and tactics for, 22–23, 74
training in, 12, 71
vehicle fires v., 4
Hoff, Bob, 256
hoist extraction, by helicopter, 310
hoist insertion, by helicopter, 303, 305
hoisting machine, 84, 85
hoist-ways, elevator, 32, 83, 85–86
    blind, 86, 102
    buffers for, 85
    doors, 85, 86, 97–100
    fire or water in, 89, 92, 93
    inspection of, 92–93, 98
    single/multiple, 86
    smoke in, 89, 92, 93, 254
hollow v. noisy streams, 193–194
hose packs, xx, 4–5
    assembling, storing, and carrying, 118–119, 133–134, 139–141
    checking/maintaining, 14
    Denver, 118, 139, 172, 187
    nozzles on, 139–141, 156
    selection of, 118, 162
    standpipe, user-friendly, 133–134
    straps, 133–134, 139–140, 141, 171–172, 174
    teamwork with, 118, 134, 171–174
hose stretches
    apartment, 187–188
    from gated wye, 176
    gravity and, 110, 189
    hand stretching attack lines, 37–39, 42
    with increaser, 170–174
    labeling SPG and, 167–168
    nozzle placement with, 190
    rope bags for, 41, 45
    stairwell stretch, 110–111, 135–138, 188–191, 325
    stretching from static hose bed, 37–38, 42, 45
    in T pattern, 190
    for tandem pumping, 211–213
    training with, 325, 329
hoseline
    1½-inch handline, 123–124, 125
    advancing, 137–138, 191–193
    attack line from outside tunnel, 45
    backup lines, 195–197, 200, 234
    connecting, 187

connection failure, apparatus placement and, 210
couplings, 53, 58, 139–140, 172, 176, 185, 187, 208, 210
egress by following, 37–38, 186, 233, 241
extending, with increaser, 170–174
extra, 140, 163, 190–191, 325
flaking, 55, 135
friction points and, 110, 137–138, 139, 183, 188, 190–192, 326
in horseshoes, 133, 139, 173
kinks in, 54, 55, 135, 168, 169, 185, 188, 189–190, 214, 325
multiple/backup, 194–197, 200
primary attack, 170, 171, 174, 195–196, 200, 201, 234
for standpipe systems, 50–52
unglamorous operating positions with, 137–138, 191, 326
hoseline, 1¾-inch handline, 123, 124, 141
2½-inch and, 37, 126, 156–159, 183, 184
with combination fog nozzles, 155–156
extended with additional 1 3/4-inch hose, 37
guidelines for using, 126
hydraulic calculations for, 127–128
realistic flow and, 125–126, 128
with smoothbore nozzles, 126
hoseline, 2-inch handline, 123, 128–129
advantages of, 129
hydraulic calculations for, 128–129
hoseline, 2½-inch handline, 123–124, 130–139, 141
1¾-inch and, 37, 126, 156–159, 183, 184
ADULTS, 130–131
flow from, 125
hydraulic calculations for, 131–133
positions of six engine companies with, 200–201
with smoothbore nozzles, 68, 70, 71, 132
hoseline, big line, 130–139
seven keys to success with, 133–139, 141
hoseline selection, 123–141, 123–142
friction loss and, 125, 126, 141
GPM v. Btu's and, 124–125, 183
realistic flow and, 125–126
size, 123–124, 141
hoseline, supply lines
3-inch, 125
5-inch, 54, 77
adapters/reducers for, 214

for bypassing fire department connection, 213–214
protecting, 215–216
securing to fire department connection, 210–211
hospitals, 42–43
HOT (standpipe hands-on training), 56–57, 141, 155–159, 192, 199, 329, 335–336
HVAC systems
air movement and, 250–251, 258
concerns regarding, 27, 28, 30
fire dampers in, 32
monitoring, 27
in second generation high-rise buildings, 25, 30
in third generation high-rise buildings, 26, 30
hydration, importance of, 120, 121, 201, 289, 328

I

ICS (incident command system), 232, 275–294
identification cards, 280, 281
IFSTA Manual, 144, 155
immediately dangerous to health and life (iDLH), 231, 263, 270, 273, 285
incident command system (iCS), 232, 275–294
branches of, 292
organization sections of, 292–294
other operational components of, 289–291
incident commander (IC)
with elevator operations, 89, 93, 96, 97
with HHRT, 300–303, 304, 308
initial, 277
responsibilities of, 27–28, 114, 275–280, 283–287, 292–294
incident stabilization, as fireground priority, 154, 155
increaser, 1½-inch to 2½-inch, 50–52, 166, 170–175
creating second 2½-inch outlet with, 174–175
extending hoseline with, 170–174
Indianapolis Athletic Club (IAC), 175
injuries, xix, 30
burns, 145, 146, 148
from complacency, 1–3, 5, 11
elevators and, 82, 93, 95–96, 103

from falling debris, 215–216
from steam, 145, 146, 148, 153
insurance industry, 153–154
irons, 91–92, 102, 224, 226, 326. *See also* Tools, forcible entry/exit

J

Jahnke, Jay, 8, 37–38

K

keys
    for elevators, 90, 91, 96, 98–100, 102, 103, 266
    master, 224–226, 266, 326
    for opening windows, 247
knox box, 225

L

La Salle Bank Building, Chicago, 121
ladder company, 220. *See also* Truck company operations
ladders
    aerial, 19–21, 22–23, 75, 261
    for elevator rescues, 102
    ground, 19–21, 216, 261
    high-rise buildings and, 19–21, 22–23, 75, 105, 221–222
    at low-rise buildings, 220–221
    telesquirt, 19
    tower, 19–21, 22–23, 248
    training on, 221
    truck companies and, 221–222
    usable length of, 19–21
landings
    insertion, helicopters and, 304
    stairs and, 107, 108, 120, 135–136, 189–191, 257
    standpipe hose connection valves on, 55
lapping, 32
latch strap, 237
Layman, Lloyd, 153–155, 159

leadership, management v., 276–277, 294
learning curve, fire service, 12–13
life hazard, 40, 41, 42–43
life safety
    as priority, 154, 155, 243
    rescue and, 226
    systems, 27, 115
light-duty, physical fitness and, 12
lighting
    fluorescent, 25
    natural, 25
    tools, 92, 93, 101, 103, 266
    and ventilation for elevators, 84, 102
lobbies
    command and control and, 277–283, 287, 290, 291, 293, 294
    fires in upper floors v., 74–75
    of high-rise buildings, 74–75, 86, 90, 97
    sky, 86, 282, 284
lobby control officer (LCO), 182, 277, 279, 281, 324
lobby control/systems, 181, 182, 270, 277, 279–283, 287, 293, 294
lobby pull station, for elevator recall, 90
lock box, 224, 225, 326
lockout, tag out procedure, for elevators, 96, 97
logistics
    defined, 74
    exposures and, 73–78, 182
    in high-rise buildings, 32–33, 73–78, 105–106, 119
    in low-rise buildings, 36, 38
    standpipe systems and, 77–78
logistics section, of ICS, 281, 292–293
    chief, 279, 282, 287, 290
"Lone Rangers," 237
Longworth House Office Building, Washington, D.C., 25
lookouts, communications, escape routes, and safety zones (LCES), 300
LOVERSU (Laddering/Overhaul/Ventilation/Entry (Forcible Entry)/Rescue (Search and Rescue)/Salvage/Utilities), 221–228
low-rise buildings, 35–45
    defined, 21, 35
    elevators in, 86, 91, 95
    exposures in, 36, 38
    garden level with, 36–37
    high-rise v., 17, 22, 35–36
    ladders at, 220–221

logistics in, 36, 38
master stream appliances in, 197–199
multiple dwellings, 36–37, 223
priority in, 221
RIT at, 260, 272–273
sprinkler systems in, 36, 42
stairs in, 91
standpipe systems and, 35–45

M

Manning, Bill, 148
markers, colored, 166, 179, 238–239
master stream appliances
    flow from, 125
    offensive use of, 197–199
    with One Meridian Plaza, 23
    tools for countering nozzle reaction of, 198
material safety data sheets (MSDS), 261
McCormick, Jim, 14
McDonnell, Kevin, 329–330
McGrail, Pat, xx, 6, 146–147
mechanical air movement, 250–251
mechanical levels, 27
mechanical pressure-reducing/restricting devices, 58–59
mechanical rooms, 106, 291, 307, 308
medical aircraft/helicopter (air ambulance), 296–297, 299
medical facilities, 42–43
medical problems, 95, 289–290, 309
medical unit, 290, 293
mega-high-rise buildings
    construction features of, 26
    defined, 18, 21
    high-pressure pumpers for, 211–212
mentors, xix, 6, 13, 88, 143, 146–148, 335
MGM Grand Hotel and Casino, Las Vegas, 78
Mile High Stadium, Denver, 40
military aircraft/helicopter, 298–299
Miller, Mike, 248
Miller, Phil, 198–199, 329
mind-set, firefighting, 1–16
    attitude and, 4, 9, 10–11
    belief and, 4, 10
    building proper, 9–16
    complacency and, 1–3, 4–5, 10–11, 336
    conversational, second gear as, 10
    daydreaming, first gear as, 10

discipline and, 9, 10, 11, 12
life-or-death, fourth gear as, 10
preparation and, xxi, 4, 10–16, 71, 76, 105, 121, 333
vehicle analogy for, 9–10
mobile ventilation unit (MVU), 256, 257
Monadnock Building, Chicago, 24
Montoya, Rich, 331

N

nasty hazards, 261–262
National Fire Incident Reporting System (NFIRS), 6
National Incident Management System (NIMS), 276
National Institute for Occupational Safety and Health (NIOSH), 156
National Institute of Standards and Technology (NIST), 256–257
national standard hose (NST) thread, 175
natural gas, 30, 227, 228, 261, 290, 297
neutral pressure plane (NPP), 246, 247–248
news media aircraft/helicopter, 297–298, 299
NFPA (National Fire Protection Association)
    101, Life Safety Code, 18
    704, Hazardous Material placard, 43, 261
    Alert Bulletin, on pressure-regulating devices, 56, 61
    Fire Investigation Report of 1993 World Trade Center bombing, 18
NFPA 14 (National Fire Protection Association)
    on hoses and nozzles, 156
    on pressure-regulating devices, 57, 61, 62
    on PRVs, 60, 61
    Standard for the Installation of Standpipe and Hose Systems, 57
    on standpipe systems, 49, 52, 54–55, 70
    on water pressure, 54–55, 199
911 system
    abuse of, 7
    reverse, 230, 231, 318
Norman, John, 181, 221, 222
nozzle selection, 143–163
    burns and, 145, 146, 148
    cost, durability, maintenance and, 161
    direct v. indirect attack and, 143–144, 148, 153–154

flow tests for weapons and, 156–159
great debate over, 143–145
nozzle weight and, 160, 161
one Meridian Plaza and, 155–157
personal experience with, 145–148
proper, 155–163
steam and, 145, 148, 150, 152, 153, 159
water damage and, 145, 147, 153–155
nozzle team, 191–194
    advancing hoseline, 191–193
    on hollow v. noisy streams, 193–194
    sweeping floor prior to advancement, 193
nozzleman, 119, 137–138
nozzles
    break-apart, 162, 174
    cellar (Bresnan distributor), 175
    inadvertent closure of, 174
    low-pressure, 136–137, 157, 162–163
    primary, 166, 170
    spare, 163, 166, 170
    tips, 160, 161, 170–174
nozzles, combination fog
    automatic, 5, 57, 163
    automatic, reaction force formula for, 132
    ball valve shutoff with, 170
    in center core high-rise, 151–152, 196–197
    debris passing with, 160, 163, 170
    dual pressure, 157
    protection myth with, 149–151
    smoothbore v., 136–137, 143–149, 156, 158–163, 170, 194
    in standpipe equipment kit, 170
    treating disease v. symptom with, 150, 151, 155, 194
nozzles, smooth-bore
    ball valve shutoff with, 162, 166, 170
    combination fog v., 136–137, 143–149, 156, 158–163, 170, 194
    debris passing with, 160
    hoseline, 2½-inch with, 68, 70, 71, 132
    reaction force formula for, 132
    recommended, 162–163
nuisance alarms, 1–2, 43, 333

O

obesity, 11–12
off-duty hours, mind-set and, 10
One Meridian Plaza, Philadelphia, 105
    access stairs at, 112–113
    helicopter at, 297
    master stream strategy with, 23
    nozzle selection and, 155–157
    PRVs and, 56–57, 61, 66, 68, 77, 155
one-skid insertion, 304
operations section, of ICS, 292–293
    chief, 283, 284, 285, 286, 292–293
orifice plate, 58, 62
"Original Men in Black," 123
Otis, Elisha Graves, 17, 82
outside stem and yoke valves (OS&Y), 55
outside survey, 261
overhaul, 222–224
    at commercial buildings, 222–223
    post-control, 224
    precontrol, 49, 222–224
    at residential buildings, 223–224

P

panic attacks, 95
paramedics, 289, 290, 309
parking garages, 20, 41, 88
Patti, Dennis, 38
Pearl Harbor, 333
penciling upper atmosphere, 194
penthouse level, 27
personal protective equipment (PPE), 9, 14, 91, 92, 116–121, 139, 145
physical fitness, xxi, 5, 10–12, 76. *See also* Training
pigtail, 175
pike poles, 91, 101
pipes and tubes, with standpipe systems, 54–55
plane crash, 297
planning section, of ICS, 292–294
plenum space, opening, 193, 222, 261
pliers, needle-nose, 205, 207
police department aircraft/helicopter, 296
poling across, in elevator operations, 100–101
poling down, in elevator operations, 100, 101
poling tools, 101–102
Polo Club fire, Denver, 105, 248
positive pressure ventilation (PPV) blowers, 245, 256–257
post indicator valves (PIVs), 55, 207
Powell, Colin, 141

PPE (personal protective equipment), 9, 14, 91, 92, 116–121, 139, 145
PRDs (pressure-restricting devices), 58–59, 60, 61, 62, 63, 65, 68–69
pre-fire planning, 28
    adapters and, 175
    pressure-regulating devices and, 66–68
    stairs and, 108, 115, 121
    weapon selection with, 182–183
preparation mind-set, xxi, 4, 10–16
    mental and physical, xxi, 5, 10–12, 71, 76, 105, 121, 333
    third gear as, 10
    training/development and, xxi, 12–14
preparing for battle, xix, 321–337
pressure-reducing devices, mechanical, 58–59
pressure-regulating devices, 56–65, 328
    defined, 57, 62
    false information regarding, 69–70
    identifying, 28, 68–70
    location of, 57
    at standpipe hose valve outlets, 57, 214
    types of, 57–65
    weapon selection and, 183
pressure-restricting devices (PRDs), 58–59
    adapters for, 58, 59
    Allen wrench for removing limiting device on, 59, 60
    automatic, 63
    field adjustable, 59
    fireground adjustable, 59, 65, 68
    fireground removable, 65
    identifying, 68–69
    limiting, 59, 60
    mechanical, 58–59
    orifice plate, 58, 62
    removing pin and collar from, 59, 61
    stop, 59
property conservation, as fireground priority, 154–155
protect in place strategy, 43, 49, 226, 230–231, 232–233, 317
PRVs (pressure-regulating/reducing valves), 60–65
    characteristics of, 60–61, 328
    drills with, 12
    Elkhart, 62–63, 68, 69
    factory pre-set, 42, 61, 62–63, 69
    field adjustable, 57, 63–65, 66
    fireground adjustable, 57
    Giacomini, 66–68, 69
    One Meridian Plaza and, 56–57, 61, 66, 68, 77
    pilot-operated, 60, 61
    Powhatan, 62–63, 69
    removing bonnet portion of, 64, 65
    terminology for, 61–62
    tools for, 63–68
    Zurn, 63, 64, 65, 68, 69, 70
public address (PA) system, 227, 317–318
    for evacuation, 76–77, 317–318
    during search and rescue, 230, 231
public hallways, fire spread to, 149–150, 171, 223, 228, 232–233
pull stations
    lobby, for elevator recall, 90
    manual, 28
    phone jacks near, 316, 317
pumpers
    apparatus placement and, 209–210, 211–213, 216
    deck gun on, 199
    discharge outlets, elbows on, 169
    handlines from standpipe systems v., 37, 45
    high-pressure, 211–212
    pump panel of, 166, 167, 210, 212, 216
pumping
    relay, 331–333
    tandem, 209–213
pyramids, of Egypt, 17

R

radios, 314–315
    channels, non-repeated, 314
    communication with, 179, 242, 313–314
    elevator operations and, 91, 96, 98
    listening to, 14–15, 260–261
    mobile repeaters and, 314
    in stairway operations, 121
ram or piston, with hydraulic elevators, 88
rapid Assent Teams, 236
rapid intervention crew (RIC), 259
rapid intervention team (RIT) operations, 259–273
    basics of, 265–269
    command post, 262
    extreme, 267–269
    fire attack, 262–263, 269
    general strategies for, 269–273

HHRT as, 302, 309
at high-rise buildings, 260, 262–271, 273
history of, 259
improvised carrying devices with, 268
at low-rise buildings, 260, 272–273
RECON, 260–262
resources needed for, 266
roof, 264, 271
at single family dwellings, 259, 260, 262, 273
staging location for, 263
Superman carry with, 268–269
tools for, 265–266, 327
upper search and evacuation (USE), 263–264, 269–270, 271
rappelling, HHRT and, 303, 304–305
Raynor, Tracy, 12
RECON (Rescue of firefighters/Egress for firefighters/Construction of the fire building/Outside survey (360 degrees)/Nasty hazards (hazardous materials), 260–262
reconnaissance, 27, 31–32, 95, 300–302, 308
reducers, 51, 162, 166, 175, 214
reflex time, 77, 182, 183
REHAB (rehabilitation support unit), 290, 293, 327
rehabilitation support unit (REHAB), 289, 293, 327
air cylinders and, 201, 289
medical unit and, 290
staging and, 201, 288, 289
removal, 231
procedures, for elevators, 95–104
rescue v., 95, 96, 102
smoke, systems, 251
from stairwells, 114–115
rescue, 181, 231. *See also* Search and rescue operations
companies, in elevator operations, 95
of firefighters, 260–261, 266–269, 327–328
rescue litter, 310
residential buildings. *See also* Apartment fires
balconies on, 32, 235
commercial v., 19–21, 29, 33, 88–89, 94, 107, 193, 224, 230, 232, 262
compartmentalization in, 130, 271
fire spread to public hallway at, 149–150, 171, 223, 228, 232–233
overhaul at, 223–224
search and rescue in, 232–233, 235
residential high-rise buildings

backup lines in, 195
fire doors in, 48
forcible entry/exit in, 225–226
RIT at, 271
resource status and situation status (re-stat/sit-stat), 288
"retirement home," 6
retrofitting, of buildings, 36
riding lists, 281
risk, xix
RIT (Rapid intervention team) operations, 259–273
Roeper, Sean, 69–70, 109, 328
Rogers, Jack, 147
roof
air support from, 22–23, 28, 75, 76, 105, 264, 271
construction of, 28
doors, 255, 306–307
obstructions on, 28, 303, 304, 306, 307
stairs to, 106
wind conditions on, 28, 307
roof hatch door, in elevator rescues, 91, 93, 102–103
Roosevelt, Franklin D., 333
rope bags, for hose stretches, 41, 45
ropes
hoisting, in elevators, 83, 84–85, 87
for search and rescue operations, 22, 234, 241–243
tag line, 266
tether, 241–242
runners, 319

S

safe zones, 301, 308
safety and survival, 3, 12, 186, 330
safety lift, 17, 29, 82
salvage group, 227, 291
salvage operations, 30
IFSTA Manual on, 155
overhaul and, 222, 224
salvage covers/tarps, 155, 216, 227, 291
truck companies and, 226–227
water damage and, 226–227, 291
SCBA (self-contained breathing apparatus)
checking/maintaining, 14
elevator operations and, 91, 92

face-piece-mounted voice amplifiers, 314, 315
improper use of, 1
long-duration, 265
stairway operations and, 14, 116–121
training with, 14, 139
schools, 42
sealed buildings, 30
search and rescue operations, 22–23, 229–243. *See also* evacuation; protect in place strategy; removal
via aircraft, from roof, 22–23
in commercial buildings, 230, 233–234, 235, 239–243
communication during, 230, 231–232, 242
equipment for, 236–241
final, 236
fire floor and, 232–234
forcible entry/exit in, 224
with HHRT, 307–308
identifying who needs to be rescued, 229–231
improved methods for, 239–240
marking systems for, 166, 178, 236–239
primary, 22, 48, 150, 154, 220, 234–235, 237, 285, 308
prioritizing areas for, 231–232, 249
refuge areas and, 231, 235
residential buildings in, 232–233, 235
ropes for, 22, 234, 241–243
secondary, 236, 237
speed of, 33
three floor rule for, 230–231
three-phase, 234–236
training in, 22, 234
truck companies for, 195, 226
search markers/door straps, 166, 178, 236–238, 239, 266–267
Sears Tower, Chicago, 21, 26, 29
security personnel, 77, 94, 224, 314
self-contained breathing apparatus (SCBA) packs, 1, 14, 91, 92, 116–121, 139, 265, 314, 315
September 11, 2001, xix, xx, xxi, 318. *See also* World Trade Center buildings
falling debris/bodies and, 22, 216
firefighters lost on, 336
impacts of, 23–24, 334
logistics and, 105–106
towers' collapse on, 8, 78, 278, 301
Shapiro, Jeffrey M., 50, 52, 58

Shepherd, Mike, 328
ships, 45, 153
shopping malls, 37–38, 186
short-haul extraction, by helicopter, 310
siamese connection, 53, 182, 208–209, 214
single family dwellings, 73–74, 77, 152, 182
RIT at, 259, 260, 262, 273
search and rescue at, 226, 236
SKED rescue sled, 268
skyscrapers, 18
smoke, 18. *See also* fire/smoke towers
air movement and, 246, 248–249, 250
carbon monoxide in, 249
deaths from, 30, 249
disorientation from, 37, 231
elevators and, 89, 91, 92, 93, 95–96, 254
at floor of alarm, 1–3
mushrooming of, 249, 255
removal systems, 251
spread, fire extension and, 30–33
in stairwells, 26, 29, 49, 111–113, 254–255, 257, 308
stratification of, 248–249
from vehicle fires, 41
ventilation and, 26, 30, 32, 246, 248–258
smoke detectors, 9, 28
smoke ejectors, 245, 255
smoke management systems, 112–113, 227, 250–252
activated with alarms, 112, 251–252, 254
codes on, 251
fans v., 255–257
shutting down/controlling, 252
solid gasoline, 8, 40
space shuttle disasters, 333–334
spalling (breaking up of concrete), 23
spandrel distance, between floors, 32
SPG (standpipe inline pressure gauge), 56, 166–169, 176
sports venues, 39–41, 186
sprinkler systems
automatic, xxi, 36, 47–48
codes on, 36
in high-rise buildings, 28–29, 49
in low-rise buildings, 36, 42
out of service, 29
at schools, 42
standpipes and, 47–49, 52
types of, 28–29
stack effect, 26, 246–247, 254, 255
wind and, 114, 152, 196–197, 248

staging area, 74, 90, 120
   air cylinders at, 286, 327
   exterior, 286
   Level II, 286
   manager, 286
   REHAB and, 201, 288, 289
   RIT, 263
   training at 27th-floor, 325–326
stairs
   access, 29, 75–77, 111–113, 193, 252
   assessment of, 29
   attack, 26, 29, 49, 76, 93, 113–115, 232, 234, 236, 254–255, 284–285, 307–308, 318
   codes on, 109–110
   convenience, 111
   elevators and, 3, 90–91, 92–93, 100, 105–106, 116, 119, 160
   evacuation, 22, 29, 76–77, 93, 113, 114, 115, 231, 235, 255, 264, 270, 284, 308, 318
   exit, 49
   landings and, 107, 108, 120, 135–136, 189–191, 257
   in low-rise buildings, 91
   pre-fire planning and, 108, 115, 121
   return, 29, 106–108, 109, 110–111
   scissor, 26, 29, 108–111, 326
   search and rescue operations via, 22
   tactical use of, 113–115
   tenant, 29, 111
   training on, 14, 116
   types of, 106–113
   U return, 106, 189
stairs, climbing, 11–12, 14, 105–106, 110–111, 116–121, 322
   with equipment, 14, 76–78, 90, 91, 106, 116–121, 134, 160, 176, 323–325, 330
   heat buildup and, 117
stairway operations, 105–121
   clothing and, 117–121
stairway support unit (SSU), 119–121, 282–283, 289, 293
stairwell stretch, 110–111, 135–138, 188–191, 325
stairwells
   assessment of, 29
   doors, locks on, 225
   exits from, 109
   labeling, 109
   open, 25, 31, 32
   open doors to, 26, 49, 115–116, 117
   pressurization of, 26, 29, 112, 115–116, 117, 256
   removal from, 114–115
   smoke in, 26, 29, 49, 111–113, 254–255, 257, 308
   standpipes in, 28, 37, 55
   telephones in, 318
   vertical ventilation via, 257, 285, 306–307
standpipe cabinets
   attack stair selection and, 114
   elbows and, 169
   on fire floor, 184–185
   on floor v. in stairwell, 37, 55, 186
   PRV tools in, 66–67
   sign regarding PRV tools in, 63–64
standpipe equipment kit (tool bag), 64, 67, 165–179
   engineer's/pump operator's, 205, 207–208, 211
standpipe Hands-on Training (HOT), 56–57, 141, 155–159, 192, 199, 329, 335–336
standpipe hose packs, user-friendly, 133–134
standpipe hose valves, 55–56
   deluge valve with, 52
   hand wheel for, 40, 55–56, 166, 175, 176–177
   on landings, 55
   open, 37, 40–41
   orifice plate mounted inside, 58
   PRVs with, 42
   at same level as fire, 37
   threaded v. smooth stem, 69
   threads, brush for cleaning, 177
   threads, compatibility of, 175
   valves, 37, 40–41, 42, 52, 55–56
standpipe inline pressure gauge (SPG), 56, 166–169, 176
   labeling, 167–168
   testing, 167
standpipe outlets, 50–52
   adapters and, 175
   attack stair selection and, 114
   brush for cleaning, 177
   drill with PRV and, 12
   elbows and access to, 169
   flush, 325
   increaser for creating second, 174–175
   pressure-regulating devices at, 57, 214
   SPG and, 166–167
standpipe risers, 50, 52
   backfilling, 213–214
   location of, 28

pressure and, 56, 174
for sprinkler and standpipe systems, 47, 48, 69
standpipe systems, 47–71
automatic dry, 52
automatic wet, 29, 52, 203–204
backup plans for, 41, 45, 77–78, 163
bridges and, 43–44
Class I, 29, 50–51, 52, 185–186
Class II, 38–39, 42, 50–51, 175
Class III, 38, 42, 51–52, 174, 185–186
classes of, 28, 50–52
codes on, 50–51, 53, 57, 157, 170, 186
cold weather and, 40
components with, 28, 53–56
debris in, 159–160, 162, 176
defined, 49–50
dry, 37, 40–41, 44–45, 52–53, 204
in educational facilities, 42
fire department connection and, 44, 47–49, 53–55
in fire department operations, 38, 42, 50–52, 70
flow rates for, 51
flow testing, 37, 68, 199
handlines from pumper v., 37, 45
hooking up on floor below, 37, 38, 40, 48, 55
horizontal, 37–38, 42, 45, 186, 272–273
hoseline for, 50–52
hydraulic calculations for supplying, 215
inadequate fire flow from, 38–39
increasers with, 50–52
inspections of, 41
location of, 28
logistical problems of, 77–78
low-rise buildings and, 35–45
manual dry, 52–53
manual wet, 53
in medical facilities, 42–43
non-automatic dry, 204
for occupant use, 38, 51–52
in parking garages, 41
pipes and tubes with, 54–55
portable, 77
pressure-regulating devices and, 56–65, 126
pumping into, 214–215
semiautomatic dry, 52
at shopping malls, 37–38, 186
at sports venues, 39–41, 186
sprinkler systems and, 47–49, 52
in stairwells, 28, 37, 55

in tunnels, 44–45
two-zone (high zone/low zone), 64
types of, 28, 52–53, 203–204
at warehouses, 38–39, 186
water pressure and, 37, 40–41, 52, 54–65, 126, 128, 214–215
water supply for, 29, 47–49, 52–53, 60
on water vessels, 45
wet, 37, 42, 45, 52–53, 203–204
steam
disruption of thermal layering and, 159
expansion rate of water to, 159
injuries/deaths from, 145, 146, 148, 152, 153
nozzle selection and, 145, 148, 150, 152, 153, 159
as nuisance alarm, 2, 43
temperature at which water turns into, 159
thermal shock of, 197
as utility, 30, 227
stokes litter, 267
storz coupling, 53, 54
strike teams, 121, 264, 287
structural instability, 303
superman carry, 268–269
superman syndrome, 240
support systems, 181–182
systems control, 277, 281–282, 290, 291, 293
officer, 27, 252, 282, 301
systems group, 27

T

Tarver, Brett, 266, 333
task forces, 121, 264, 287
telephones, 315–317, 318
cell, 316
sound-powered, 316–317
terrorism, 8, 24, 301, 310. *See also* September 11, 2001
Terry, Bob, 102
tethered basket extraction, by helicopter, 310–311
tethered basket insertion, by helicopter, 303, 305–306
theory, new age and, 194
thermal imaging cameras (TIC), 234, 240–241, 242–243, 308
Thiokol, Morton, 334

tools
- axe, 91, 118
- carrying, in stairway operations, 14, 118–121
- checking/maintaining, 14
- Halligan, 91, 102, 118
- hand, 102, 106, 118, 166, 175, 176–179
- lights, 92, 93, 101, 103, 266
- for PRVs, 63–68
- for RIT, 265–266, 327
- training with, 14

tools, forcible entry/exit (irons), 91–92, 224–226
- for countering nozzle reaction of master stream appliances, 198
- for elevator operations, 91–92
- Halligan, 91, 224
- prying tool, 91, 92, 233
- rabbit tool/Hydra-Ram, 226, 233, 265
- saws, 224, 265
- sledge hammer, 91
- striking tool, 92, 224, 233
- TNT (Denver), 91, 102

towers, fire/smoke, 26, 114, 115, 171, 254–255
Tracy, Jerry, 113, 152, 198, 335
training
- for air support operations, 296, 299, 300, 305, 327–328
- as discipline, 12, 14
- drills, 12–14, 138–139, 145, 147, 151, 321–337
- with equipment, 14, 139, 323–326
- in high-rise buildings, 12, 71
- on ladders, 221
- preparation mind-set and, xxi, 12–14
- on private property/playgrounds, 139
- programs, 56–57, 141, 155–159, 192, 199, 321–337, 335–336
- propane Christmas tree burn, 145, 151
- quarterly district/battalion, 330–331
- in search and rescue, 22, 234
- on stairs, 14, 116

triage, 290, 291, 309
truck company operations, 219–228
- apparatus placement and, 209, 221–222
- basic functions of, 220–228
- elevators and, 95, 96
- engine company and, 219–220
- forcible entry/exit and, 224–226
- laddering and, 221–222
- location of fire by, 184
- LOVERSU, 221–228
- overhaul and, 222–224
- salvage and, 226–227
- for search and rescue, 195, 226
- training assignment with, 327–328
- utilities and, 227–228
- ventilation and, 221, 224

Trujillo, Mark, 102
tuberculated pipe, 160
tunnels, standpipe systems in, 44–45
"two in, two out," 259

U

United States Fire Administration (USFA), 23
upper atmosphere
- opening line into overhead to cool, 48
- penciling, 194

upper search and evacuation (USE), 232, 284–285, 292–293
- RIT, 263–264, 269–270, 271

urban search and rescue (USAR) units, 238, 300, 302
utilities, 30
- control of, 227–228, 290
- emergency, 228
- natural gas as, 30, 227, 228
- poke-through construction for, 30–31, 222–223
- protection from water damage, 226–227
- steam as, 30, 227
- truck company operations and, 227–228

V

valves. *See also* Standpipe hose valves
- adjustment rods, 66–68
- check, 55, 60
- clapper, 54, 208–209, 214
- clapper, broken/capped, 208–209
- defined, 62
- deluge, 52
- flow testing, 62, 68
- four-way hydrant, 331–333
- globe, 55–56
- indicator-type, 55

isolation, 55
post indicator valves (PIVs), 55, 207
PRV, 12, 42, 56–57, 60–71, 77, 328
sectional control, 55
standard hose, 61, 62, 66, 69
outside stem and yoke, 207
wall indicator, 207
vehicle analogy, for mind-set, 9–10
vehicle fires, 4, 41, 123, 149
ventilation
air movement and, 26, 30, 32, 224, 245–258
via attack stair, 49
bridges and, 40
group, 290, 291
horizontal, 221, 257
lighting and, for elevators, 84, 102
operational guidelines for, 257
in second generation high-rise construction, 25
smoke/fire and, 26, 30, 32, 246, 248–258
at sports venues, 40
in third generation high-rise construction, 26
topside, 219
truck companies and, 221, 224
vertical, 155, 255, 257, 285, 306–307

W

war years, 6, 124, 153–154
warehouses, 17, 38–39, 186
master stream appliances in, 197–199
RIT at, 273
water damage
elevators and, 89, 92, 93, 119
nozzle selection and, 145, 147, 153–155
salvage operations and, 226–227, 291
during training, avoiding, 325–326
water pressure
with 5-inch supply line, 54
friction loss and, 37, 52, 57, 68, 125, 126, 141, 215
low, problems with, 37, 40–41, 49, 69
NFPA 14 on, 54–55, 199
proper operating, 136
at sports venues, 40–41
standpipes and, 37, 40–41, 52, 54–65, 126, 128, 214–215
static v. flow, 168–169

water supply, 203–217, 290
backup, 212–214, 331–333
broken pipes and, 41, 48, 49
city water, 29, 49, 55
engine company operations and, 182–194, 203–217
engineer/pump operator and, 47–49, 54, 61, 166, 204–216
equipment, 29
fire department connection and, 29, 53–55, 205–215
fire protection systems and, 29, 47–49
fireground operations and, 29, 204–205
general pumping considerations for, 211–216
via master streams, 23
priority, 47–49
for standpipe systems, 29, 47–49, 52–53, 60
training with, 328, 329
water vessels, 45
weapons
flow tests for, 156–159
package, one-size-fits-all, 4–5
selecting appropriate, xxi, 71, 124–125, 126, 182–183
Web site, fire service, 12
webbing, with carabiners, 205, 211
Wesseldine, Mark, 8
"white shirt syndrome," 12
wide-rise buildings, 17, 38–39, 272
wildland firefighting, 121, 123, 264, 287, 299
wind
conditions on roof, 28, 307
donut effect and, 152
impacts of, 28, 185, 196–197, 248, 254, 256
stack effect and, 114, 152, 196–197, 248
windowless buildings, 26, 30
windows
breaking, 247–248, 253–254
broken, 30, 32, 152, 194, 215–216, 235, 247–248, 307
for egress, 106, 235
fire blanket for, 253
grates/bars on, 261
keys for opening, 247
opening, 247, 257
wood door wedges (door chocks), 49, 116, 166, 177–178
World Trade Center buildings
1993 bombing, 18, 105–106, 334
collapse of, 8, 23–24, 78, 278, 301, 303

  construction of, 9, 23–24, 26, 28
  people jumping from, 22
  protect in place strategy with, 230
  Seven, 8, 27
wrenches
  Allen, 59, 60
  hand wheel, 166, 177
  pipe, 166, 176, 177, 205
  spanner, 166, 176, 205
ws, four (who, what, where, when), 281

Y

Young, Mike, 328

    construction of, 9, 23–24, 26, 28
    people jumping from, 22
    protect in place strategy with, 230
    Seven, 8, 27
wrenches
    Allen, 59, 60
    hand wheel, 166, 177
    pipe, 166, 176, 177, 205
    spanner, 166, 176, 205
ws, four (who, what, where, when), 281

Y

Young, Mike, 328

isolation, 55
post indicator valves (PIVs), 55, 207
PRV, 12, 42, 56–57, 60–71, 77, 328
sectional control, 55
standard hose, 61, 62, 66, 69
outside stem and yoke, 207
wall indicator, 207
vehicle analogy, for mind-set, 9–10
vehicle fires, 4, 41, 123, 149
ventilation
air movement and, 26, 30, 32, 224, 245–258
via attack stair, 49
bridges and, 40
group, 290, 291
horizontal, 221, 257
lighting and, for elevators, 84, 102
operational guidelines for, 257
in second generation high-rise construction, 25
smoke/fire and, 26, 30, 32, 246, 248–258
at sports venues, 40
in third generation high-rise construction, 26
topside, 219
truck companies and, 221, 224
vertical, 155, 255, 257, 285, 306–307

W

war years, 6, 124, 153–154
warehouses, 17, 38–39, 186
master stream appliances in, 197–199
RIT at, 273
water damage
elevators and, 89, 92, 93, 119
nozzle selection and, 145, 147, 153–155
salvage operations and, 226–227, 291
during training, avoiding, 325–326
water pressure
with 5-inch supply line, 54
friction loss and, 37, 52, 57, 68, 125, 126, 141, 215
low, problems with, 37, 40–41, 49, 69
NFPA 14 on, 54–55, 199
proper operating, 136
at sports venues, 40–41
standpipes and, 37, 40–41, 52, 54–65, 126, 128, 214–215
static v. flow, 168–169

water supply, 203–217, 290
backup, 212–214, 331–333
broken pipes and, 41, 48, 49
city water, 29, 49, 55
engine company operations and, 182–194, 203–217
engineer/pump operator and, 47–49, 54, 61, 166, 204–216
equipment, 29
fire department connection and, 29, 53–55, 205–215
fire protection systems and, 29, 47–49
fireground operations and, 29, 204–205
general pumping considerations for, 211–216
via master streams, 23
priority, 47–49
for standpipe systems, 29, 47–49, 52–53, 60
training with, 328, 329
water vessels, 45
weapons
flow tests for, 156–159
package, one-size-fits-all, 4–5
selecting appropriate, xxi, 71, 124–125, 126, 182–183
Web site, fire service, 12
webbing, with carabiners, 205, 211
Wesseldine, Mark, 8
"white shirt syndrome," 12
wide-rise buildings, 17, 38–39, 272
wildland firefighting, 121, 123, 264, 287, 299
wind
conditions on roof, 28, 307
donut effect and, 152
impacts of, 28, 185, 196–197, 248, 254, 256
stack effect and, 114, 152, 196–197, 248
windowless buildings, 26, 30
windows
breaking, 247–248, 253–254
broken, 30, 32, 152, 194, 215–216, 235, 247–248, 307
for egress, 106, 235
fire blanket for, 253
grates/bars on, 261
keys for opening, 247
opening, 247, 257
wood door wedges (door chocks), 49, 116, 166, 177–178
World Trade Center buildings
1993 bombing, 18, 105–106, 334
collapse of, 8, 23–24, 78, 278, 301, 303